Dielectric Materials Engineering

Dielectric Materials Engineering

Edited by **Doreen Rowe**

NY RESEARCH
P R E S S

New York

Published by NY Research Press,
23 West, 55th Street, Suite 816,
New York, NY 10019, USA
www.nyresearchpress.com

Dielectric Materials Engineering
Edited by Doreen Rowe

International Standard Book Number: 978-1-63238-507-9 (Hardback)

Printed in the United States of America.

Contents

Preface

Dielectric materials are insulators that can be polarized when placed in an electric field. These materials are used for the storage and dissipation of magnetic and electric energy. Dielectric materials engineering is mainly concerned with the design and development of dielectric materials for industrial as well as domestic usage. These materials include industrial coatings, mineral oils, electrets, piezoelectric materials, capacitors, resonators, etc. This book traces the progress of this field and highlights some of its key concepts and applications. This book is appropriate for students seeking detailed information in this area as well as for experts. It elucidates the current theories and innovative models around prospective developments with respect to dielectric materials. The aim of this text is to provide an in-depth understanding of this discipline and unravel the unexplored aspects related to it.

The researches compiled throughout the book are authentic and of high quality, combining several disciplines and from very diverse regions from around the world. Drawing on the contributions of many researchers from diverse countries, the book's objective is to provide the readers with the latest achievements in the area of research. This book will surely be a source of knowledge to all interested and researching the field.

In the end, I would like to express my deep sense of gratitude to all the authors for meeting the set deadlines in completing and submitting their research chapters. I would also like to thank the publisher for the support offered to us throughout the course of the book. Finally, I extend my sincere thanks to my family for being a constant source of inspiration and encouragement.

Editor

Dielectric Properties of Er^{3+} Doped ZnO Nanocrystals

N. K. Divya, P. U. Aparna, P. P. Pradyumnan[*]

Department of Physics, University of Calicut, Malappuram, India
Email: [*]drpradyumnan@gmail.com

Abstract

Erbium doped ZnO nanocrystals were synthesized through solid state reaction route. A detailed structural study was carried out by using X-ray diffraction (XRD) data. This work investigates the changes occurred to the hexagonal wurtzite crystal lattice of ZnO due to the incorporation of Er^{3+} based on crystalline size. The chemical composition of the samples was confirmed using Energy Dispersive Spectra (EDS) data analysis. Decrease in average crystallite size with increase in rare earth concentration was observed in XRD. The frequency dependence of dielectric constant (ε_r), dielectric loss (tanδ) and AC conductivity of pure and rare earth doped ZnO were calculated and well explained with Maxwell-Weigner model.

Keywords

Solid State Reaction, XRD, Crystallite Size, Bond Length, Dielectric Constant, AC Conductivity

1. Introduction

The nontoxicity, excellent chemical and thermal stability, specific electrical and optical properties make ZnO an attractive compound for researchers. Growth conditions, environment, the type and purity of dopants have very important role in the determination of structural and dielectric properties of ZnO. It has potential applications in electro-optic, acousto-optic, ultraviolet (UV) light emitters, chemical sensors, piezoelectric materials and high power optoelectronic devices [1]. The endless requirement of materials with high dielectric constant in the field of microelectronics makes the study on the development of new dielectric materials more relevant. Dielectric constant and Loss tangent are the most important parameters determining the dielectric behaviour of materials [2]. Dielectric loss of a dielectric material is the inherent dissipation of electromagnetic energy into heat. It can be expressed in terms of either the loss angle δ or the corresponding loss tangent tanδ. Zinc oxide possesses

---

[*]Corresponding author.

semiconductor behaviour; it can be a dielectric at low temperature and can be a conductor at high temperature [3].

Rare earth (RE^{3+}) doped ZnO has the speciality of co-existance of semiconducting, electrochemical and optical properties, which motivates the researchers to work with these materials. The improvement of dielectric properties of ZnO via doping has been reported several times. Since the ionic radius of Zn^{2+} ions is very small compared with RE^{3+} ions, the incorporation of trivalent RE^{3+} ions in ZnO makes distortions in the ZnO matrix, which leads to asymmetricity in the electronic environment of the system [4]-[18]. Thus electrical conductivity and other physical properties of the semiconductor can be controlled by the incorporation of RE^{3+} impurities into semiconductor lattices. Dielectric properties of erbium doped ZnO are rarely reported and this report opens a path for large scale production of dielectric materials with low dielectric loss.

The synthesis methods have prominent role in the properties of the materials. There are several methods like sol-gel method, co-precipitation method, electrochemical crystal growth, hydrothermal method, solid state reaction method, etc., which can be adapted for the preparation of samples. Each method has its own advantages and limitations. Compared with conventional methods, solid state reaction method facilitates the production of large quantity of high pure samples at low cost.

This report is unique in such a way that, this is first time reporting the production of perfectly doped ZnO within the solubility limit of erbium in ZnO via solid state reaction method. Thus this opens a door to the large scale production of RE doped ZnO with high dielectric constant and low dielectric loss. An important observation of this work is the frequency dependence of the dielectric constant of the samples with erbium concentration 0.6 wt% and 0.9 wt% and the frequency independent behaviour of dielectric constant of samples with 0.3 wt% and 1.2 wt% erbium incorporation. This observation will be very useful in the practical applications of these materials.

Aim of present study is (i) to synthesise Erbium doped ZnO via solid state reaction route and (ii) to study the structural and dielectric properties in detail. This work focuses on the development of dielectric material with low dielectric loss.

2. Experimental Details

2.1. Synthesis

Er^{3+} doped ZnO powders were prepared via controlled solid state reaction route at high temperature. The samples were prepared by thorough mixing of ZnO (Sigma aldrich product −99.99% purity) and Er_2O_3 (Sigma aldrich product −99.99% purity) and made in to slurry by adding $LiOH.H_2O$ (merck product −99.99% purity) and ethanol (merck product −99.99% purity). In this study LiOH act as a heat transfer medium. The slurry then dried at 100°C in an oven. Then obtained powder sample was well ground for 1hr using a mortar and pestle and made in to pellets using hydraulic pelletizer at a pressure of 8 Torr. The pellets were packed in a quartz boat and fired at 900°C. The heat treated pellets were again ground for 15 minutes and used as the sample. Five batches of samples were prepared by varying the wt% of erbium as 0.3 wt%, 0.6 wt%, 0.9 wt% and 1.2 wt% of total amount of the powder sample.

2.2. Characterization

The phase purity and structural informations were obtained by using Rigaku Miniflex 600 machine with Cu Kα radiation (1.5406 Å). The powder X, PDXL: Integrated powder X-ray diffraction software and ICDD data base were used for phase determination, peak reflection assignments and to analyse variations in peak positions and intensities. The purity of samples was confirmed with Energy Dispersive X-ray Analysis (EDAX) equipment of Hitatchi. The dielectric studies were done with Hioki 3532-50 LCR HiTESTER.

3. Results and Discussion

3.1. X-Ray Diffraction Analysis

The crystallinity and structure of the synthesised samples were analysed by X-ray diffraction data recorded. The intensity data was collected over a 2θ range of 10° - 90°. Powder X-ray diffraction patterns of undoped and Er-doped ZnO samples plotted in the **Figure 1**. All patterns are well matched with the standard X-ray diffraction pattern indicating hexagonal wurtzite structure of bulk ZnO. Diffraction peaks corresponding to reflections from planes (100), (002), (101), (102), (110), (103), (112) of Wurtzite ZnO structure, which are consistent with the standard ICDD Card no.01-074-9940. When the concentration of erbium was increased further, extra phase related to Er_2O_3 arises and is observed in **Figure 2**. It indicates that at this concentration the solubility limit of erbium in ZnO has exceeded. And thus multiphase formation occurs at this concentration.

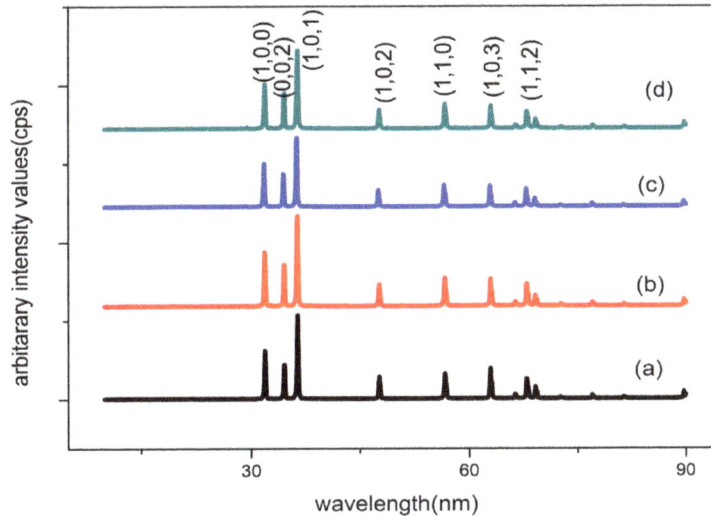

Figure 1. XRD pattern of samples (a) heat treated pure ZnO (b) 0.3 wt% Er doped ZnO (c) 0.6 wt% Er doped ZnO (d) 0.9 wt% Er doped ZnO (e) 1.2 wt% Er doped ZnO.

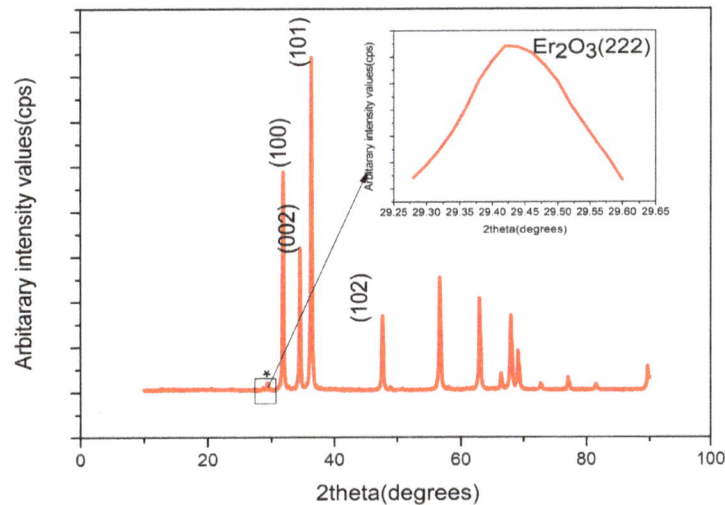

Figure 2. XRD pattern of 1.2 wt% Er doped ZnO sample.

Using Scherrer formula, $D = 0.9\lambda/\beta\cos\theta$, where λ is the wavelength of CuKα radiation (1.514 Å), θ is the peak position and β is the FWHM, the average crystallite size of the samples was estimated [5]. It is found that average crystallite size decreases with increase in doping concentration. The average crystallite size of pure ZnO nanoparticle was 57.71 nm. It is decreased to 45.26 nm with 1.2 wt% doping of erbium. The decrease in crystallite size is due to the distortion of host ZnO lattice by RE^{3+} ions, which actually reduces the nucleation and subsequent growth rate of ZnO crystals.

3.2. Chemical Compositional Confirmation

In order to confirm the chemical composition of the samples, Energy Dispersive Spectra were taken. The sam-ples were cleaned by Ar+ ion beam sputtering to remove surface contamination if any. **Figure 3(a)** is the energy dispersive spectra of undoped ZnO. **Figure 3(b)** is the energy dispersive spectra of doped (Er^{3+} −1.2 wt%) ZnO ceramics. From **Figure 3(a)**, the purity of parent sample is confirmed. There are no impurities in the parent ZnO sample. The incorporation of Er^{3+} ions in ZnO crystal lattice are confirmed from the **Figure 3(b)**.

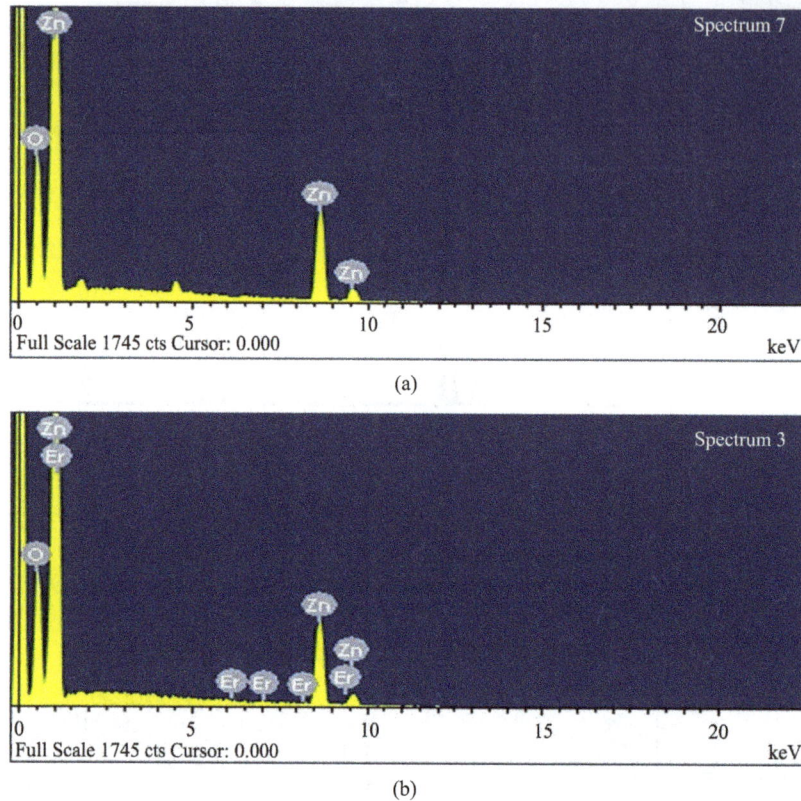

(a)

(b)

Figure 3. (a) (b) EDS spectra of pure and Er doped ZnO respectively.

3.3. Dielectric Properties

Dielectric measurements of the undoped and erbium doped ZnO were carried out as a function of frequency at room temperature using an LCR meter. The samples were pelletized by using a hydraulic press and this pellet was placed between the electrodes. The dielectric constant (ε_r) is calculated using the relation $\varepsilon_r = Cd/\varepsilon_0 A$ and the ac conductivity is calculated by the relation, $\sigma_{ac} = \varepsilon_0 \varepsilon_r \omega \tan(\Delta)$, where C is the capacitance, d is the thickness, A is the area of cross section of the pellet and $\tan(\Delta)$ is the dielectric relaxation of the sample. It is observed that the dielectric constant decreases with increasing frequency for all samples. This is evident in the low frequency region and is shown in **Figure 4**. At high frequencies the dielectric constant becomes frequency independent. Maxwell-Weigner model can be used to explain the dielectric properties of the homogeneous double structure systems. According to this model the dielectric medium is made of well conducting grains with poorly conducting or resistive grain boundaries. Thus by the application of external electric field, the charge carriers easily moves, but will be accumulated at the grain boundaries. This process causes large polarization and high dielectric constant produced by the sample. At low frequency, the small conductivity of grain boundary contributes to the high value of dielectric constant. The interfacial/space charge polarization due to inhomogeneous dielectric structure can also be considered as the reason for high dielectric constant. Beyond certain level of frequency of external field, the hopping between different metal ions (Zn^{2+}, Er^{3+}) can not follow the alternating field [2]. This is why the dielectric constant becomes frequency independent beyond certain limit. The decrease in dielectric constant with increase in frequency in the lower frequency range can be considered as a natural phenomenon because any species contributing to polarizability is found to show lagging behind the applied field at higher and higher frequencies.

In order to investigate further the effect of erbium doping, the dielectric constant of material is re plotted in

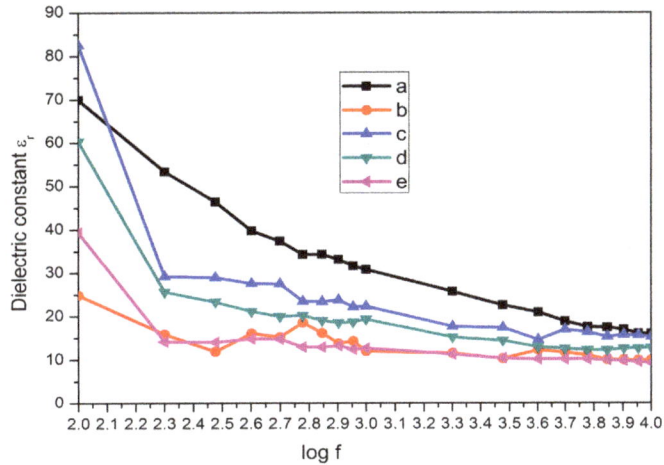

Figure 4. Figure of Dielectric constant vs logf of the samples (a) pure ZnO (b) 0.3 wt% Er doped ZnO (c) 0.6 wt% Er doped ZnO (d) 0.9 wt% Er doped ZnO (e) 1.2 wt% Er doped ZnO.

Figure 5 as a function of concentration of erbium. The dielectric constant is found to be decreased first with small amount of erbium incorporation (0.3 wt%). It is striking to see that the dielectric constant increases dramatically while the concentration of erbium is increased to 0.6 wt%. Further increase in erbium concentration decreases the dielectric constant of ZnO. The incorporation of small amount of Erbium ions produces distortion and an abrupt unstability in the crystal lattice of ZnO, which results in the decrease of dielectric constant of the material. But when the concentration is increased to an optimum concentration of 0.6 wt%, the system gets stabilized and results in the reduction of the number of oxygen vacancies [18]. This leads to an increase in the dielectric constant. Further increase of erbium content would result in the unit cell contraction and hence the free volume available for the displacement of ions gets reduced. This results in the decrease of dielectric polarization and dielectric constant.

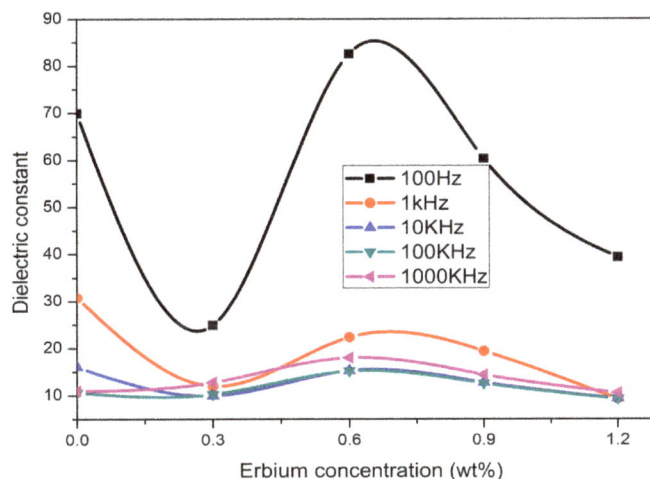

Figure 5. Dielectric constant of erbium doped ZnO at selected frequencies.

An important observation from **Figure 5** is the frequency dependence of the dielectric constant of the samples with erbium concentration 0.6 wt% and 0.9 wt% and the frequency independent behaviour of dielectric constant of samples with 0.3 wt% and 1.2 wt% erbium incorporation. This observation will be very useful in the practical applications of these materials. It is interesting to note that the sample with 0.6 wt% erbium incorporation appears to be the best dielectric material than the parent ZnO at 100 Hz, 10 KHz, 100 KHz and 1000 KHz. This helps us to optimize the concentration of erbium in ZnO for making good dielectric material.

Figure 6 shows the variation of ac conductivity of undoped and Erbium doped samples with frequency. The ac conductivity of all the samples gets increased with increase in frequency. But the maximum ac conductivity is seen for undoped ZnO sample and which decreases with the erbium incorporation. The production of defects like zinc interstitials in the ZnO host system by erbium ions can be considered as the reason for this reduction in ac conductivity. It is very relevant that the doping percentage plays important role in tuning the electrical property of the sample. Because the 0.6 wt% erbium incorporation show maximum ac conductivity as compared to other doped samples. After this optimum doping percentage the ac conductivity is found to be decreased with increase in doping percentage. This decrease may be due to the defects produced by erbium ions in ZnO host matrix. As doping concentration increases, the defect ions which facilitates the formation of grain boundary defect barrier, leading to blockage to the flow of charge carriers. This in turn decreases the conductivity of the system on doping.

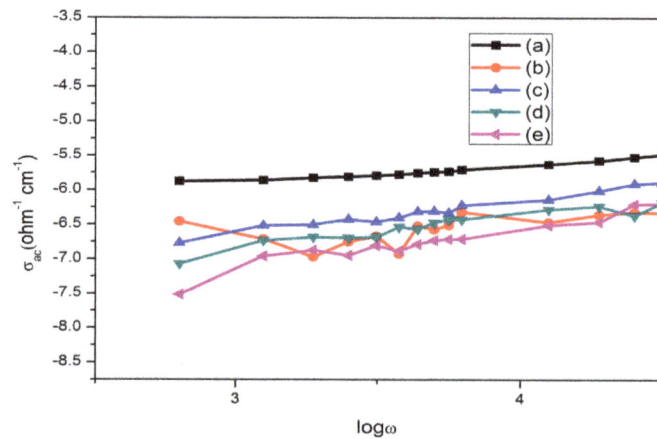

Figure 6. Figure of log σ_{ac} vslogω of the samples (a) pure ZnO (b) 0.3 wt% Er doped ZnO (c) 0.6 wt% Er doped ZnO (d) 0.9 wt% Er doped ZnO (e) 1.2 wt% Er doped ZnO.

Figure 7 shows the variation of dielectric loss values tan(δ) with frequency of the un doped and erbium doped ZnO samples. Loss tangent *i.e.* tan(δ) is the energy dissipation in the dielectric system. It is found to be maximum for undoped ZnO samples. As erbium doping increases, this loss factor is decreased to a large extend. The decrease of tan(δ) with increase in frequency seen in erbium doped samples is due to the space charge polarization. This peculiar behaviour as well as the very low loss factor compared to the parent ZnO makes the prepared samples suitable for high frequency device applications.

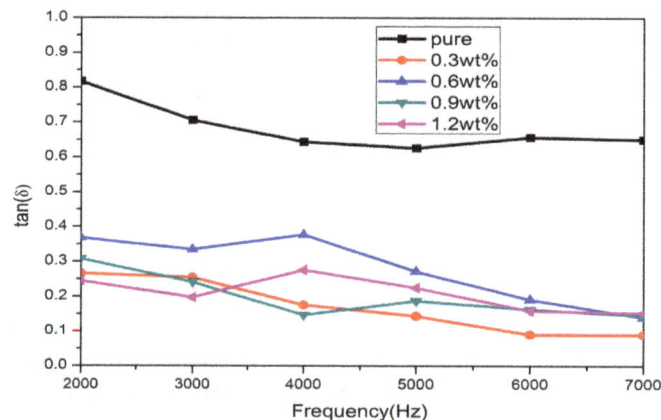

Figure 7. Plot of tan(δ) Vs Frequency of the samples (a) pure ZnO (b) 0.3 wt% Er doped ZnO (c) 0.6 wt% Er doped ZnO (d) 0.9 wt% Er doped ZnO (e) 1.2 wt% Er doped ZnO.

4. Conclusion

The structural and dielectric studies of Erbium doped ZnO synthesised via simple solid state reaction route were investigated. Since rare earth elements have larger ionic radii compared with Zinc, the incorporation of trivalent erbium ions in ZnO host lattice can cause a significant distortion in ZnO crystal lattice. This is first time reporting the production of perfectly doped ZnO within the solubility limit of erbium in ZnO via solid state reaction method. Thus, this report opens a path for the production of RE doped ZnO with high dielectric constant. The crystallite size is calculated using Scherrer formula and is found to be decreased with erbium doping. The average crystallite size of pure ZnO nanoparticle was 57.71 nm. It is decreased to 45.26 nm with 1.2 wt% doping of erbium. The incorporation of Er^{3+} ion was confirmed using EDS. The AC conductivity of the erbium doped samples is found to be decreased with frequency. Dielectric constant of 0.6 wt% erbium incorporated ZnO is found to be maximum compared with other undoped and doped samples. The decrease in loss factor with erbium incorporation as well as its decrease with increase in frequency is very relevant results as far the application of material concerned. Sample with 0.6 wt% erbium incorporation is showing maximum dielectric constant and low loss compared with other samples. Thus it can be concluded that 0.6 wt% erbium doped ZnO is a very good dielectric material than the parent ZnO. These results make the prepared samples as promising candidates for high frequency device applications.

Acknowledgements

Supporting of this investigation by UGC-SAP, BSR is gratefully acknowledged. Author (P.P.Pradyumnan) thanks DST-SERB, DST-FIST, Department of Physics, University of Calicut for financial assistance and equipment facility.

References

[1]   Ishizum, Y. and Kanemitsu, I. (2005) Structural and Luminescence Properties of Eu-Doped ZnO Nanorods Fabricated by a Microemulsion Method. *Applied Physics Letters*, **86**, Article ID: 253106. http://dx.doi.org/10.1063/1.1952576

[2]   Zamiri, R., Kaushal, A., Rebelo, A. and Ferreira, J.M.F. (2014) Er-Doped ZnO Nanocplate: Synthesis, Optical and Dielectric Properties. *Ceramics International*, **40**, 1635-1639. http://dx.doi.org/10.1016/j.ceramint.2013.07.054

[3]   Omar, K., Jhohan Ooi, M.D. and Hassin, M.M. (2009) Investigation on Dielectric Constant of Zinc Oxide. *Modern Physics*, **3**,110-116.

[4]   Jadwisienczak, W.M., Lozyokowski, H.J., Xu, A. and Patel, B. (2002) Visible Emission from ZnO Doped with Rare-Earth Ions. *Journal of Electronic Materials*, **31**, 776-784. http://dx.doi.org/10.1007/s11664-002-0235-z

[5]   Chen, G.Y., LIU, H.C., Somesfalean, G., Sheng, Y.Q., Liang, H.J., Zhang, Z.G., Sun, Q. and Wang, F.P. (2008) Enhancement of the Upconversion Radiation in Y_2O_3:Er^{3+} Nanocrystals by Codoping with Li^+ Ions. *Applied Physics Letters*, **92**, Article ID: 113114. http://dx.doi.org/10.1063/1.2901039

[6]   Rodnyi, P.A. and Khodyuk, I.V. (2011) Optical and Luminescence Properties of Zinc Oxide. *Optics and Spectroscopy*, **111**, 776-785. http://dx.doi.org/10.1134/S0030400X11120216

[7]   Willander, M., Nur, O., Sadaf, J.R., Qadir, M.I., Zaman, S., Zainelabdin, A., Bano, N. and Hussain, I. (2010) Luminescence from Zinc Oxide Nanostructures and Polymers and Their Hybrid Devices. *Materials*, **3**, 2643-2667.

[8]   Li, H.L., Zhang, Z., Huang, J.Z., Liu, R.X. and Wang, Q.B. (2013) Optical and Structural Analysis of Rare Earth and Li Co-Doped ZnO Nanoparticles. *Journal of Alloys and Compounds*, **550**, 526-530. http://dx.doi.org/10.1016/j.jallcom.2012.10.080

[9]   John, R. and Rajakumari, R. (2012) Synthesis and Characterization of Rare Earth Ion Doped Nano ZnO. *Nano-Micro Letters*, **4**, 65-72. http://dx.doi.org/10.1007/BF03353694

[10] Kaid, M.A. and Ashour, A. (2007) Preparation of ZnO-Doped Al Films by Spray Pyrolysis Technique. *Applied Surface Science*, **253**, 3029-3033. http://dx.doi.org/10.1016/j.apsusc.2006.06.045

[11] Binitha, M.P. and Pradyumnan, P.P. (2013) Structural, Thermal and Electrical Characterization on Gel Grown Copper Succinate Dihydrate Single Crystals. *Physica Scripta*, **87**, Article ID: 065603. http://dx.doi.org/10.1088/0031-8949/87/06/065603

[12] Rahman, Md.T., Vargas, M. and Ramana, C.V. (2014) Structural Characteristics, Electrical Conduction and Dielectric Properties of Gadolinium Substituted Cobalt Ferrite. *Journal of Alloys and Compounds*, **617**, 547-562. http://dx.doi.org/10.1016/j.jallcom.2014.07.182

[13] Li, J.-L., Chenn, G.-H. and Yuan, C.-L. (2013) Microstructure and Electrical Properties of Rare Earth Doped ZnO-

Based Varistor Ceramics. *Ceramics International*, **39**, 2231-2237.
http://dx.doi.org/10.1016/j.ceramint.2012.08.067

[14] Pathak, D., Wagner, T., Adhikari, T. and Nunzi, J.M. (2015) Photovoltaic Performance of AgInSe$_2$-Conjugated polymer Hybrid System Bulk Heterojunction Solar Cells. *Synthetic Metals*, **199**, 87-92.
http://dx.doi.org/10.1016/j.synthmet.2014.11.015

[15] Pathak, D., Bedi, R.K. and Kaur, D. (2010) Growth of Heteroepitaxial AgInSe$_2$ Layers on Si (100) Substrates by Hot Wall Method. *Optoelectronics and Advanced Materials—Rapid Communications*, **4**, 657-661.

[16] Pathak, D., Bedi, R.K., Kaur, D. and Kumar, R. (2011) 200 Mev Ag+ Ion Beam Induced Modifications In AgInSe$_2$ Films Deposited By Hot Wall Vacuum Evaporation Method. *Chalcogenide Letters*, **8**, 213-222.

[17] Divya, N.K. and Pradyumnan, P.P. (2014) Variation in Morphology and Crystallinity of ZTO Ceramics. *Research Journal of Recent Sciences*, **3**, 71-74.

[18] Chang, F.G., Song, G.L., Fang, K. and Zeng, Q.J. (2006) Effect of Gadolinium Sbstitution on Dielectric Properties of Bismuth Ferrite. *Journal of Rare Earths*, **24**, 273-276. http://dx.doi.org/10.1016/S1002-0721(07)60379-2

The Effect of γ-Radiations on Dielectric Properties of Composite Materials PE + x vol% TlGaSe$_2$

E. M. Gojayev, A. G. Hasanova

Azerbaijan Technical University, Baku, Azerbaijan
Email: geldar-04@mail.ru

Abstract

The purpose of the paper is to study the effect of gamma irradiation on the temperature dependence of the dielectric constant (ε) and dielectric loss (tanδ) of composite materials PE + x vol% ($0 \leq x \leq 10$). Measurements are carried out with an alternating current at a frequency of 1 kHz using the measuring bridge E-20. Measurements are carried out at temperature range 300 - 450 K, irradiated at doses of 50, 100 and 150 kGy. It is revealed that in all irradiated samples with increasing volumetric filler content increase the dielectric characteristics of composites PE + x vol%. TlGaSe$_2$. Temperature variation of the dielectric parameters, after gamma irradiation are the result of occurring in the electron-ion and polarization at the interface of the matrix polymer with filler of TlGaSe$_2$.

Keywords

Gamma Irradiation, Dielectric Properties, Composite Materials

1. Introduction

It is known that there are several ways for changing the properties of polymers. One of them is to add the low molecular weight disperse fillers to the volume of the polymer. This composite materials take an entirely new electro-physical, electret, strength, thermal, optical and other properties [1]-[5].

Usually, additives change supramolecular structure that largely determines physical, chemical and mechanical properties of polymers and composites based on them. This is due to the fact that when injecting small amount of fillers into polymers, the fillers act as artificial crystallization nuclei that leads to a change in the material properties. Thus, by changing the kind of fillers, the properties of polymer-based composite materials may be controlled. At present, there exists a rather large amount of works on investigation of fundamental properties of composites. However, there are no perfect models that explain the processes responding for change of electro-physical properties of polymers when introducing into them fillers, and also the action of ionizing radiations on these properties. In this connection, investigations of the effect of disperse fillers and also irradiation with γ-rays on electro-physical properties of polymers and composites based on them are very actual [6]-[8].

The purpose of work is to study the dielectric properties of γ-irradiated composites with semiconducting fillers as PE + x vol% TlGaSe$_2$ dimensions 100 nm.

2. Experimental Technique

Composite samples are obtained by mechanical mixing of TlGaSe$_2$ powder with the PE powder in porcelain mortar. The mixing is continued until we have obtained a homogeneous mixture. The mixture is kept for some time at the melting temperature of the polymer under pressure of 5 MPa. In the same temperature by compacting the homogeneous mixture and the pressure slowly increases to 15 MPa. At this temperature, the sample is maintained for 5 minutes and then is quickly cooled in water. The sizes of the samples are: thickness is about 80 - 120×10^{-3} mm, diameter of the obtained samples is 35 mm. Provide reliable electric contact between the samples and electrodes made of stainless steel, the electrodes made of thin aluminum foil of thickness 7×10^{-3} mm was pressed to both working surfaces of the sample.

The measurement of dielectric constant and dielectric loss PE + x vol% TlGaSe$_2$ carried out in the range 296 - 520 K with a linear rise in temperature at the rate of 2.5 K/min. Block diagram of the setup is shown in **Figure 1**. Measurement of ε and tgδ realized with alternating current frequency at 1 kHz. Sample (2) is kept between the two electrodes of the cell for measurement. Then the sample is heated in the cell via the heater (4) that was designed integrated in the laboratory with constant speed 2.5 K/min. The temperature of sample is registered by the thermocouple (3) by the system that attaches the heater 4. Resistance of the sample is measured by the tera-ohmmeter (5) (E6-13A), while dielectric constant and dielectric loss measuring by the bridge E-20. The sample's temperature is measured by the temperature meter (6) (M-64). For the power supply of the heater used three-auto-regulated system (7).

3. Results and Discussion

The results of dielectric constant (ε) and dielectric loss of composites PE + x vol% TlGaSe$_2$ at different tem-

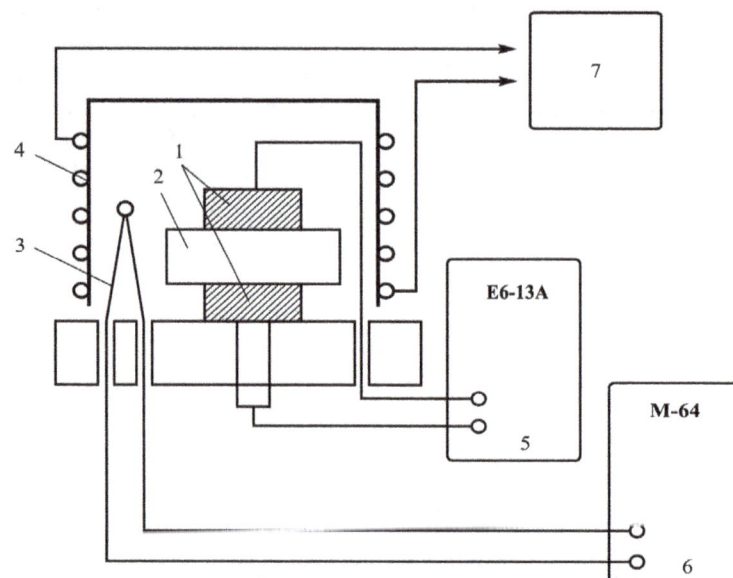

Figure 1. Block diagram of the apparatus for measuring temperature dependences of dielectric constant and dielectric losses: 1-measuring cell, 2-sample, 3-thermocouple, 4-heater, 5-immittance meter, 6-termometer, 7-three-auto-regulated system.

peratures up to γ-radiations is shown in **Figure 2**. The investigations were carried out at temperature range 300 - 450 K for the composites PE + x vol% TlGaSe$_2$, where x = 0; 3; 5; 7 and 10. TlGaSe$_2$. As is evident in **Figure 2**. in dependences ε(T) and tgδ(T) at temperatures 320 - 326 K there appears one low temperature maximum. With increasing the content of the filler TlGaSe$_2$ the values of ε(T) and tgδ(T) increase.

At room temperature the values of dielectric constant of the composites with fillers x = 0; 3; 5; 7 and 10 vol% TlGaSe$_2$ is 2.35, 2.7, 3, 3.40 and 3.65, while at temperature 320 K, 2.35, 3.1, 3.55, 4.15 and 4.5, respectively. In particular, for the composite 3 vol% TlGaSe$_2$ with increasing the temperature from the room temperature to 320 K, ε(T) increases by 17% (curve 2), for the composite with x = 5 (curve 3) at the same temperature range ε(T) increases by 18%, for the composite with x = 7 (curve 4) by 22%, for the composite with x = 10 (curve 5) by 23%. For the compositions under investigation, with increase of the filler content there happens increase of the quantity of ε(T) and tgδ(T), in particular, for the composites with x = 0; 3; 5; 7 and 10 x vol% TlGaSe$_2$ at room temperature tgδ(T) becomes 0.013, 0.027, 0.043, 0.070 and 0.10, while for 330 K - 0.013, 0.043, 0.073 0.113, 0.24, respectively. With increasing the filler content at temperature 330 K the diffuse maximums are observed. With change of temperature from the room temperature to 330 K, for the composite 3 vol% TlGaSe$_2$ the amplitude of tgδ(T) increases for the composites with fillers 5; 7 and 10 vol% TlGaSe$_2$ the increase of tgδ(T) at the indicated temperature range is 17%, 16% and 24% respectively. For this composites with increasing the filler content, the value of ε(T) and tanδ(T) in the entire investigated temperature range is increased.

The effect of γ-irradiations on dielectric properties of the composites PE + x vol% TlGaSe$_2$ is investigated. The studies were carried out on the composites irradiated at 50, 100, 150 kGy. The results are given in **Figure 3**. Temperature dependence of dielectric constant of composites PE + x vol% TlGaSe$_2$, irradiated at dose of 50 kGy is given in **Figure 3(a)**. As it follows in **Figure 3(a)**, with increasing the filler's content, on the whole of temperature range there happens increase of ε, and on the curves ε(T) the maximums appear. However, in the irradiated samples the maximums are shifted to the area of higher temperatures (360 - 370 K) and become weakly expressed. In the composites irradiated at dose of 50 kGy dependencies of maximums in tanδ(T) were not revealed, monotone decrease of tanδ(T) (**Figure 3(b)**) are observed.

The results of investigations of temperature dependence of dielectric constant of composites PE + x vol% TlGaSe$_2$, irradiated at dose of 100 kGy are given in **Figure 3(c)**. As is seen from **Figure 3(c)**, ε increases due to increase in the filler content. In particular, at room temperature, ε for the composites with admixtures x = 3; 5; 7; and 10 vol% TlGaSe$_2$ is 3.6, 4, 4.4 and 4.66, respectively, and at temperature 360 K, ε of these composites was 3.87; 5.2; 5.87 and 7.47, respectively. Thus, in irradiated 100 kGy composites, at low temperatures with increasing the temperature from the room temperature to 370 K, ε increases, and with a further increase in temperature it decreases. Dielectric losses of these irradiated composites typical for all samples increase to 380 K, and at temperature range 380 - 450 K it decreases (**Figure 3(d)**). The results of investigations of temperature dependences of dielectric constant of composites PE + x vol% TlGaSe$_2$ irradiated at dose 150 kGy are given in **Figure 3(e)**, from which it follows that in ε(T) dependence with the exception of pure polyethylene for all irradiated samples, with increasing temperature at wide range, dielectric constant increases. With increasing the filler content, the appropriate maximums of composites shift to the area of higher temperatures. Temperature dependence

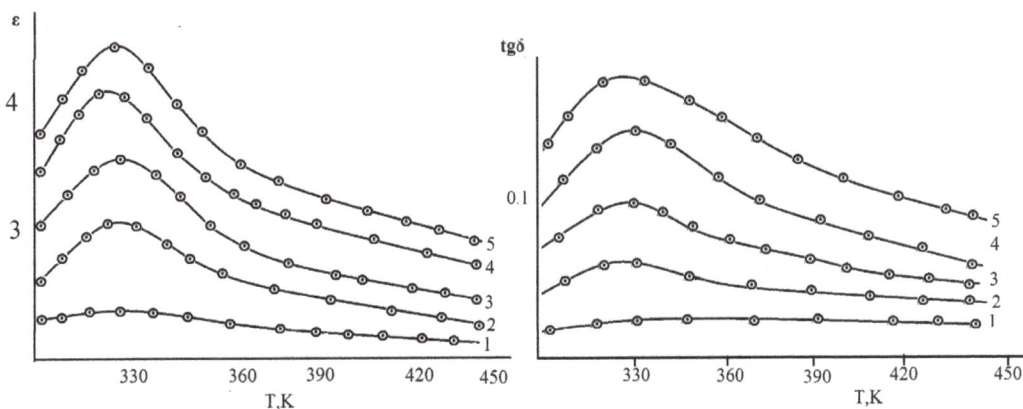

Figure 2. Temperature dependences of dielectric constant (ε) and dielectric loss of composites PE + x vol% TlGaSe$_2$, where 1-x = 0; 2-x = 3; 3-x = 5; 4-x = 7; 5-x = 10.

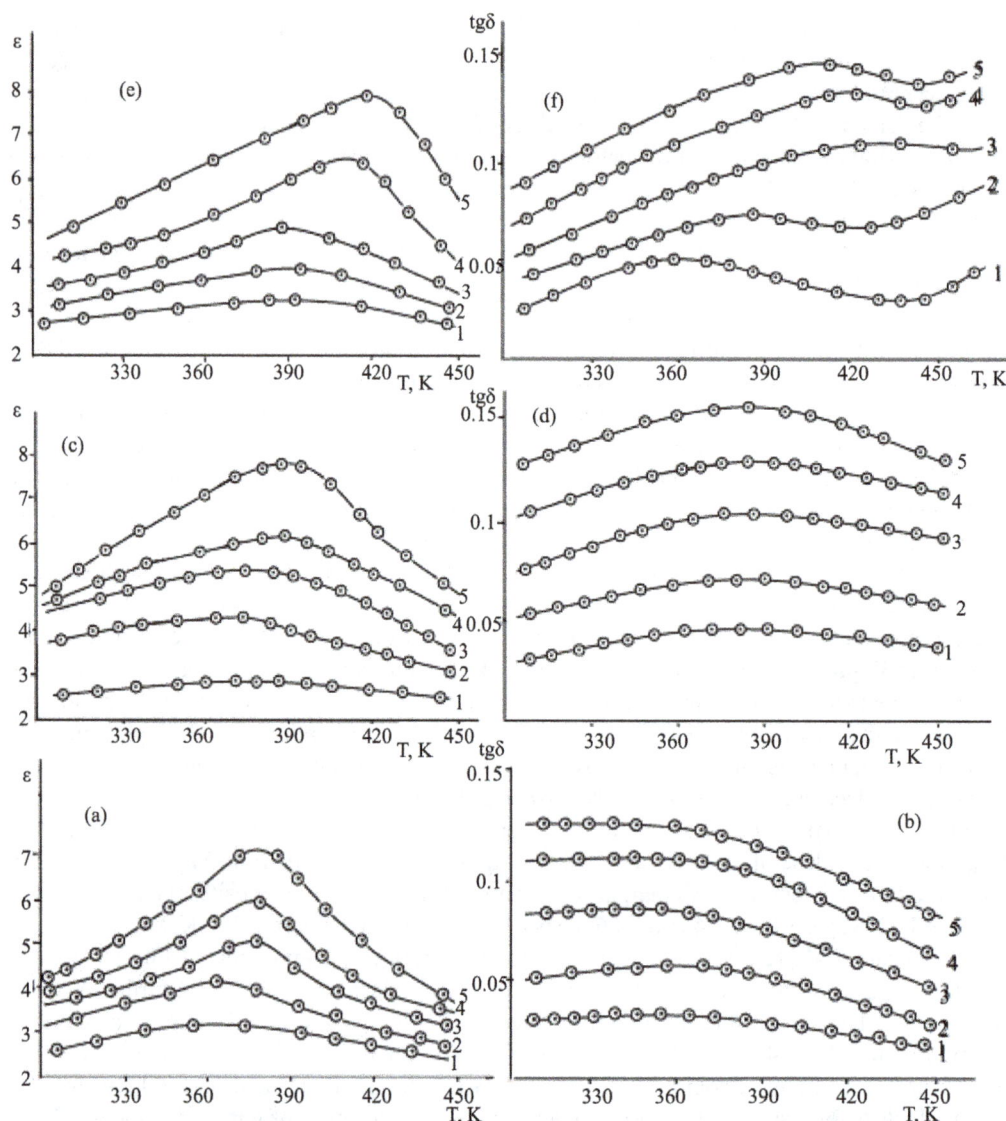

Figure 3. Temperature dependences of dielectic constant (ε) and dielectric loss of composites PE + x vol% TlGaSe$_2$, under irradiated 50 kGy (a) (b); 100 kGy (c) (d); and 150 kGy (e) (f) where 1-x = 0; 2-x = 3; 3-x = 5; 4-x = 7; 5-x = 10.

of dielectric loss of composites radiated at the dose 150 kGy is given in **Figure 3(f)**. As it follows from **Figure 3(f)**, the temperature dependence tg(T) of composites, irradiated at dose of 150 kGy is relatively complicated, *i.e.* at lower temperatures tanδ increases, at the certain interval it decreases, and at 420 K increases. Note that for the composites irradiated by γ-rays at dose of 150 kGy the value of tgδ increases due to increase of the filler content.

Analysis of the results of investigation of dielectric properties of irradiated composites PE + x vol% TlGaSe$_2$ shows that on the temperature dependence ε(T) before and after irradiation one single maximum is observed. Apparently, appearance of this maximum in connected with participation in polarization process deeper traps emerging after γ-radiation.

Effect of γ-radiation on composites PE + x vol% TlGaSe$_2$ reduces to significant alternation in the dependence ε(T) and tanδ (**Figure 2**). So, after γ-radiation of composites in the interval 50 - 150 kGy, at temperature dependences ε(T) one distinct maximum is observed. With increasing the dose of γ-radiation, the maximums in ε(T) dependences shift to area of higher temperatures. Probably, appearance of maximum is connected with participation of deeper traps in the polarization process, and shift of the maximums is the result of change of the character of interphase boundary layer parameters after irradiation effect.

Analysis of dependences tanδ before and after γ-irradiations (**Figure 2**, **Figure 3**) shows that effect of γ-irradiation on the composites PE + x vol% TlGaSe$_2$ reduces to significant change of dielectric losses. It is seen from tanδ dependences that at lower doses (50 kGy) of radiation, the observed decrease of tanδ is connected with the cross-linking processes that occur both in polymer matrix and in interphase layer of the polymer with TlGaSe$_2$ filler. In the samples radiated at doses of 100 kGy, the observed relative stabilization of the value of tanδ is the result of balancing of cross linking and destruction processes. In the samples irradiated with dose of 150 kGy the changes of tanδ are observed. Apparently, this is connected with prevalence of destruction processes.

Thus, changes in the dielectric characteristics of the composites based on polyethylene with TlGaSe$_2$ filler after γ-radiation is the result occurring in the electron-ion and polarization processes in the polymer matrix and the interface of the polymer with the TlGaSe$_2$ filler. These processes manifest themselves in increasing the concentration of low-energy surface states with increasing the filler content and the appearance of deep traps after γ-irradiation.

4. Conclusion

In this paper, we studied the effect of different doses of γ-irradiation on composites PE + x vol% TlGaSe$_2$. The temperature dependence of the dielectric constant and dielectric loss of composites revealed that by changing the temperature and fillers content and the radiation dose could be obtained composite material with desired dielectric parameters

References

[1]   Yabmekov, M.Y. and Gelman, A.B. (2011) Electret Properties of Polypropylene-Based Composite Materials. *International Scientific Mechanical Conference*, 14-17 November 2011, 78-80. (In Russian)

[2]   Gojayev, E.M., Maherramov, A.M., Zeynalovsh, A., Osmanova, S.S. and Allahyarov, E.A. (2010) Coronoelectrets Based on High Density Polyethylene Composites with Semiconductor Filler TlGaSe$_2$. *Electronnaya Obrabotka Materialov*, **46**, 91-96. (In Russian)

[3]   Mamedov, G.A., Gojayev, E.M., Magerramov, A.M. and Zeynalovsh, A. (2011) Investigation the Microrelyef of Surface by Atomic Force Microscop and Dielectric Properties of the Composition High Density Polyethylene with Additives TlGaSe$_2$. *Electronnaya Obrabotka Materialov*, **47**, 94-98.

[4]   Radwan, R.M., Aly, S.S. and El Aal, S.A. (2008) Preparation, Characterization and Effect of Electron Beam Irradiation on the Structure and Dielectric Properties of BatiO$_3$/PVDF Composite Films. *Journal of Radiation Research and Applied Sciences*, **11**, 9-16.

[5]   Shevchenko, V.G. (2010) Fundamentals of Physics of Polymer Composite Materials: Textbook/M. Moscow State University of M.V. Lomonosov, Moscow, 99. (In Russian)

[6]   Maharramov, A.M., Quliyev, M.M. and Ismayilova, R.S. (2012) The Effect of Gamma-Irradiation on the Dielectric Behavior in Polypropylene—CdS/ZnS Composites Section II. *Radiation Physics of Condensed Matter International Conference on Nuclear Science and Its Application*, Samarkand, 25-28 September 2012, 192-194.

[7]   Gafurov, U. and Fazilova, Z. (2007) Influence of Gamma-Irradiation on Dielectric Properties of Recycled Polyethylene Composites. *MRS Proceedings*, **1038**, 61-69.

[8]   Maharramov, A.M. (2001) Structural and Radiation Modification of Electret and Piezoelectret Properties of Polymer Composites. Baku, Elm, 327 p.

3

Dielectric Properties of Fishbone Kutum and Composite Materials with Its Participation

E. M. Gojaev*, Sh. V. Alieva, K. C. Gulmammadov, S. S. Osmanova

Azerbaijan Technical University, Baku, Azerbaijan
Email: *geldar-04@mail.ru

Abstract

This paper presents the results of studying the frequency dependence of permittivity and dielectric loss of fishbone (Fb) Kutum and polyethylene composites filled with nanoparticles fishbone Kutum and the influence of aluminum(Al) nanoparticles with dimensions 80-nm on the dielectric properties of polymer composites. Studies were carried out at the frequency range of 0 - 1 MHz. It was found that the variation of the volume of filler content of fishbone and aluminum nanoparticles might be prepared composite materials with the required dielectric parameters.

Keywords

Fishbone, Polymer Composites, the Dielectric Constant, the Dielectric Loss, PE + x vol% (Fb), PE + x vol% (Fb) + 1% Al

1. Introduction

It is known that the electroactive polymer materials are widely used in various technical fields, in particular in the electret microphone, radiation monitors, pressure sensors, air filters, and electromechanical transducer. Gradually, expanding the scope of application of the modified polymeric composite is of interest to the preparation of such materials with a particular combination of properties. To modify the properties of polymers, the dispersed fillers are added to the volume of the polymer. After that, the composite material becomes quite different electro physically, electret, strength, thermal and other properties [1]-[9]. The properties of the composites are largely determined by, among other parameters, the structural state of the surface and the intensity of intermolecular interactions between the matrix material and filler material and the size of the nanoparticles. Since the

---

*Corresponding author.

nanoparticles have a size less than 100 nm, their higher specific surface in comparison with fillers with larger particles allows significantly reduce the degree of filling composite. The transition to organic and nanosized filler, when optimizing the synthesis parameters, allows not only to reduce the specific consumption of this materials, but also to obtain materials with higher performance.

It is known that biological entities are interesting objects of study. The results show that the dielectric constant of tissues, cells and animal bones is very high at low frequencies and decreases with increasing frequency. Studies of the frequency dependence of the dielectric constant of objects of biological origin can determine such important physical constants as coefficients of absorption, reflection, refraction, and establish the scope of their practical application [10]-[12]. New materials obtained based on polymers with fillers of biological origin can certainly open up new scientific-practical possibilities. Selection of fish bones in this research is due to the fact that this stuff retains its structure, properties, and resistant to external influences for a long time. Therefore, for the purchase of high quality, stable to external factors, composites are used as filler particles from fish bones. Investigation of the influence of metal nanoparticles and with the change of the composition, size and volume content of the filler and nanoparticles allowed us to obtain new class of nanocomposite materials whose properties could be controlled and improved in the right direction. Thus, the receipt and investigation of the frequency dependence of the dielectric constant and dielectric loss of the composites Al + LDPE + Fb are the undoubted scientific practical interest.

2. Experimental Technique

Composite samples are obtained by mechanical mixing of fishbone Kutum powder with the PE powder in porcelain mortar. The mixing is continued until we have obtained a homogeneous mixture. The mixture is kept for some time at the melting temperature (443 K) of the polymer under pressure of 5 MPa. In the same temperature by compacting the homogeneous mixture and the pressure slowly increases to 15 MPa. At this temperature, the sample is maintained for 5 minutes and then is quickly cooled in water. The sizes of the samples are: thickness is about $80 - 120 \times 10^{-3}$ mm, diameter of the obtained samples is 35 mm. Provide reliable electric contact between the samples and electrodes made of stainless steel, the electrodes made of thin aluminum foil of thickness 7×10^{-3} mm was pressed to both working surfaces of the sample.

Block diagram of the setup is shown in **Figure 1**. Measurement of ε and $\tan\delta$ realized with alternating current frequency at 1 kHz. Sample (2) is kept between the two electrodes of the cell for measurement. Then the sample is heated in the cell via the heater (4) that was designed integrated in the laboratory with constant speed 2, 5 K/min. The temperature of sample is registered by the thermocouple (3) by the system that attaches the heater (4). Resistance of the sample is measured by the tera-ohmmeter (5) (E6 - 13 A), while dielectric constant and dielectric loss measuring by the bridge E-20. If necessary the temperature of sample is measured by the thermometer (6) (M-64). For the power supply of the heater used three-auto-regulated system (7).

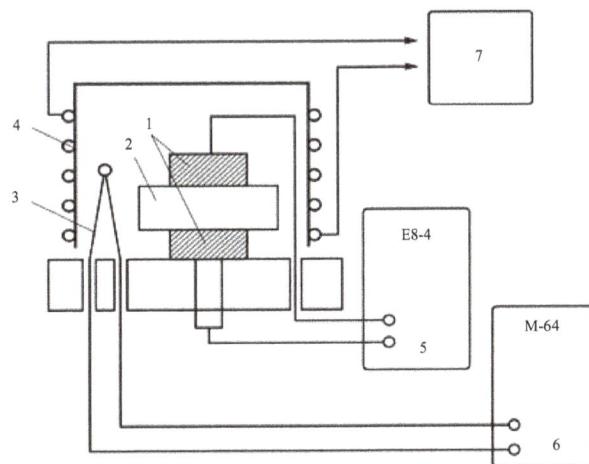

Figure 1. Block diagram of the apparatus for measuring temperature dependences of dielectric constant and dielectric losses: 1—measuring cell, 2—sample, 3—thermocouple, 4—heater, 5—immittance meter, 6—thermometer, 7—three-phase auto-regulated system.

3. Results and Discussion

The results of experimental studies of the frequency dependence of the dielectric constant and dielectric loss of fishbone and composites of the PE + x vol% (Fb) and PE + x vol% (Fb) + <Al> are shown in **Figures 2-4**.

The dependence of the dielectric constant and dielectric loss versus frequency for fish bones (Kutum) are shown in **Figure 2**. As shown in **Figure 2** at very low (0 - 1 kHz) frequencies place a significant decrease of in the dielectric constant from 34.72 to 10.97. In the frequency range 1 - 20 kHz revealed a deep minimum at the frequency of 9.6 and a maximum at the frequency of 10 kHz, and later, with change in the frequency to 10^3 kHz -ε remains constant. A similar variation of dielectric constant versus frequency detected for the second bone. The minimum is observed at a frequency of 2.8 kHz and a maximum at a frequency 5, 1 kHz. The dependence of the dielectric loss of fish bones were also investigated in the frequency range 0 - 1000 kHz. As follows from **Figure 2(a)** at low frequencies for both fishbone is a strong decrease. However, for one of bones at a frequency of 10 kHz is observed variance. In the future, with increasing frequency to 500 kHz comes moderate decrease in dielectric loss. Since the frequency of 500 kHz, an increase in the dielectric loses to 1000 kHz.

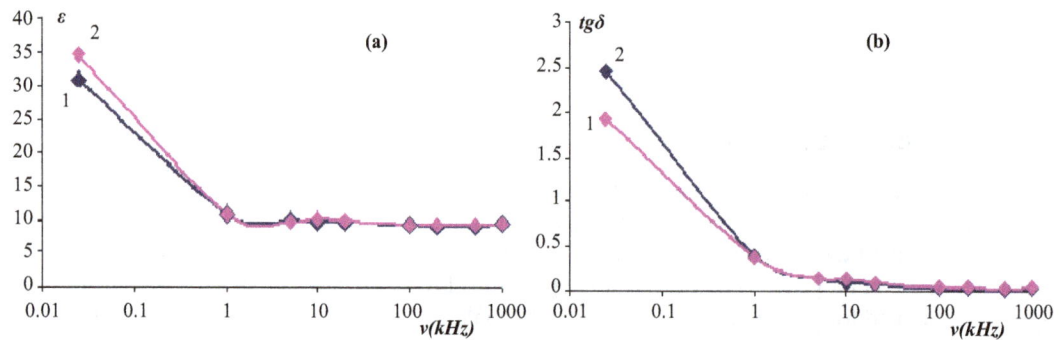

Figure 2. The dependence of the permittivity (a) and the dielectric loss (b) versus frequency for fishbone Kutum.

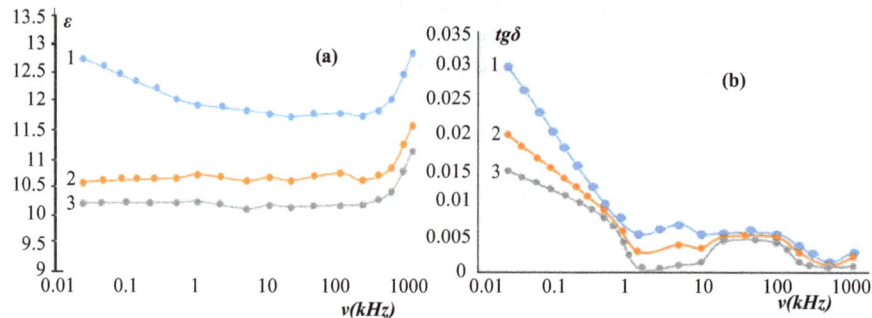

Figure 3. The dependence of the dielectric constant (a) and the dielectric loss (b) versus frequency for composites PE + x vol. % fishbone Kutumx = 3, 5, 7.

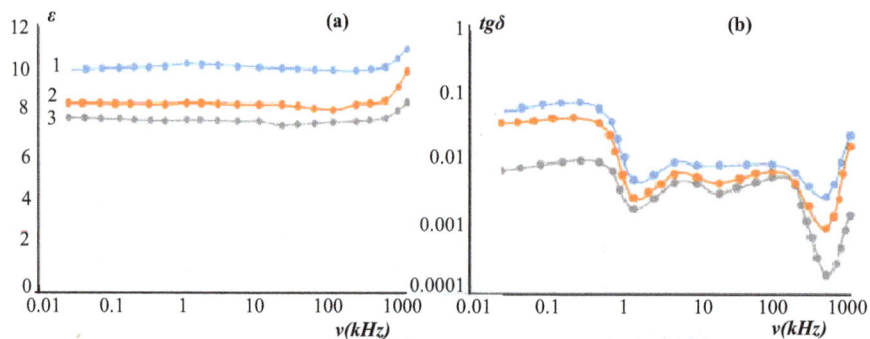

Figure 4. The dependence of the permittivity (a) and the dielectric loss (b) versus frequency for composites PE + vol. % fishbone Kutum + 1vol% Al, x = 3, 5, 7.

Studies were conducted in the frequency range 0 - 1000 kHz in composites with additions x = 3%, 5%, 7% (Fb). As follows from **Figure 3(a)** for composite PE + 3 vol% Fb in the investigated frequency range (0 - 20 kHz) dielectric constant decreases from 12.795 to 11.77. In the same frequency band to decrease ε composite additive PE + 5 vol% (Fb) does not change, ranging from 10.66 to 10.64, and for the composite with the addition 7 vol% (Fb) there was a slight decrease from 10.236 to ε 10.175. In the frequency range for all types of 20 - 100 kHz the composites there is a slight increase since for 3 vol% (Fb) from 11.77 to 11.83 to 10.64 5vol% (Fb) to 10.77, while 7 vol% (Fb), 17 to 10.21. In the frequency range of 100 - 200 kHz ε is reduced to about 3 vol% (Fb) 11.83 to 11.79, for a composite 5 vol% (Fb) of 10.77 to 10.64, and the composite of 7 vol% (Fb) from 10.205 to 10.216. Since the frequency of 200 kHz in the range 200 - 1000 kHz, the dielectric constant for all type of composites increases as the composite filled about 3 vol% (Fb) from 11.79 to 12.9 for the composite 5 vol% (Fb) 10.65 to 11.62, and the composite of 7 vol% (Fb) 10.216 to 11.16, respectively (Curve 3).

Results of the study of the frequency dependence of the dielectric loss (tanδ) are shown in **Figure 3(b)**. As the 3b for composite additive 3 vol% (Fb) in the frequency range (0 - 1 kHz), a decrease tanδ, and in the frequency range of 1 - 10 kHz tanδ it is from 0.0055 to 0.007. With a further increase in the frequency of 10 - 100 kHz tanδ it remains practically constant, and in the frequency range of 100 - 500 kHz, a decrease of 0.0056 - 0.0015, in the frequency range 500 - 1000 kHz increase from 0.0015 to 0.003.

For composite PE + 5 vol% (Fb), the frequency dependence tanδ similarly changed. In the frequency range of 0 - 10 kHz is reduced from 0.02 to 0.0035, in the frequency range of 10 - 20 kHz, the dielectric loss (tanδ) increases from 0.0035 to 0.0051, and in the frequency range 20 - 100 kHz tanδ remains constant. With further increase in the frequency 100 - 500 kHz is reduced from 0.0051 to 0.001, and in the frequency range 500 - 1000 kHz increases from 0.001 to 0.0022. For composite PE + 7 vol% (Fb), the frequency dependence tanδ similarly changed. In the frequency range of 0 - 1 kHz is reduced from 0.0148 to 0.00041, in the frequency range of 1 - 10 kHz, a dielectric loss (tanδ) increases from 0.00041 to 0.0015, and in the frequency range of 10 - 20 kHz is increased by tanδ 0.0015 to 0.0045 kHz. In the frequency range 20 - 100 kHz tanδ remains constant. With a further increase in the frequency of 100 - 500 kHz decreases from 0.0045 to 0.0015, and in the frequency range 500 - 1000kHz increased from 0.00015 to 0,0009.

We investigated the frequency characteristics of the dielectric constant and dielectric loss of nanocomposites with the general formula PE + x vol% (Fb) + <Al>. Results of the study of the frequency dependence of the dielectric constant of these composites are shown in **Figure 4(a)**. As follows from the composite 3 vol% (Fb) + 1% Al dielectric permittivity in the frequency range 0 - 500 kHz increased from 10.078 to 10.236, and the range 500 - 1000 kHz ε increases from 10.236 to 11.055 (Curve 1). For a composite with 5 vol% (Fb) + 1% Al in the frequency range 0 - 500 kHz ε increases from 8.54 to 8.68 and a frequency range 500 - 1000 kHz from 8.68 to 10.03 (curve 2). A similar change is observed for the composite with the addition of 7 vol% (Fb) +1% Al. For this composite in a wide frequency range ε increases slightly from 7.88 to 7.89. In range 500 - 1000 kHz increases from 7.89 to 8.6 (curve 3).

Results of the study of the frequency dependence of the dielectric loss of the composites PE + vol% (Fb) + 1% <Al> are shown in **Figure 4(b)**. As follows from **Figure 3(b)** for the composite with the addition of 3 vol% (Fb) + 1% Al in the frequency range 0 - 0.5 kHz tanδ increased from 0.06 to 0.066 0.5 - 2 kHz sharply decreases from 0.066 to 0.005. In the frequency range 2 - 5 kHz increases from 0.0055 to 0.01, 5 - 20 kHz tanδ reduced from 0.01 to 0.009, in the frequency range 20 - 100 kHz slightly increased from 0.009 to 0.0094, and in the frequency range 100-tanδ is reduced from 500 kHz to 0.0094, 0.00316, and a frequency range 500 - 1000 kHz increased from 0.00316 to 0.026 (curve 1). For the composite with 5 vol% (Fb) + 1% Al in the frequency range of 0 - 0.5 kHz tanδ increases from 0.04 to 0.045, and in the frequency range 0.5 - 2 kHz sharply decreases from 0.045 to 0.003. In the frequency range 2 - 5 kHz increased from 0.003 to 0.0068, and in the range 5 - 20 kHz tanδ decreases from 0.0068 to 0.005, in the frequency range 20 - 100 kHz slightly increased from 0.005 to 0.0073, and in the frequency range 100 - 500 kHz tanδ decreases from 0.0073 to 0.001, and in the frequency range 500 - 1000 kHz increases from 0.001 to 0.018 (curve 2). For the composite with additives 7 vol% (Fb) + 1% Al in the frequency range of 0 - 0.5 kHz tanδ increases from 0.0074 to 0.01, and in the frequency range 0.5 - 2 kHz sharply reduced from 0.01 to 0.002. In the frequency range 2 - 5 kHz increases from 0.002 to 0.005, and in the range 5 - 20 kHz tanδ reduced from 0.005 to 0.0035 in the frequency range 20 - 100 kHz slightly increases from 0.0035 to 0.0046 in the frequency range 100 - 500 kHz tanδ decreases from 0.0046 to 0.0002, and in the frequency range 500 - 1000 kHz increases from 0.0002 to 0.016 (curve 3).

Increasing the volume content of the filler leads to an increase in the number of dust particles fishbone Kutum

(Fb) in the total thickness of the sample, which can be regarded as a resistance connected between the electrodes. Since the particles (Fb) are highly compared with PE conductive, it can be assumed that the resistance of the composite will be mainly determined by the contact between the particles (Fb). At the boundaries of clusters (clusters are surrounded by thin layers of polyethylene, having a low ε in an alternating electric field of the accumulation and redistribution of free electric charges, which distort the original internal electric field. At low frequencies, the internal electric fields are distributed accordingly conductivity. Therefore, the change in dielectric parameters frequency increases can be explained by the emergence of a relatively strongly embedded in the filler clusters [12].

4. Conclusion

The researches of the frequency dependence of the dielectric constant and dielectric loss of the composites LDPE + Fb and nanocomposites LDPE + Fb + Al in the frequency range 0 - 1000 kHz revealed that a change in volume content of fish bones and aluminum nanoparticles could determine their optimal parameters, allowing to solve specific technical problems.

References

[1]   Shik, A., Rudo, H. and Sargent, E.H. (2000) Photoelectric Phenomena in Polymer-Based Composites. *Journal of Applied Physics*, **88**, 3448-3453. http://dx.doi.org/10.1063/1.1289228

[2]   Tashilkov, A.M., Hasanli, Sh.M. and Bayramov, H.B. (2007) The Nonlinear Resistor Composites Based on the Polymer-Ceramic. *JSF*, **77**, 127-130.

[3]   Aleksandrov, E.L., Lebedev, E.A., Konstantinova, N.N. and Aleshin, A.N. (2010) The Effect of Switching in Composite Films Based on a Conjugated Polymer Polyfluorene and Nanoparticulate ZnO. *Fiz*, **52**, 393-396.

[4]   Gojayev, E.M., Nabiyev, N.S., Zeynalov, Sh.A. and Osmanov, S. (2013) A Study of Fluorescence Spectra and Dielectric Properties of Composites PEHD + x vol TlGaSe$_2$. *Electronic Processing of Materials*, **3**, 14-18.

[5]   Yablokov, M.Y., Gilman, A.B. and Ozerin, A.N. (2011) Electronic Properties of Nanocomposite Materials Based on Polypropylene. *Nanotechnics*, **2**, 86-88.

[6]   Hippel, A.R. (1960) Dielectrics and Waves. M., K III, 351.

[7]   Muradyan, V.E., Sokolov, E.A. Babenko, S.D. and Morsevsky, A.P. (2010) The Dielectric Properties of the Composites Modified Carbon Nanostructures, in the Microwave Range. *Technical Physics*, **80**, 83-87.

[8]   Gojayev, E.M., Maharramov, A.M., Safarova, S.I., Nuriyev, M.M. and Ragimov, R.S. (2008) The Dielectric Properties of Polymer Composites with a Semiconductor Filler TlInSe$_2$. *Electronic Processing of Materials*, **44**. 66-71.

[9]   Tupik, A.B. and Garmashov, S.I. (2011) Dielectric Loss in Statistical Mixtures. *FTT*, **53**, 1129-1132.

[10]   Shrivastava, B.D., Barde, R., Mishra, A. and Phadke, S. (2014) Dielectric Behavior of Biomaterials at Different Frequencies on Room Temperature International Conference on Recent Trends in Physics (ICRTP 2014). *Journal of Physics: Conference Series*, **534**, Article ID: 012063. http://dx.doi.org/10.1088/1742-6596/534/1/012063

[11]   Rizhenkov, A.V., Klassen, N.V. and Masalit, V.M. (2013) Features of the Structure and Properties of Composites of Biopolymers with Inorganic Nanoparticles. *Science and Technology of Materials*, **53**.

[12]   Guliyev, M.M. and Ismayilova, R.S. (2012) Influence of Inorganic Filler on the Properties of High Density Polyethylene. *Plastic Weight*, No. 4, 10-13.

4

Studies on AC Electrical Conductivity and Dielectric Properties of PVA/NH$_4$NO$_3$ Solid Polymer Electrolyte Films

Alabur Manjunath*, Tegginakeri Deepa, Naraganahalli Karibasappa Supreetha, Mohammed Irfan

Department of P.G. Studies in Physics, Government Science College, Chitradurga, India
Email: *manjugsc@yahoo.com

Abstract

Solid polymer electrolytes have been extensively studied due to wide applications in various electrochemical devices [1]-[3]. Most of the solid polymer electrolytes consist of polymer as a host material to provide strength and good mechanical stability and an inorganic salt that supplies ionic carriers to cause electrical conductivity. In our studies, we prepared PVA/NH$_4$NO$_3$ polymer electrolyte films by solution casting method. The prepared films with varied Ammonium Nitrate concentration from 0 - 20 wt% are characterized by XRD & FTIR spectroscopy. XRD results show that amorphous nature increases as the amount of the Ammonium Nitrate salt in PVA is increased. IR-spectra confirm the polymer salt complexes in the range of 3700 - 712 cm^{-1}. Conductance analysis reveals that polymer electrolyte films containing 20 wt% of NH$_4$NO$_3$ exhibit the highest ionic conductivity of 1.01×10^{-7} S/cm while pure PVA films give the lowest ionic conductivity of 2.10×10^{-11} S/cm. It was evident from this study that the increase of ionic conductivity depended on the Ammonium Nitrate salt concentration. The dielectric constant exponentially decreases with increase of frequency for pure PVA and NH$_4$NO$_3$ doped PVA film composites. The temperature dependent studies of AC conductivity and dielectric constant also included to understand the conducting property. The results obtained by these studies are reported in this work.

Keywords

Polymer Electrolytes, XRD, FTIR, Electrical Conductivity

1. Introduction

Polymers form a very important class of materials without which life seems very difficult. Polymeric materials

---

*Corresponding author.

also have better mechanical properties for the construction of all practical solid-state electrochemical cells. Generally, the addition of inorganic salts into a polymer matrix can improve its conductivity. PVA is one of the most important polymer materials and it has many applications in industry, it exhibits good film forming property and enhances the conductivity. NH_4NO_3 is a white crystalline solid at room temperature and pressure. Commonly, used in agriculture as fertilizer. The present study is focused on the preparation and characterization of PVA-NH_4NO_3 polymer electrolyte films [4]-[6].

2. Experimental Part

2.1. Materials and Preparation of Polymer Electrolyte Films

The chemicals used for the preparation are AR grade Polyvinyl alcohol (PVA) and Ammonium Nitrate (NH_4NO_3). Different compositions of Polyvinyl alcohol (PVA) and Ammonium Nitrate (NH_4NO_3) films have been prepared by solution casting method , using different weight ratios of NH_4NO_3 (0, 5, 10, 15, 20) wt%. The solution of PVA andNH_4NO_3 is obtained by dissolving them in distilled water at 350 K, and the solution is stirred well using magnetic stirrer for about one hour, until highly homogenous polymer solution was formed. These homogenous solutions were casted in a glass dish (diameter of 5 cm). The whole assembly was placed in a dust free chamber and the solvent was allowed to evaporate slowly in open air at room temperature for a week and films are peeled off from the glass dish. The thicknesses of the films were in the range of (0.03 - 0.18) mm [7].

2.2. Instrumentation

In order to investigate the structure of polymer electrolytes, XRD studies of the films were carried out with an instrument RigakuMiniflex II X-ray diffractometer with CuK$_\alpha$ radiation of $\lambda = 1.5406$ Å in the range of $2\theta = 5°$ to $30°$. The FTIR spectrum has been recorded by using IRspectraphotometer in the range of 500 - 4000 cm^{-1} at a resolution of 4 cm^{-1}.

3. Results and Discussion

3.1. XRD Analysis

Figure 1 shows the XRD pattern of NH_4NO_3 doped polymer electrolyte. A broad peak around 19.29° is observed for pure PVA and has been found to be shifted in the complex systems. It is also observed for different concentrations of NH_4NO_3 added polymer films. The increase in the broadness of the peak reveals the amorphous nature of the complexed system. Peaks corresponding to NH_4NO_3 have been found to be absent in the complexes indicating the complete disassociation of salt in polymer matrix. Thus XRD analysis reveals the complex formation between the polymer and the salt [8]-[12].

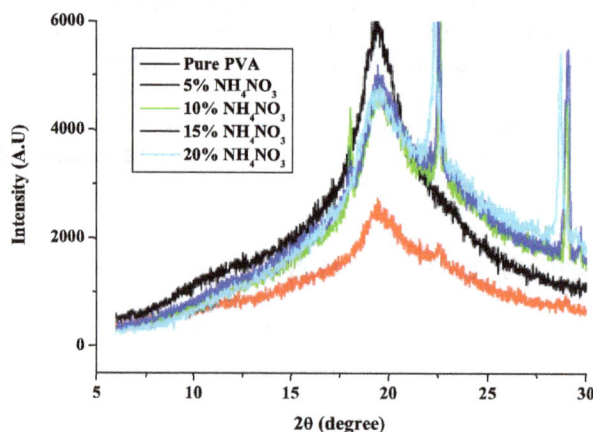

Figure 1. XRD Spectra of pure PVA and PVA/NH_4NO_3 polymer electrolyte films.

3.2. FTIR Studies

FTIR absorption spectra of pure PVA film and with different concentrations of NH_4NO_3 are shown in **Figure 2**.

The broad band observed between 3287.48 cm^{-1} - 3216.57 cm^{-1} are refers to the intermolecular hydrogen bonding and O-H stretching vibration (region I). The vibrational band observed between 2939.56 cm^{-1} - 3054.05 cm^{-1} is associated with C-H stretching from alkyl groups (region II), and the absorption peaks observed between 1731.07 cm^{-1} - 1753.91 cm^{-1} (region III) are due to the stretching C=O and C-O from acetate group.

The C-H stretching band of pure PVA has been shifted to higher wave number in doped polymer electrolytes [13].

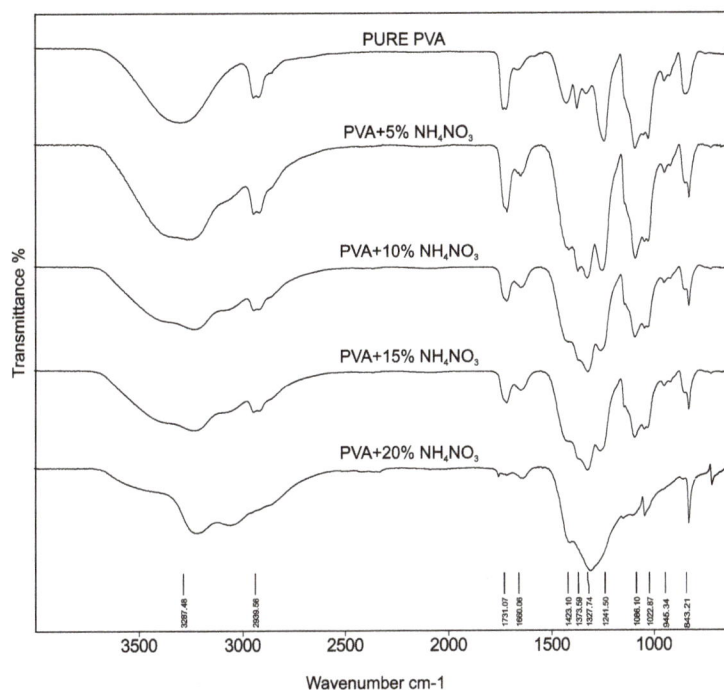

Figure 2. FTIR Spectra of pure PVA and PVA/NH₄NO₃ polymer electrolyte films.

Following **Table 1** shows the peak assignments of pure PVA and various concentrations of NH_4NO_3 polymer films.

Table 1. Values of peaks, peak assignments of pure PVA and various concentrations of NH_4NO_3 polymer films.

Peaks (cm^{-1})	Groups	Peak assignments
3500 - 3200	Alcohols, phenols	O0 O-HStretch, H-bonded
3000 - 2800	Alkanes	C-H (stretching mode)
1760 - 1690	Carboxylic acids	C=O stretch
1660 - 1633	Alkenes	–C=C– stretch

3.3. AC Conductivity Studies

Complex impedance spectroscopy gives information on electrical properties of materials and their interface with electronically conducting electrodes. The Solid Polymer Electrolyte (SPE) films were cut into small discs (2 mm diameter) and sandwiched between two stainless steel electrodes under spring pressure. The AC conductivity of these films was measured in the frequency range from 50 Hz to 5 MHz using the HIOKI 3532-50 LCR Hi-tester which was interfaced to a computer. Measurements were also made at temperatures ranging from 305K to 373K. The measured conductance $G(\omega)$ from 50 Hz to 5 MHz was used to calculate the AC conductivity of the sample

using the relation,

$$\sigma = \frac{G(\omega)d}{A}$$

where, $G(\omega)$ is the measured conductance, A is the area of the sample and d is the thickness of the sample.

The variation of AC conductivity as a function of frequency for PVA composition of (PVA + NH$_4$NO$_3$) polymer electrolyte is shown in **Figure 3**. From **Figure 3**, it is observed that conductivity increases with increase in frequency. The increase in AC conductivity is due to increase in the composition of the salt in polymer matrix resulting in relatively more number of free ions. This will increase the mobile charge carriers as observed in **Table 2**. These charge carriers move in the amorphous polymer matrix and hence the conductivity increases. Thus there is a relation between the amorphous nature of the polymer film and the conductivity. In general, conductivity increases as the degree of crystallinity decreases, as observed above, which is the compliment of increase in amorphous nature [14].

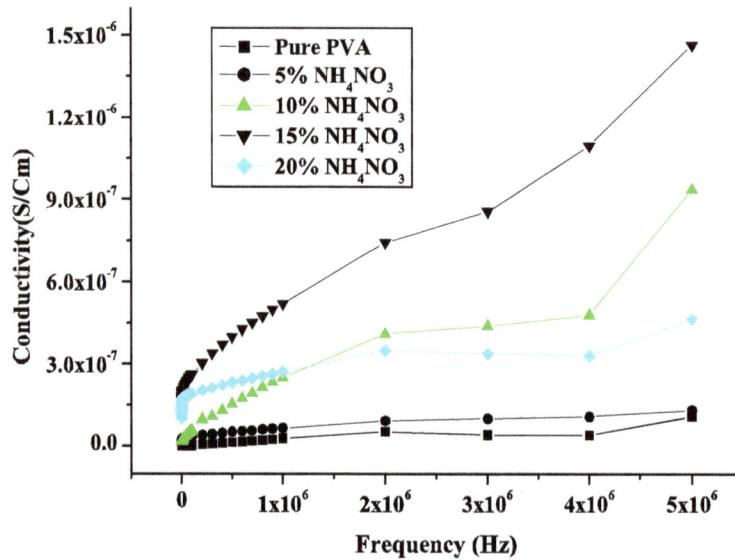

Figure 3. Variation of conductivity with frequency of pure PVA and NH$_4$NO$_3$ doped PVA polymer electrolyte films.

Table 2. Conductivity of pure PVA and NH$_4$NO$_3$doped PVA polymer electrolyte films.

Sample	Conductivity (S·cm^{-1})
Pure PVA	2.1E−10
PVA + 5%NH$_4$NO$_3$	1.95E−8
PVA + 10%NH$_4$NO$_3$	2.51E−8
PVA + 15%NH$_4$NO$_3$	1.46E−7
PVA + 20%NH$_4$NO$_3$	1.01E−7

3.4. Dielectric Studies

Figure 4 shows the variation of dielectric constant (ε^{l}) with frequency for different PVA concentration at room temperature. The permittivity (ε^{l}) was calculated using the relation,

$$\varepsilon^{l} = \frac{Cd}{A\varepsilon_o}$$

where, d is the thickness of the polymer film, A is the area of the electrolyte film, C is the capacitance of the cell with sample and $\varepsilon_o = 8.85 \times 10^{-12}$ F/m is the permittivity of free space.

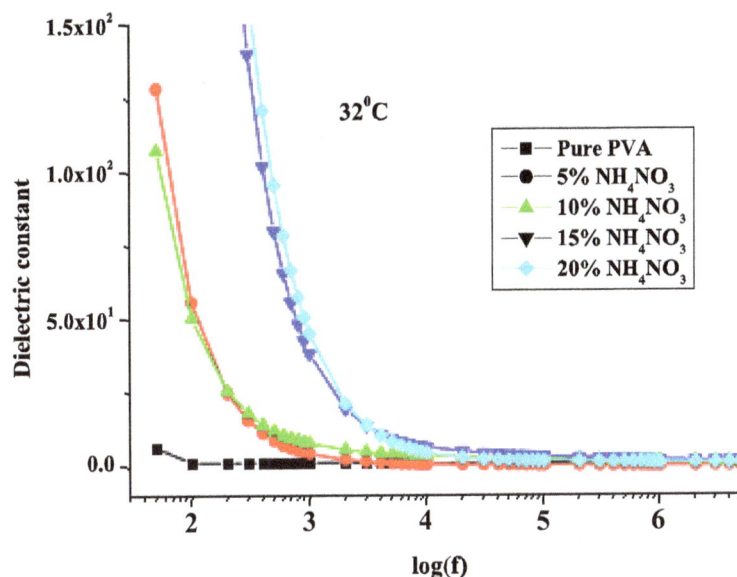

Figure 4. Variation of Dielectric constant with frequency of pure PVA and NH$_4$NO$_3$ doped PVA polymer electrolyte films.

It is observed from the **Figure 4** that $\varepsilon^1(\omega)$ decreases with increase infrequency and attain a constant value at higher frequency range similar to that for polar materials [15] [16]. The initial value of permittivity is high, but with increasing the frequency, this value begins to decrease, which could be due to the dipole not being able to follow the field variation at higher frequencies [17] and due to polarization effects [18]. Further in the case of crystalline and semi-crystalline polymeric materials, the crystalline phase dissolves progressively into amorphous phase with increase in temperature. This in turn influences the polymer dynamics and thus the dielectric behavior. The low frequency region appears due to the contribution of charge accumulation at the electrode-electrolyte interface. At higher frequencies, the periodic reversal of the electric field occurs so fast that there is no excess ion diffusion in the direction of the field. It has been observed that dielectric permittivity decreases with increasing frequency in all the samples of PVA based polymer electrolyte films.

4. Conclusions

1) PVA based polymer electrolyte films with different concentrations of Ammonium Nitrate has been prepared by solution casting technique. The obtained film composites are **solid, glassy to touch, transparent materials in appearance** as shown in **Figure 5**.

2) The temperature dependence **AC conductivity** implies that for pure PVA and NH$_4$NO$_3$ doped PVA film composites, the **AC conductivity increases with increase in frequency**. It may be due to the **increase of the mobility of charge carriers** in the composite film. As the temperature increases from **40°C to 70°C in steps of 10°C**, it is observed that **Dielectric constant exponentially decreases with increase of frequency** as shown in **Figure 6**.

3) In XRD, a broad peak around **19.29°** is observed for pure PVA and has been found to be shifted in the complex systems. It is also observed that for different concentrations of NH$_4$NO$_3$ added polymer films, **the intensity of the peak decreases, and the full width at half maximum increases**. This increase in the broadness of the peak reveals the **amorphous nature** of the complexed system. Peaks corresponding to NH$_4$NO$_3$ have been found to be absent in the complexes indicating the **complete disassociation of salt in polymer matrix**. Thus XRD analysis reveals the complex formation between the polymer and the salt.

FTIR Studies of pure PVA and NH$_4$NO$_3$ doped PVA exhibit formation of several bonds. The **O-H (hydroxyl) stretching bond** is the most characteristic stretching indicates the presence of hydroxyl group IR bond of alcohols. The free vibrations occur as a sharp peak at **3287.48 cm^{-1}** in pure PVA, this peak is broadened in the doped PVA at **3264.70 cm^{-1}**, which is due to **hydrogen bond formation**. The wave number at **2939.56 cm^{-1}** indicates an asymmetric stretching of **C-H group** which is shifted to **2942.22 cm^{-1}** for NH$_4$NO$_3$ doped PVA film compo-

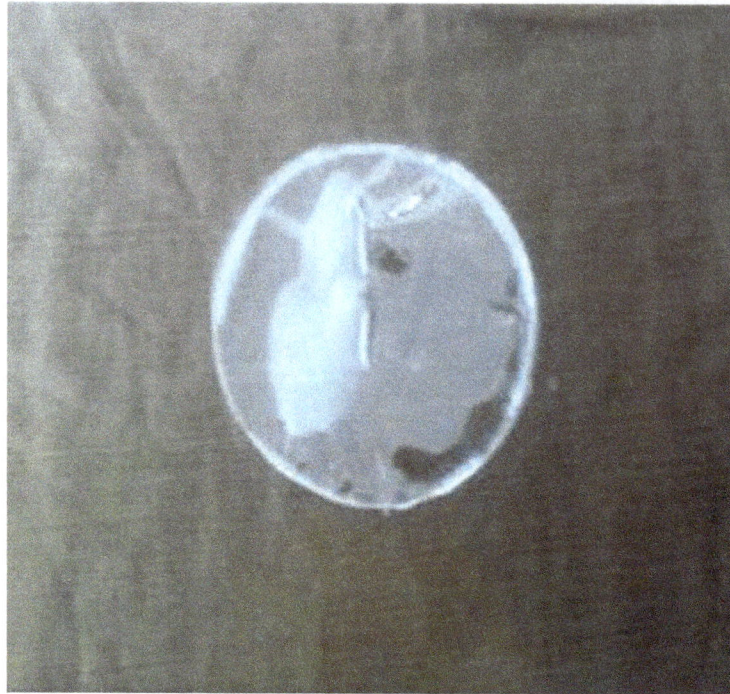

Figure 5. PVA + 5% NH_4NO_3.

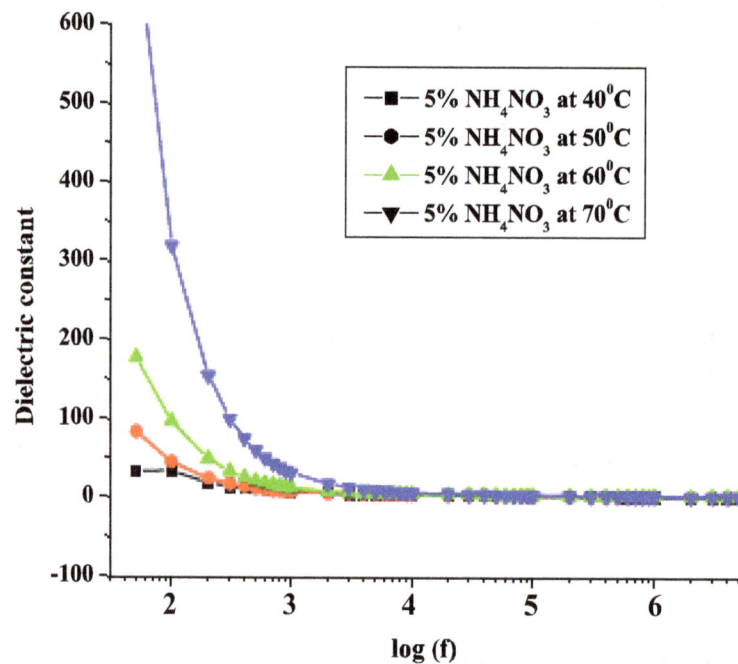

Figure 6. Temperature dependent Dielectric constant with frequency for PVA/ 5% NH_4NO_3 doped polymer electrolyte films.

site. An absorption peak in pure PVA at **1731.07 cm⁻¹** attributes to the **C=O stretching mode** and remains unaltered for the doped samples also. Similarly the absorption peak observed in pure PVA at **1660.06 cm⁻¹** has been assigned to **C-O in stretching mode** in which it is shifted to **1633.97 cm⁻¹** in doped samples. The FTIR studies indicate that NH_4NO_3 doped polymer film composites forms ionic bonds, molecular bonds weak anti bonding which may be indicated that **unstable and stable complex structure**.

References

[1]   Qiao, J., Fu, J., Lin, R., Ma, J. and Liu, J. (2010) Alkaline Solid Polymer Electrolyte Membranes Based on Structurally Modified PVA/PVP with Improved Alkali Stability. *Polymer*, **51**, 4850-4859. http://dx.doi.org/10.1016/j.polymer.2010.08.018

[2]   Hema, M., Selvasekerapandian, S., Hirankumar, G., Sakunthala, A., Arunkumar, D. and Nithya, H. (2009) Structural and Thermal Studies of PVA: NH_4I. *Journal of Physics and Chemistry of Solids*, **70**, 1098-1103. http://dx.doi.org/10.1016/j.jpcs.2009.06.005

[3]   Benedict, T.J., Banumathi, S., Veluchamy, A., Gangadharan, R., Ahamad, A.Z. and Rajendran, S. (1998) Characterization of Plasticized Solid Polymer Electrolyte by XRD and AC Impedance Methods. *Journal of Power Sources*, **75**, 171-174. http://dx.doi.org/10.1016/S0378-7753(98)00063-9

[4]   Leones, R., Sentanin, F., Rodrigues, L.C., Marrucho, I.M., Esperança, J.M.S.S., Pawlicka, A. and Silva, M.M. (2012) Investigation of Polymer Electrolytes Based on Agar and Ionic Liquids. *Polymer Letters*, **6**, 1007-1016. http://dx.doi.org/10.3144/expresspolymlett.2012.106

[5]   Selvasekarapandian, S., Hema, M., Kawamura, J., Kamishima, O. and Baskaran, R. (2010) Characterization of PVA—NH_4NO_3 Polymer Electrolyte and Its Application in Rechargeable Proton Battery. *Journal of the Physical Society of Japan*, **79**, 163-168. http://dx.doi.org/10.1143/JPSJS.79SA.163

[6]   Kadir, M.F.Z., Aspanut, Z., Majid, S.R. and Arof, A.K. (2010) FTIR Studies of Plasticized Poly(Vinyl Alcohol)-Chitosanblend Doped with NH_4NO_3 Polymer Electrolyte Membrane. *Spectrochimica Acta Part A*, **78**, 1068-1074. http://dx.doi:10.1016/j.saa.2010.12.051

[7]   Abdullah, O.Gh., Aziz, B.K. and Hussen, S.A. (2013) Optical Characterization of Polyvinyl Alcohol—Ammonium Nitrate Polymer Electrolytes Films. *Chemistry and Materials Research*, **3**. http://www.iiste.org/Journals/index.php/CMR/article/view/7045

[8]   Stephen, A.M., Saito, Y., Muniyandi, N., Ranganathan, N.G., Kalyanasundaram, S. and Elizabeth, R.N. (2002) *Solid State Ionics*, **148**, 467. http://dx.doi.org/10.1016/S0167-2738(02)00089-9

[9]   Fernandez, A., Torrecilla, J.S., Garcia, J. and Rodriguez, F. (2007) Thermophysical Properties of 1-Ethyl-3-methylimidazolium Ethylsulfate and 1-Butyl-3-methylimidazolium Methylsulfate Ionic Liquids. *Journal of Chemical & Engineering Data*, **52**, 1979-1983. http://pubs.acs.org/doi/abs/10.1021/je7002786 http://dx.doi.org/10.1021/je7002786

[10] Singh, M.P., Singh, R.K. and Chandra, S. (2010) Effect of Ultrasonic Irradiation on Preparation and Properties of Ionogels. *Journal of Physics D: Applied Physics*, **43**, 4.

[11] Bhargav, P.B., Mohan, V.M., Sharma, A.K. and Rao, V.V.R.N. (2009) Investigations on Electrical Properties of (PVA:NaF) Polymer Electrolytes for Electrochemical Cell Applications. *Current Applied Physics*, **9**, 165-171. http://dx.doi.org/10.1016/j.cap.2008.01.006

[12] Kurumova, M., Lopez, D., Benavente, R., Mijangos, C. and Pevena, J.M. (2000) Effect of Crosslinking on the Mechanical and Thermal Properties of Poly(Vinyl Alcohol). *Polymer*, **41**, 9265-9272. http://dx.doi.org/10.1016/S0032-3861(00)00287-1

[13] El-Hefian, E.A., Nasef, M.M. and Yahaya, A.H. (2010) The Preparation and Characterization of Chitosan Poly(vinyl alcohol) Blended Films. *E-Journal of Chemistry*, **7**, 1212-1219. http://dx.doi.org/10.1155/2010/626235

[14] Hou, W.H., Chen, C.Y. and Wang, C.C. (2004) Conductivity, DSC, and Solid-State NMR Studies of Comb-Like Polymer Electrolyte with a Chelating Functional Group. *Solid State Ionics*, **166**, 397-405. http://www.sciencedirect.com/science/article/pii/S0167273803005307 http://dx.doi.org/10.1016/j.ssi.2003.09.021

[15] Ramesh, S. and Chai, M.F. (2007) Conductivity, Dielectric Behavior and FTIR Studies of High Molecular Weight Poly(vinylchloride)—Lithium Triflate Polymer Electrolytes. *Materials Science and Engineering: B*, **139**, 240-245. https://scholar.google.co.in/citations?view_op=view_citation&hl=en&user=Xx64OREAAAAJ&cstart=140&sortby=pubdate&citation_for_view=Xx64OREAAAAJ:blknAaTinKkC http://dx.doi.org/10.1016/j.mseb.2007.03.003

[16] Ramya, C.S., Savitha, T., Selvasekharapandian, S. and Hiran Kumar, G. (2005) Transport Mechanism of Cu-Ion Conducting PVA Based Solid-Polymer Electrolyte. *Ionics*, **11**, 436-441. http://dx.doi.org/10.1007/BF02430262

[17] Ramesh, S., Yahana, A.H. and Arof, A.K. (2002) Dielectric Behaviour of PVC-Based Polymer Electrolytes. *Solid State Ionics*, **152-153**, 291-294. http://dx.doi.org/10.1016/S0167-2738(02)00311-9

[18] Tareev, B. (1979) Physics of Dielectric Materials. MIR Publications, Moscow.

5

Correlations of Materials Surface Properties with Biological Responses

Robert E. Baier

Industry/University Center for Biosurfaces, 110 Parker Hall, State University of New York at Buffalo, Buffalo, NY, USA
Email: baier@buffalo.edu

Abstract

More than 50 years have passed since it was first recognized that the surface properties, and predominantly the surface energies of materials controlled their interactions with all biological phases via their spontaneous acquisition of proteinaceous "conditioning films" of differing degrees of denaturation but usually of the same substances within any given system. This led to the understanding that useful engineering control of such interactions could thus be manifested through adjustments to those surface properties, giving significant control and utility to the biomaterials developer without requiring detailed discovery of the biological specifications of the components involved. Thus, effective selection of adhesive versus abhesive (non-stick, non-retention) outcomes for such useful appliances as dental implants versus substitute blood vessels, or water-resistant bonded structures versus clean, nontoxic ship bottoms is now facilitated with little biological background required. A historical overview is presented, followed by a brief survey of the forces involved and most useful analyses applied. Utility for blood-contacting materials is described in contrast to utility for bone- and tissue-contacting materials, demonstrating practical uses in controlling cell-surface interactions and preventing biofouling. New research directions being explored are noted, urging applications of this prior knowledge to replace the use of toxicants.

Keywords

Surface Property, Biological Response, Adhesive, Blood-Contacting Materials

1. Introduction

Biological responses to non-physiological surfaces are, usually, mediated by spontaneous deposits of organic films and particulate matter from all biological fluids. The earliest events follow a common pattern. Such interactions result in differing degrees of bioadhesion and can be effectively correlated usually, even controlled by the surface properties, (especially surface energies) of the substrata involved [1].

There are convincing proofs of the bioengineering utility of surface property modification to minimize biological fouling [2]. Such proofs are found across the broad range of biomaterials development, from the successful implantation of substitute human blood vessels and total artificial hearts, to the fabrication of entirely nontoxic fouling-release coatings that can replace poison paints on ship bottoms. Surfaces contacting blood have received the most careful scrutiny, and first revealed the general features of biological response later found in other circumstances [3], like milk fouling of pasteurizers [4], oceanic fouling of heat exchangers [5], cell culture propagation [6] and dental plaque formation [7]. Control of the surface properties of biomaterials is a unifying approach to control of interactions between all that which is alive and all that which is not.

2. Historical Overview

It has been a successful strategy to tailor the surface properties of biomedical organ substitutes to exhibit desired degrees of biological response. As an example, general approval of the US Food and Drug Administration was granted as long ago as January, 1979 for human peripheral vascular reconstruction using tanned umbilical cord veins meeting specific surface properties and standards [8]. The natural pavements of endothelial cells of the original living blood vessels were lost during their processing from fresh or frozen umbilical cords [9]. What was preserved was mainly the surface quality of the sub-endothelial lining, a physiologically tolerable layer, along with a basement membrane and elastic internal lamina. The primary quality control criterion utilized was the "critical surface tension" parameter [10], adjusting processing conditions to maintain this value in the mid-20 dynes/cm. The international experience included thousands of successful human implantations, saving numerous limbs from hitherto inevitable amputation [11].

In the occasional circumstances when properly surface-controlled materials do become coated with clotted blood, usually as a result of constriction of tubes at their connections or stagnation in the outflow tract, post-implantation analyses show the clots to reside innocently in the lumens of the vessels, exhibiting little to no adhesion to "biocompatible" walls [12]. Similar findings, and understanding, regarding materials surface properties in other settings will allow prediction and control of biological responses to materials in food processing units [13], in the sea [14], in tissue culture [15], in the womb [16], in the eye [17], and in the oral cavity [18].

3. Background Concepts

3.1. Control of the Forces

All inquiries about adhesive strengths and properties of biological deposits on various materials must include close control of the exposure geometries, flow fields and shear forces. The boundary hydrodynamics during both the deposition and detachment processes are crucial. It is not acceptable in biomaterials research, for example, to terminate a biological experiment (e.g. for the purpose of noting the accumulating microfouling films) by "rinsing" a sample in an unspecified or non-quantifiable manner [19]. Neither is it acceptable to equate the forces needed to detach or distract a biological deposit from a solid substratum with the thermodynamic strength of adhesion of the attached elements [20]. Cell and film separations from solids are not simply the reverse of the first attachment events. Joint failure upon separation is usually of the mixed-mode (cohesive and adhesive) type. Special devices for flow or shear-rate control should be employed during both the first exposure of test surfaces to biological media and in the final rinsing steps required to remove loosely held debris, co-adsorbed interfering salts, or entrained substances [21].

One can also incorporate transducers and electrodes into the circuits to record dynamic electrical events accompanying biological film deposition and adhesion of fouling elements. The technique of streaming potential measurement, for instance, allows direct calculation of the important boundary phase parameter of the plane of shear called zeta potential [22], cited in many theoretical models of biological responses to material samples.

3.2. Selection of Surface Analyses

Many techniques for rapid (and, in most cases, nondestructive) instrumental analysis of both synthetic material and biological surfaces are now available and in routine use. The sensitive procedure of multiple attenuated internal reflection spectroscopy [23] produces infrared and/or UV/visible absorption spectra of the exterior molecular layers of matter defining biological interfaces with connecting solids. The acquired spectra reveal compositions, rates of accumulation, and modifications of such important boundary layers.

There are several sensitive, simultaneous or sequential analytical techniques immediately applicable to probing the structure and composition of thin biological deposits without requiring their manipulation or removal from the substrata of interest [24]. These techniques are: a) as described above, acquisition of infrared absorption spectra by the internal reflection technique; b) determination of film thicknesses and refractive indexes by the method of reflected polarized light (called ellipsometry); c) measurement of surface electrical states and contact potentials; d) inference of wettabilities, strengths of adhesion, operational surface free energies and critical surface tensions by comprehensive contact angle analysis (not just measurements of water contact angles!); e) morphological inspection of such layers by direct scanning electron microscopy without requiring over-coating the samples with obscuring metallic conducting films [25]; f) simultaneous analysis by energy-dispersive X-ray techniques of the presence and relative abundances and locations of elements of atomic number higher than sodium [26], and g) identification of actual crystalline forms present, by glancing angle X-ray diffraction [27]. These methods have contributed significant information to the development of dental restorative composites, surgical adhesives, prosthetic implants, and extracorporeal circuits, as well as provided basic data on the initial events of blood clotting and marine fouling. The early addition of ESCA (Electron Spectroscopy for Chemical Analysis), also known as XPS (X-ray Photoelectron Spectroscopy) to the battery of methods was widely accepted [28].

4. Basic Precepts

4.1. Blood-Contacting Materials

The preferred materials of construction for pulsatile blood-handling devices are hardy elastomers of the polyether types of polyurethane, preferably fabricated to exhibit surface enrichment of "biocompatible" methyl groups [29]. These groups can be provided by direct alloying or admixture of polydimethylsiloxane (PDMS) in the original polymer blend [30]. All blood-contacting surfaces are preferably fabricated to be smooth and free of entrapped air prior to their first exposure to blood. Even with stringent preparation and precaution, the best synthetic materials for blood contact applications will still acquire spontaneously deposited plasma protein layers (usually dominated by fibrinogen)and support modest, though temporary, cellular adhesion [31].

Once this "conditioning" film is in place on mid-Critical Surface Tension (20 - 30 dynes/cm) materials with "passive" surface properties, arriving blood platelets that do attach typically maintain their natural discoid shapes [32]. They retain their granules and/or clotting factors. When in contact with higher or lower surface energy "active" materials, the platelets become distressed, undergoing a "viscous metamorphosis" that triggers both thrombosis and activation of the coagulation cycle. In all cases, segmented polymorphonuclear leucocytes, also called neutrophils, then stream toward the platelet- or platelet-debris-strewn foreign surfaces. On "active" incompatible surfaces, these perform as phagocytic units while white cells attach without any beneficial effect. Rather, the white cells add to the growing adherent cellular mass and accelerate the thrombotic episodes.

On "passive" compatible materials, selected by virtue of their less-retentive surface properties [33], the white cells function as a "clean-up squad", using fibrinolytic enzymes to break up the original fibrinogen films. Interfacial shear forces are then sufficient to detach the original bound mass, seldom to form again. Long-term, dynamic equilibrium with fresh flowing blood, then, is obtained through a more "native" plasma protein film acting as a "passivation" layer.

4.2. Bone- and Tissue-Contacting Materials

Basically, the operating rules for materials choices for bone- or tissue-contact devices are just the opposite of those for blood-contact devices, with a few interesting exceptions [34].

Biological mineral attachment to various materials has been demonstrated by the surface analytical techniques described earlier-including X-ray diffraction-to be dominated by calcium phosphate microcrystals of hydroxyapatite form. This basic mineral phase of bones and teeth has been grown, inadvertently, on flexible elastomeric and tissue materials used in artificial hearts and substitute heart valves [35]. Although this presents difficulties in tissue-valve and artificial heart development programs, where unwanted mineralization must be limited or overcome, one might seize upon this observation and turn adversity into virtue for use of similar materials in orthopedic or dental applications [36].

In the ongoing pursuit of improved surface conditions for dental implants of all types, including complicated structures that must interface with bone on one aspect, and overlying tissue on another, and protrude through the tissue interface into the oral cavity with an infection-free stable seal, coatings of the same elastomeric materials used in artificial hearts might be applied. Additional roughening and surface-activation would enhance cellular in-growth processes and stimulate osteogenesis [37].

The more conventional approach, using intrinsically high surface energy or bioactive materials, has meanwhile served quite well. This involved either the initial choice of self-cleansing (through surface dissolution or erosion in the host site) glass compositions [38], or calcium phosphate minerals, or the scrupulous prior cleansing (from metallic implants) of all organic debris and polishing agents. Their sterilization in the same process, and resultant activation for adhesion, is readily accomplished by exposure to radio-frequency-initiated glow discharges [39]. Apparatus to achieve such surface activation and cleansing is becoming generally available, and protocols require only a few minutes to execute [40]. Useful improvements in dental implant immobilization, encouragement of direct tissue binding, and elimination of fibrous encapsulation do result. Similar apparatus, scaled to handle larger implants such as artificial hips, can provide excellent surface cleaning and activation of these prostheses to enhance their cementation into prepared bone sites.

5. Practical Use

5.1. Controlling Cell-Surface Interactions

One of the interesting findings of microbiology has been that many hetero typically bound bacteria, isolated from oceanic films as well as from human dental plaque, often have filamentous tufts at one end specialized for preferential adhesion [41]. Internal reflection infrared spectra identifies the polar tuft material as being of mainly glycoprotein composition, at variance with the general hypothesis that lipoteichoic acid (a poly glycerol phosphate polymer) dominates the adhesive sites of those microorganisms.

Even such organisms, arguably specialized by evolution for colonization of any "foreign surface", can be denied successful adhesion (that is, there is no retention after gentle removal forces are applied). Proof of this requires only that one employ substrata exhibiting surface properties in the bioabhesive, or adhesion-resistant "biocompatible", zone identified long ago in studies dealing with mammalian cells [42]. The most utilitarian" surface energy conversion" coating of the sort required is that formed by covalent binding of methyl-silanes (often to create poly-dim ethylsiloxane layers) onto glass, silica, other mineral or metallic surfaces. Standard laboratory "siliconization" of glassware or metals can be excellent in this regard, but only when exposing new exterior chemical arrays of closely packed methyl groups to the colonizing bio-environment. Silicones not dominated by methyl group side chains do not resist bioadhesion [43]! Closely packed methyl group "lawns" exhibit a composite critical surface tension or apparent surface free energy of about 22 dynes per centimeter, in the minimally adhesive zone of the surface energy scale. The prevention of adhesion is mediated in these cases, nevertheless, by native structure persistence of deposited glycol proteinaceous films as discussed earlier [44].

Assuring specific biological responses to materials on the basis of their relative surface properties eventually will allow us to overcome still-troublesome biomaterials-centered infections. Early examples of the latter problem were those of pelvic inflammatory disease (PID) associated with Intrauterine Contraceptive Devices (IUDs) [45] having multi-filamentous nylon or frayed, micro-fibrillated polypropylene tail-strings. These structures provided sites for attachment and migration of infective microorganisms from the vaginal canal to the normally infection-free uterus, where they colonized as "biofilms" on other intrinsically bioadhesive materials used for the IUD bodies. Multimillion dollar legal judgments were levied against the major pharmaceutical firms that introduced these inappropriate biomaterials to the human reproductive system.

Unfortunately, the infection-related biomaterials problems diminished attention to beneficial features of earlier, more benign intrauterine contraceptive devices. The spontaneous coating of such devices by adsorbed glycoproteinaceous layers from the cervical mucus fluid supported speculation that such coatings may prevent the "capacitation" of sperm transiting the IUD locale. Sperm that do not experience such required changes in their initial surface (adhesive) properties do not successfully engage in another form of heterotypic adhesion, that of sperm-to-ovum, upon which fertilization of mammalian oocytes is premised. Again, the practical application of biomaterials with better-selected surface properties can make IUD regulation of fertility a useful, safe and effective option for world-wide population control.

It is abundantly clear that "strength" of biological adhesion (more accurately, "resistance to detachment" or "strength of retention") is associated with defined ranges of surface energies, or critical surface tensions, for solid substrata in extremely diverse biological circumstances [46]. Other examples abound: demonstration that the minimum binding strength (resistance to detachment) for liver cells in tissue culture is sharply in the zone between 20 and 30 dynes/cm [47]; statistically sound observations, on solid substrata ranging in surface free energy from about 10 to over 50 dynes/cm (ergs per square centimeter), that the spread areas and associated degrees of distortion of settled human cells [48] and blood platelets, freshly obtained from plasmapheres is, are minimized in the critical surface tension zone between 20 and 30dynes/cm; observation of minimal forces required to pull musselbyssus discs from solid materials having these same critical surface tensions; and more. Even the induction of hemolytic damage to circulating red blood cells is minimized when the critical surface tensions of the walls of the shearing device are adjusted to the mid-20's dynes/cm range [49]. This last observation strongly implies that biological macromolecules that arrive at, and then are displaced from, solid surfaces of varying surface properties carry away with them (back into solution or suspension) varying "messages" based on their differing degrees of surface-contact-induced "denaturation" from their original solution states [50].

5.2. Preventing Biofouling

Biofouling refers to those unwanted deposits that frequently occur on contact lenses, dentures, periscope windows, and ship bottoms, to name just a few cases. A continuing difficulty in all food processing operations is the deposition of organic matter and mineral layers on the surfaces of heat exchangers and membranes. Dairy products are near the top of the list of problem-makers in this regard [51]. After many years of study, the predominant compositions of the first most strongly bound layers are generally not known. Differences of opinion still exist, for example, on whether the earliest deposits from raw milk are mainly proteins or minerals such as the calcium-phosphate-rich "milk stone". This is a fertile territory for biomaterials specialists to enter and enrich, using the same concepts of materials-related control of bioadhesion already successful in medical-device development.

For example, special flow cells may be designed to allow control of the surface shear rates and stresses, investigating the earliest fouling events for heated surfaces in contact with homogenized whole milk [52]. Preliminary results already available endorse the concept that milk protein adsorption is the first, essentially irreversible, event in surface fouling by dairy products, with mineral deposition occurring much later.

Useful flow cell devices have already been constructed [53]. Some apply "voltage clamping" circuitry so that the primary events of biofilm deposition and secondary events of microbial attachment, polymer exudation, and mineralization can be observed on substrata surfaces controlled electronically at any desired surface potential/charge condition. Such flow cells helped in documentation of the early events of microbiological fouling of model heat exchange surfaces in the warm subtropical waters of the Gulf of Mexico. Acquisition of that fundamental knowledge on biological fouling of heat exchange devices in warm seawater was of critical importance to international efforts to extract stored solar energy from the tropical oceans, using the principles of "ocean thermal energy conversion". The Gulf of Mexico findings were compelling in their demonstration of slow deposition of humic-like conditioning films on the heat exchange surfaces. These adsorbed films were colonized in less than 3 days by pioneer bacteria of both flagellated and un-flagellated types. Within 6 days of exposure, growing deposits of polymer exudates were revealed around the pioneer microorganisms by using a special adaptation of scanning electron microscopy, eliminating the need for obscuring electro-conductive layers of sputtered metals. At the same exposure time, deposition of calcareous smatter, usually associated with diatoms and other algal forms, was noted. Patches of extreme microbiological diversity, with ecological succession already in progress, were observed on many test surfaces in the period between 3 and 6 days. It is certainly clear from these, and considerable supporting data, that biomaterials uses within that 70% of the earth's surface that is" wet" will require attention to (and control of) such biofouling phenomena [54].

Already gathered experimental data on critical inter facial layers of biological films in marine environments suggest nontoxic mechanisms by which their adhesion-or lack thereof to practical materials—can be controlled [55]. Recognizing serious drag-enhancing penalties associated with bacterial/slime fouling layers on even toxic marine paint surfaces, intriguing early results suggest that biofilms, or their synthetic analogues, maybe created to eliminate the drag effects of more-usual fouling layers and perhaps even to provide significant drag reduction

[56]. Proof of-principle testing, using natural low-drag skin of living porpoises and killer whales [57], has shown the best surfaces have low-critical-surface-tension, protein-dominated characters inconsonance with similar findings for fouling-resistant layers of human oral mucosa and blood vessel endothelium [58]. Since synthetic materials for numerous biomedical devices have been successful without the need for toxicants to prevent fouling by even concentrated biological fluids, it should not be surprising that field tests endorsed similar material surface modifications for service in seawater. Results indicated that adjustment of a material's critical surface tension to the zone between 20 and 30 dynes/cm correlated with the most facile detachment of fouling debris [59]. The coincidence of this finding with results from tests of biomedical devices is encouraging in suggesting a strong conservatism among natural mechanisms promoting or preventing biological adhesion.

It even has been shown that algae produce base attachments of significantly greater area in order to resist separation from substrata having low critical surface tensions [60]. Diatom colonization success has also been demonstrated to be dependent on the initial surface energy of test substrata [61]. Our major conclusions must be that understanding, prediction, and control of surface properties of materials in all biological settings will be crucial to achieving improved performance in practical cases. Interestingly, here is a situation where large-scale (and large volume) commercial benefits may arise from the preceding "small science" (and small volume) of biomaterials development for artificial internal organs.

6. New Research Directions

Many authors still cite the needs for further study of the events of "fouling" in medical equipment, so it is only necessary to add some new discussion of the areas of biological adhesion noted here [62]. It is especially important to note the striking similarities in the fouling processes that occur in the bloodstream and the oral cavity, as similar to events in cooling water structures and processing equipment for various dairy products.

A need certainly still exists for improved understanding, on a basic level, of the deposition of macromolecules and of living cells at solid and semisolid (hydrogel, tissue) surfaces. Lack of the required detailed knowledge of these fundamental processes limits practical control of biological adhesion on a more general basis. Specific major areas for continuing study include the effects of different material surface properties (texture, charge, chemistry, energy) on binding of deposited macromolecules, seeking evidence for selective retention of particular components or sub fractions, determining the actual degree of coverage of the "cleaned" original surfaces, assessing the orientation of the adsorbed molecular entities, and defining the longer term modifications to the original surfaces' properties and to the attached molecules. Further, it is important to learn more about the cascade of specific cells arriving at and attaching to indwelling engineering materials, particularly noting cellular exudation of bio reactants or enzyme catalysts concentrated at the interfaces [63]. There is a known tendency for attached cells to produce polymeric exudates that both permanently bind them to the "pellicles" first acquired and engulf the growing, metabolizing units in "coats of slime". These events must be better understood and controlled if biomaterials-centered infections are to be successfully combatted [64].

It is also extremely important to extend modern surface analytical methods to the problems of identifying, more completely, the nature of cellular exopolymers produced, addressing as well their mode of production and changes with time and conditions. This is a crucial topic in many emerging sub-disciplines of what is generally called "biotechnology". The reactions between "exported" cellular products and the adsorbed, usually glycol-proteinaceous, films that provide early binding of the cells to the starting surfaces must be ascertained. Improved measures of the actual strengths of adhesion (retention) between the cellular polymers and the films coating solid surfaces of differing physical chemical states must be developed. A practical focus of these efforts could be toward the identification of any "weak links" in the chain of cellular colonization, growth, and binding processes, that may be subject to direct interruption or strengthening [65].

Other biological responses to materials' surface properties are of equally urgent concern, specifically as they influence the events occurring at the initial attachment interfaces. It is recognized that cellular migration as well as other transport processes in tissue or fluid phases may limit the rate of cellular attachment to different engineering surfaces. The compositions, adhesive and cohesive strengths, and densities of the attached cell layers, as well as the geometries and hydrodynamic features at specific sites, are clearly important. These factors bear heavily on issues of human health and disease through their influence on the re-suspension and removal of biological deposits from the interfacial zones. Based upon the criteria reviewed here, one important lesson for fouling limitation in natural fluids may be that complete prevention of film formation and early cellular adhesion is not necessary: appropriate adjustments of material surface properties can control-indeed enhance or limit-the rate and reliability of re-entrainment of original deposits into the adjacent aqueous phases [66]. It is not yet known whether similar adjustments in materials' surface properties can modulate the biological responses of still-attached organisms to systemic (volume) agents for therapy or antisepsis.

Another identified research priority is for improvement and introduction of uniformity in testing methods regarding biological film deposition and cellular attachment, so that more direct comparisons of different materials in different circumstances can be made [67]. What seems to be required are small standardized units that can be adjusted to experience known flow rates, shear forces, nutrient conditions and so on, the purpose being to provide bio response indices with comparable meanings under different conditions at different times and at different sites. Availability and regular use of inexpensive, standardized test units would provide a continually expanding data base concerning the effects of material surface properties and treatments on rates of deposit formation, for example, under a variety of physical, chemical and biological conditions. More effective means of cleaning and sterilizing biomaterials surfaces are also required since it is obvious that current techniques, with the possible exception of still-emerging glow-discharge-plasma processes, are not likely to completely remove cellular and adsorbed film layers closest to the material surfaces [68]. When incomplete cleanliness is accepted, despite sterility, subsequent buildup of secondary fouling deposits occurs at rates much more rapid than that observed on truly clean substrata. Obviously, immunologic, antigenic, pyrogenic responses can still be triggered by remnants of even sterile biological debris. Thus, research directed at more complete removal of adherent deposits and their anchoring organic films, by both mechanical and chemical cleaning techniques, concomitant with or as a precursor to sterilization, is essential [69].

7. Concluding Remarks

Application of ambient environment as well as high-vacuum surface analytical techniques to biomaterials reveals fundamental similarities in the primary events of bioadhesion to them that are well correlated with the initial materials' surface properties. Initial fouling film formation in milk processing equipment exhibits a similar pattern to process in the oralcavity and in subtropical ocean water heat exchangers, including subsequent microorganism attachments, polymer exudation, and mineralization in most such circumstances. Implications of the identification of preferred surface energy ranges favoring or inhibiting permanent biological adhesion are clear in results from medical device trials including the artificial heart, dental implants, artificial hips, and substitute blood vessels. Calcification of blood-contacting surfaces of flexing elastomeric heart-assist-sacs indicates directions for similar methods to promote bone formation on dental or orthopedic fixtures [70]. Research priorities in the field of biological responses to materials surfaces should include additional attention to the effects of different surfaces on macromolecular retention, to the properties of various cells colonizing immersed surfaces, to the nature of exopolymers from cells attached to surfaces, to reactions between cellular products and surface "conditioning" films, to strengths of adhesion between cellular polymers and pre-adsorbed films, to transport processes in the bulk tissue or fluid phases, to the geometries and hydrodynamics of specific systems, and to selection of surface properties that will enhance the utility of chemical and mechanical techniques for removal of fouling films, while obtaining sterilization.

References

[1]   Baier, R.E. (2006) Surface Behaviour of Biomaterials: The *Theta Surface* for Biocompatibility. *Journal of Materials Science: Materials in Medicine*, **17**, 1057-1062. http://dx.doi.org/10.1007/s10856-006-0444-8

[2]   Baier, R.E. (1982) Conditioning Surfaces to Suit the Biomedical Environment: Recent Progress. *Journal of Biomechanical Engineering*, **104**, 257-271. http://dx.doi.org/10.1115/1.3138358

[3]   Baier, R.E. and Dutton, R.C. (1969) Initial Events in Interactions of Blood with Foreign Surfaces. *Journal of Biomedical Materials Research*, **3**, 191-206. http://dx.doi.org/10.1002/jbm.820030115

[4]   Baier, R.E. (1981) Modification of Surfaces to Reduce Fouling and/or Improve Cleaning. *Proceedings of Fundmentals and Applications of Surface Phenomena Associated with Fouling and Cleaning in Food Processing*, Tylosand, 6-9 April 1981, 1-22.

[5]   Baier, R.E. (1981) Early Events of Micro-Biofouling of All Heat Transfer Equipment. In: Somerscales, E.F.C. and Knudsen, J.G., Eds., *Fouling of Heat Transfer Equipment*, Hemisphere Publishing Corp, Washington, DC, 293-304.

[6]   Baier, R.E. (1985) Cell Seeding: Biomaterial Surface Preparation. *ASAIO Journal*, **8**, 104-108.

[7]   Baier, R.E. (1973) Occurrence, Nature, and Extent of Cohesive and Adhesive Forces in Dental Integuments. In: Lasslo A. and Quintana, R.P., Eds., *Surface Chemistry and Dental Integuments*, Charles C. Thomas Publisher, Springfield, 337-391.

[8]   Baier, R.E., Akers, C.K., Perlmutter, S., Dardik, H., Dardik, I. and Wodka, M. (1976) Processed Human Umbilical Cord Veins for Vascular Reconstructive Surgery. *Transactions—American Society for Artificial Internal Organs*, **22**, 514-524.

[9]   Baier, R.E. (1978) Physical Chemistry of the Vascular Interface: Composition, Texture, and Adhesive Quality. In: Sawyer, P.N. and Kaplitt, M.J., Eds., *Vascular Grafts*, Appleton-Century-Crofts, New York, 76-107.

[10] Baier, R.E. and Loeb, G.I. (1971) Multiple Parameters Characterizing Interfacial Films of a Protein Analogue, Poly-methylglutamate. In: Craver, C.D., Ed., *Polymer Characterization: Interdisciplinary Approaches*, Plenum Press, New York, 79-96.

[11] Dardik, H., Baier, R.E., Meenaghan, M., Natiella, J., Weinberg, S., Turner, R., Sussman, B., Kahn, M., Ibrahim, I. and Dardik, I.I. (1982) Morphologic and Biophysical Assessment of Long Term Human Umbilical Cord Vein Implants Used as Vascular Conduits. *Surgery, Gynecology & Obstetrics*, **154**, 17-26.

[12] Baier, R.E. and Abbott, W.M. (1978) Comparative Biophysical Properties of the Flow Surfaces of Contemporary Vascular Grafts. In: Dardik, H., Ed., *Grafts Materials in Vascular Surgery*, Symposia Specialists, Inc., Miami, 70-103.

[13] Baier, R.E. and Meyer, A.E. (1985) Surface Chemical Approaches to Decontamination and Disinfection. In: Lund, D., Plett, E. and Sandu, C., Eds., *Fouling & Cleaning in Food Processing*, University of Wisconsin-Madison, Madison, 336-339.

[14] Baier, R.E. (1984) Initial Events in Microbial Film Formation. In: Costlow, J.D. and Tipper, R.C., Eds., *Marine Biodeterioration: An Interdisciplinary Study*, Naval Institute Press, Annapolis, 57-62. http://dx.doi.org/10.1007/978-1-4615-9720-9_8

[15] Baier, R.E. and Weiss, L. (1975) Demonstration of the Involvement of Adsorbed Proteins in Cell Adhesion and Cell Growth on Solid Surfaces. In: *Applied Chemistry at Protein Interfaces*, Advances in Chemistry Series, Vol. 145, American Chemical Society, Washington DC, 300-307.

[16] Baier, R.E. and Lippes, J. (1975) Glycoprotein Adsorption in Intrauterine Foreign Bodies. In: *Applied Chemistry at Protein Interfaces*, Advances in Chemistry Series, Vol. 145, American Chemical Society, Washington DC, 308-318.

[17] Baier, R.E. and Thomas, E.B. (1996) The Ocean: The Eye of the Earth. Contact Lens Spectrum, 37-44.

[18] Baier, R.E., Meyer, A.E., Natiella, J.R. and Carter, J.M. (1984) Surface Properties Determine Bioadhesive Outcomes: Methods and Results. *Journal of Biomedical Materials Research*, **18**, 337-355. http://dx.doi.org/10.1002/jbm.820180404

[19] DePalma, V.A. and Baier, R.E. (1978) Microfouling of Metallic and Coated Metallic Flow Surfaces in Model Heat Exchange Cells. *Proceedings of the Ocean Thermal Energy Conversion (OTEC) Biofouling and Corrosion Symposium*, U.S. Department of Energy, PNL-SA-7115, Washington DC, 89-106.

[20] Baier, R.E. (1982) Comments on Cell Adhesion to Biomaterial Surfaces: Conflicts and Concerns. *Journal of Biomedical Materials Research*, **16**, 173-175. http://dx.doi.org/10.1002/jbm.820160210

[21] Baier, R.E. and DePalma, V.A. (1979) Flow Cell and Method for Continuously Monitoring Deposits on Flow Surfaces. 8 Claims. U.S. Patent No. 4, 175, 233.

[22] Working Group on Physicochemical Characterization of Biomaterials, National Heart, Lung, and Blood Institute, National Institutes of Health, Leading to Publication of "Guidelines for Physico-Chemical Characterization of Biomaterials", NIH Publication No. 80-2186, September 1980.

[23] Baier, R.E. and Zisman, W.A. (1970) Wettability and Multiple Attenuated Internal Reflection Infrared Spectroscopy of Solvent-Cast Thin Films of Polyamides. *Macromolecules*, **3**, 462-468. http://dx.doi.org/10.1021/ma60016a017

[24] Baier, R.E., Shafrin, E.G. and Zisman, W.A. (1968) Adhesion: Mechanisms that Assist or Impede It. *Science*, **162**, 1360-1368. http://dx.doi.org/10.1126/science.162.3860.1360

[25] Baier, R.E., Meyer, A.E., DePalma, V.A., King, R.W. and Fornalik, M.S. (1983) Surface Microfouling during the Induction Period. *Journal of Heat Transfer*, **105**, 618-624. http://dx.doi.org/10.1115/1.3245630

[26] Baier, R.E., Forsberg, R.L., Meyer, A.E. and Lundquist, D.C. (2014) Ballast Tank Biofilms Resist Water Exchange but Distribute Dominant Species. *Management of Biological Invasions*, **5**, 241-244. (Special ICAIS Issue) http://dx.doi.org/10.3391/mbi.2014.5.3.07

[27] Baier, R.E., Mack, E.J., Rogers, C.W., Pilie, R.J. and DePalma, V.A. (1981) Source Assessment of Atmospheric Aerosols: Spectroscopic Data from a Rapid Field Technique. *Optical Engineering*, **20**, 866-872. http://dx.doi.org/10.1117/12.7972828

[28] Vargo, T.G., Hook, D.J., Gardella, J.A., Eberhardt, M.A., Meyer, A.E. and Baier, R.E. (1991) A Multitechnique Sur-

face Analytical Study of a Segmented Block Copolymer Poly (Ether-Urethane) Modified through an H_2O Radio Frequency Glow Discharge. *Journal of Polymer Science Part A: Polymer Chemistry*, **29**, 535-545. http://dx.doi.org/10.1002/pola.1991.080290410

[29] Boretos, J.W., Pierce, W.S., Baier, R.E., Leroy, A.F. and Donachy, H.J. (1975) Surface and Bulk Characteristics of a Polyether Urethane for Artificial Hearts. *Journal of Biomedical Materials Research*, **9**, 237-340. http://dx.doi.org/10.1002/jbm.820090308

[30] Pierce, W.S., Donachy, J.H., Rosenberg, G. and Baier, R.E. (1980) Calcification inside Artificial Hearts: Inhibition by Warfarin-Sodium. *Science*, **208**, 601-603. http://dx.doi.org/10.1126/science.7367883

[31] Baier, R.E. and Kurusz, M. (2012) Understanding Blood/Material Interactions: Contributions from the Columbia University Biomaterials Seminar. *ASAIO Journal*, **58**, 450-454. http://dx.doi.org/10.1097/MAT.0b013e3182631e3e

[32] Baier, R.E. (1987) Selected Methods of Investigation for Blood-Contact Surfaces. In: Leonard, E.F., Turitto, V.T. and Vroman, L., Eds., *Blood in Contact with Natural and Artificial Surfaces, Annals of the New York Academy of Sciences*, **516**, 68-77. http://dx.doi.org/10.1111/j.1749-6632.1987.tb33031.x

[33] Baier, R.E., DePalma, V.A., Goupil, D.W. and Cohen, E. (1985) Human Platelet Spreading on Substrata of Known Surface Chemistry. *Journal of Biomedical Materials Research*, **19**, 1157-1167. http://dx.doi.org/10.1002/jbm.820190922

[34] Baier, R.E., Meyer, A.E. and Natiella, J.R. (1992) Implant Surface Physics and Chemistry: Improvements and Impediments to Bioadhesion. In: Laney, W.R. and Tolman, D.E., Eds., *Tissue Integration in Oral, Orthopedic, and Maxillofacial Reconstruction*, Quintessence Publishing Co., Inc., Chicago, 240-249.

[35] Banas, M.D. and Baier, R.E. (2000) Accelerated Mineralization of Prosthetic Heart Valves. *Molecular Crystals and Liquid Crystals Science and Technology*, **354**, 249-267. http://dx.doi.org/10.1080/10587250008023619

[36] Sendax, V.I. and Baier, R.E. (1992) Improved Integration Potential for Calcium-Phosphate-Coated Implants after Glow Discharge and Water Storage. *Dental Clinics of North America*, **36**, 221-224.

[37] Baier, R.E. (1981) Catheter for Long-Term Emplacement. 8 Claims. U.S. Patent No. 4, 266, 999.

[38] Baier, R.E. (2002) A Challenging Anomaly—Glass that Does Not Clot Blood! *The Glass Researcher*, **12**, 23-24.

[39] Baier, R.E. and Meyer, A.E. (1988) Implant Surface Preparation. *International Journal of Oral & Maxillofacial Implants*, **3**, 9-20.

[40] Baier, R.E., Carter, J.M., Sorenson, S.E., Meyer, A.E., McGown, B.D. and Kasprzak, S.A. (1992) Radiofrequency Gas Plasma (Glow Discharge) Disinfection of Dental Operative Instruments, Including Handpieces. *Journal of Oral Implantology*, **18**, 236-242.

[41] Glantz, P.O., Baier, R.E. and Christersson, C.E. (1996) Biochemical and Physiological Considerations for Modeling Biofilms in the Oral Cavity: A Review. *Dental Materials*, **12**, 208-214. http://dx.doi.org/10.1016/S0109-5641(96)80024-8

[42] Baier, R.E. and DePalma, V.A. (1971) The Relation of the Internal Surface of Grafts to Thrombosis. In: Dale, W.A., Ed., *Management of Arterial Occlusive Disease*, Year Book Medical Publishers, Inc., Chicago, 147-163.

[43] Gould, J.A., Liebler, B., Baier, R., Benson, J., Boretos, J., Callahan, T., Canty, E., Compton, R., Marlowe, D., O'Holla, R., Page, B., Paulson, J. and Swanson, C. (1993) Biomaterials Availability: Development of a Characterization Strategy for Interchanging Silicone Polymers in Implantable Medical Devices. *Journal of Applied Biomaterials*, **4**, 355-358. http://dx.doi.org/10.1002/jab.770040410

[44] Baier, R.E., Loeb, G.I. and Wallace, G.T. (1971) Role of an Artificial Boundary in Modifying Blood Proteins. *Federation Proceedings*, Federation of AmerSoc for Experimental Biol, Bethesda, Vol. 30, 1523-1538.

[45] Tietze, C. (1966) Contraception with Intrauterine Devices. *American Journal of Obstetrics & Gynecology*, **96**, 1043-1054.

[46] Glantz, P.O., Arnebrant, T., Nylander, T. and Baier, R.E. (1999) Bioadhesion—A Phenomenon with Multiple Dimensions. *Acta Odontologica Scandinavica*, **57**, 238-241. http://dx.doi.org/10.1080/000163599428634

[47] Baier, R.E. (1980) Substrata Influences on the Adhesion of Microorganisms and Their Resultant New Surface Properties. In: Bitton, G. and Marshall, K.S., Eds., *Adsorption of Microorganisms*, Wiley-Interscience Publishers, Hoboken, 59-104.

[48] Baier, R.E. (1970) Surface Properties Influencing Biological Adhesion. In: Manly, R.S., Ed., *Adhesion in Biological Systems*, Academic Press, New York, 15-48. http://dx.doi.org/10.1016/B978-0-12-469050-9.50007-7

[49] Baier, R.E., Dutton, R.C. and Gott, V.L. (1970) Surface Chemical Features of Blood Vessel Walls and of Synthetic Materials Exhibiting Thromboresistance. In: Blank, M., Ed., *Surface Chemistry of Biological Systems*, Plenum Press, New York, 235-260. http://dx.doi.org/10.1007/978-1-4615-9005-7_14

[50] Baier, R.E. (1975) Blood Compatibility of Synthetic Polymers: Perspective and Problems. In: Kronenthal, R.L., Oser,

Z. and Martin, E., Eds., *Polymers in Medicine and Surgery*, Plenum Press, New York, 139-159. http://dx.doi.org/10.1007/978-1-4684-7744-3_10

[51] Baier, R.E., DePalma, V.A., Meyer, A.E., King, R.W. and Fornalik, M.S. (1981) Control of Heat Exchange Surface Microfouling by Material and Process Variations. In: Chenoweth, J.M. and Impagliazzo, M., Eds., *Fouling in Heat Exchange Equipment*, HTD-Vol. 17, AmerSoc Mechanical Engineers, New York, 97-103.

[52] King, R.W., Meyer, A.E., Ziegler, R.C. and Baier, R.E. (1981) New Flow Cell Technology for Assessing Primary Biofouling in Oceanic Heat Exchangers. *Proceedings of the 8th Ocean Energy Conference*, U.S. Department of Energy, Washington DC, 431-436.

[53] Baier, R.E., Meyer, A.E. and King, R.W. (1988) Improved Flow-Cell Techniques for Assessing Marine Microfouling and Corrosion. In: Thompson, M.F., Sarojini, R. and Nagabhushanam, R., Eds., *Marine Biodeterioration*, Oxford & IBH Publishing Co., Ltd., New Delhi, 385-394.

[54] Forsberg, R.L., Baier, R.E. and Meyer, A.E. (2014) Sampling and Experiments with Biofilms in the Environment: Part 2, Sampling from Large Structures Such as Ballast Tanks. In: Dobretsov, S., Thomason, J.C. and Williams, D.N., Eds., *Biofouling Methods*, Wiley-Blackwell, Oxford.

[55] Baier, R.E., Meyer, A.E. and Forsberg, R.L. (1997) Certification of Properties of Nontoxic Fouling-Release Coatings Exposed to Abrasion and Long-Term Immersion. *Naval Research Reviews*, **49**, 60-65.

[56] Baier, R.E., Meyer, A.E., Forsberg, R.L. and Ricotta, M.S. (1997) Intrinsic Drag Reduction of Biofouling-Resistant Coatings. *Proceedings, Emerging Nonmetallic Materials for the Marine Environment*, U.S.-Pacific Rim Workshop Sponsored by the U.S. Office of Naval Research, Honolulu, 1-36 through 1-40.

[57] Baier, R.E., Gucinski, H., Meenaghan, M.A., Wirth, J. and Glantz, P.O. (1984) Biophysical Studies of Mucosal Surfaces. In: Glantz, P.O., Leach, S.A. and Ericson, T., Eds., *Oral Interfacial Reactions of Bone, Soft Tissue & Saliva*, IRL Press Ltd, Oxford, 83-95.

[58] Baier, R.E. and Meyer, A.E. (1983) Surface Energetics and Biological Adhesion. In: Mittal, K.L., Ed., *Physiochemical Aspects of Polymer Surfaces*, Vol. 2, Plenum Publishing Corporation, New York, 895-909.

[59] Baier, R.E. (1973) Influence of the Initial Surface Condition of Materials on Bioadhesion. *Proceedings, Third International Congress on Marine Corrosion and Fouling*, Northwestern University Press, Evanston, 633-639.

[60] Fletcher, R.L. and Baier, R.E. (1984) Influence of Surface Energy on the Development of the Green Alga Enteromorpha. *Marine Biology Letters*, **5**, 251-254.

[61] Meyer, A., Baier, R., Wood, C.D., Stein, J., Truby, K., Holm, E., Montemarano, J., Kavanagh, C., Nedved, B., Smith, C., Swain, G. and Wiebe, D. (2006) Contact Angle Anomalies Indicate that Surface-Active Eluates from Silicone Coatings Inhibit the Adhesive Mechanisms of Fouling Organisms. *Biofouling*, **22**, 411-423. http://dx.doi.org/10.1080/08927010601025473

[62] Glantz, P.O.J., Arnebrant, T., Nylander, T. and Baier, R.E. (1999) Bioadhesion—A Phenomenon with Multiple Dimensions. *Acta Odontologica Scandinavica*, **57**, 238-241. http://dx.doi.org/10.1080/000163599428634

[63] Baier, R.E. (1992) Influence of Surface and Fluid Conditions on Thrombus Generation. *Proceedings of the Amer Acad of Cardiovascular Perfusion*, **13**, 143-146.

[64] Dutton, R.C., Webber, A.J., Johnson, S.A. and Baier, R.E. (1969) Microstructure of Initial Thrombus Formation on Foreign Materials. *Journal of Biomedical Materials Research*, **3**, 13-23. http://dx.doi.org/10.1002/jbm.820030104

[65] Nayak, S.C., Baier, R.E., Meyer, A.E. and Abuhaimed, T. (2010) Improvement of Root Canal X-Ray Imaging by Delmopinol Pretreatment-Assisted Contrast Media Infiltration. *Northeast Bioengineering Conference Proceedings*, Columbia University, New York, 26-28 March 2010, ABS-026, 39.

[66] L'Italien, G.J., Megerman, J., Hasson, J.E., Meyer, A.E., Baier, R.E. and Abbott, W.M. (1986) Compliance Changes in Glutaraldehyde-Treated Arteries. *Journal of Surgical Research*, **41**, 182-188. http://dx.doi.org/10.1016/0022-4804(86)90023-5

[67] Meyer, A.E., King, R.W., Baier, R.E. and Fornalik, M.S. (1985) A Field Study of Fouling of Test Surfaces Exposed to Flowing Brackish River Water. *Proceedings, Condenser Biofouling Control Symposium*, Electric Power Research Institute.

[68] Baier, R.E., Meyer, A.E., Akers, C.K., Natiella, J.R., Meenaghan, M.A. and Carter, J.M. (1982) Degradation Effects of Conventional Steam Sterilization on Biomaterial Surfaces. *Biomaterials*, **3**, 241-245. http://dx.doi.org/10.1016/0142-9612(82)90027-8

[69] Park, J.H., Olivares-Navarrete, R., Baier, R.E., Meyer, A.E., Tannenbaum, R., Boyan, B.D. and Schwartz, Z. (2012) Effect of Cleaning and Sterilization on Titanium Implant Surface Properties and Cellular Response. *Acta Biomaterialia*, **8**, 1966-1975. http://dx.doi.org/10.1016/j.actbio.2011.11.026

[70] White, J.A., Baier, R.E., Meyer, A.E., Burke, R.P. and Hausmann, E.M. (2000) Biomechanical and Biochemical Paths to Dystrophic Mineralization of Stented Cardiovascular Tissues. In: Vossoughi, J., Kipshidze, N. and Karanian, J.W., Eds., *Stent Graft Update*, Chapter 5, Medical and Engineering Publishers, Inc., Washington DC, 37-65.

6

Effect of Zr on Structural and Dielectrical Properties of $(Ba_{0.9}Mg_{1.0})(Zr_xTi_{1-x})O_3$ Ceramics

Sankararao Gattu[1], Venuturupalli Durga Prasadu[2], Kocharlakota Venkata Ramesh[2*]

[1]Department of Physics, MVJ College of Engineering, Bangalore, India
[2]Department of Physics, GITAM Institute of Technology, GITAM University, Visakhapatnam, India
Email: *kv_ramesh5@yahoo.co.in

Abstract

Barium titanate, $BaTiO_3$ (BTO) is the most common ferro electric material, which is used to manufacture electronic components such as multilayer capacitors, positive temperature coefficient thermistors, piezo electric transdures, and ferro electric memory. Zr doped barium magnesium titanate $(Ba_{0.9}Mg_{1.0})(Zr_xTi_{1-x})O_3$ (with x = 0.10, 0.20, 0.40 (BMZT 10, BMZT 20 and BMZT 40) perovskite is prepared by conventional solid state reaction method. The starting raw materials were $BaCO_3$, TiO_2, MgO and ZrO_2. The XRD study at room temperature suggests that these have cubic and tetragonal symmetry phases. The behavior of the measured dielectric permittivity and dielectric loss with temperature and frequency reveals that the materials undergo a diffuse para-ferroelectric phase transition and are of the relaxor type. The crystal structure, surface morphology and dielectric properties of Zr and Mg doped barium titanate ceramics were investigated. Zr^{4+} and Mg^{2+} ions have entered the unit cell maintaining the perovskite structure of solid solution without the evidence of any additional phase when Mg content is 0.1 mole% and the Zr content is 0.10, 0.20 and 0.40 mole%.

Keywords

Lead Free Ceramics, $BaTiO_3$, Dielectric Materials, Impedance, XRD, SEM

1. Introduction

Lead based ceramic has been studied more than anyone else ferroelectric because of their excellent dielectric

---

*Corresponding author.

properties [1]. However, the presence of lead in those materials is about 60% in weight [2], reconsidering its use in technical applications, due to its high toxicity of lead for the environment as well as for humans [3]-[7]. Barium titanate, $BaTiO_3$ (BTO) is the most common ferro electric material, which is used to manufacture electronic components such as multilayer capacitors, positive temperature coefficient thermistors, piezo electric transducers, and ferro electric memory, because of its excellent dielectric, piezo electric and ferro electric properties [8]-[12]. Moreover, a constant effort is being made to develop new dielectric oxides [13]-[15]. The micro level structure and dielectric properties of BTO can be modified by addition of the dopants such as La^{3+}, Ce^{2+}, Mn^{4+}, Nb^{5+}, Nd^{3+}, Cr^{3+}, Zr^{4+}, Mg^{2+}, Sr^{2+} and Si^{4+} to occupy Ba^{2+} on A sites or Ti^{4+} on B sites to form the solid solution. Numerous works have also been carried out to confirm the doping effects of rare earth oxides on the microstructure and electrical properties of $BaTiO_3$-MgO based system [16]-[32]. However, there are only few works concerned with the properties of the MgO singly doped barium titanate system. It is reported that when Ba is replaced by Mg in small quantities in the composition $Ba(Zr_xTi_{1-x})O_3$ the dielectric properties have been changed. But the transition temperature has been shifted towards lower temperature values. S. K. Rout et al. [33] reported that the transition temperature decreased with increase of substitution of Mg upto <1.5 mole%. It has been reported that [34] with 15% Zr substitution in $Ba(ZrTi)O_3$ (BZT) it exhibited three transitions rhombohedra to orthorhombic, orthorhombic to tetragonal and tetragonal to cubic. At room temperature, the doped material exhibits enhanced dielectric constant with further increase in Zr content beyond 15%. Diffuse phase transition has been observed with the decrease in one transition temperature [35] and the material showed typical relaxor like behavior in the range 25 - 45 mole% Zr substitution [7]. Unexpectedly the lead free ceramic shows the relaxor properties at low temperatures [3]. Several attempts have been made by researchers on these materials to shift the T_c to close to room temperature. It is well known that homovalent and hetrovalent substitution for barium and titanium ions gives rise to various behaviors including the shifting of the transition temperature. This inspires to work on effect of Zr on structural and dielectrical properties of barium magnesium titanate $((Ba_{0.9}Mg_{1.0})(Zr_xTi_{1-x})O_3)$ perovskite composition prepared through solid state reaction route because in this method limited formation of side products, no solvents are needed in the reaction and hence no waste disposal issues associated with the solvent need to be considered and do not require extensive purification to remove traces of solvent and impurities. The samples synthesized through solid state reaction method may be used for obtaining in bulk form with high density over other methods.

2. Experimental

The perovskite samples of Zr doped Barium Magnesium Titanate $(Ba_{0.9}Mg_{1.0})(Zr_xTi_{1-x})O_3$ (with x = 0.10, 0.20, 0.40) (BMZT 10, BMZT 20 and BMZT 40) were prepared by conventional solid state reaction method. The starting raw materials were $BaCO_3$ (Chen Chems., Chennai), TiO_2 (Loba Chem., Mumbai), MgO (Chen Chems., Chennai) and ZrO_2 (Loba Chem., Mumbai). All the powders were having more than 99% purity. The powders were taken in a suitable stachiometry for 20 gm of samples. The powders were thoroughly mixed in an agate mortar in dry and wet mixing with appropriate amount of acetone for 6 hr. After proper mixing, mixed powders were calcinated at 1300°C for 2 hr, 1400°C for 2 hr and 1500°C for 4 hr. A small amount polyvinyl alcohol was added to the calcinated powder for fabrication of pellets, which was burnt out during high temperature sintering. The circular disc shaped pellets were prepared by applying a uniaxial pressure of 4.5×10^6 N/m^2. The pellets were subsequently sintered at an optimized temperature of 1550°C for 5 hr. A preliminary study on compound formation and structural parameters was carried out using an X-ray diffraction (XRD) technique with an X-ray powder diffractometer. The XRD pattern of the calcinated powder was recorded at room temperature using PANAlytical X'pert pro with CuK_a radiation (1.5405 Å) in a wide range of Bragg's angles $2\theta (15 \leq 2\theta \leq 80°)$. Micro structures of sintered pellets were recorded by scanning electron microscope (SEM) (JEOM JSM-6380 LA). The pellets were then electrode with high purity air-drying silver paste and then dried at 500°C for 1 hr. Dielectric measurement analysis was done using Agilent E4980A Precision LCR meter with temperature (150 - 573 K) and frequency (20 Hz - 200 KHz).

3. Results and Discussion

3.1. Structural Analysis

Figure 1 shows the XRD pattern of the Zr doped BMZT (0.1, 0.2, 0.4) samples. The XRD analysis provides that

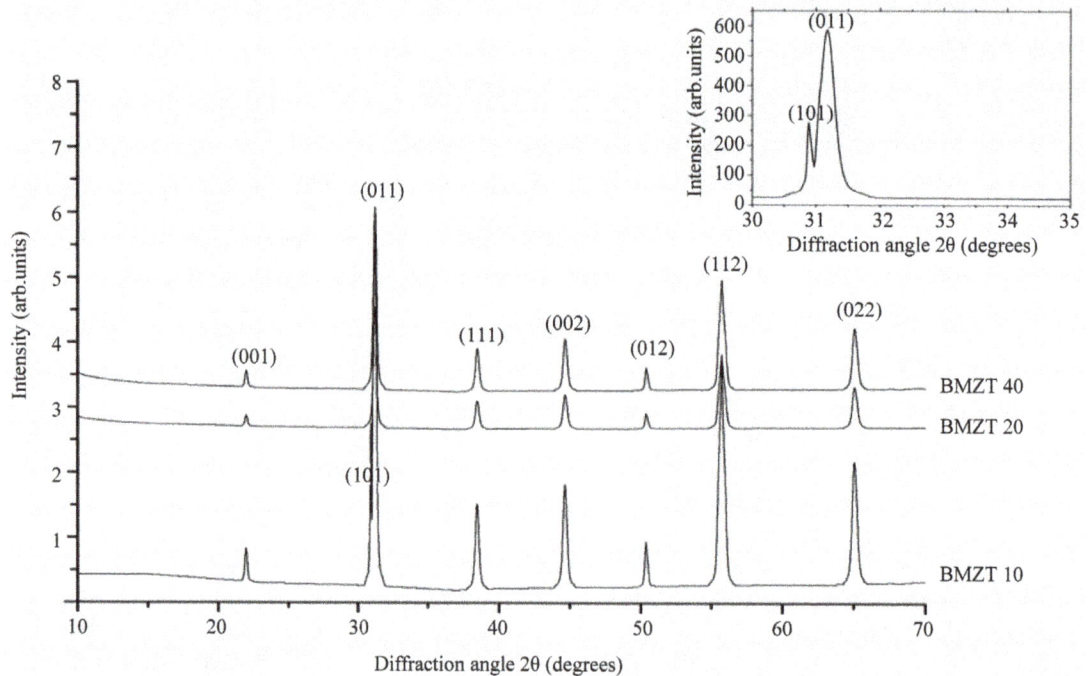

Figure 1. X-ray diffractograms of BMZT 10, BMZT 20 and BMZT 40 samples..

the samples are having single perovskite structure. BaTiO$_3$ (BTO) has the tetragonal structure at room temperature. The ionic radii of Ba^{2+} and Ti^{4+} are 1.35 Å and 0.605 Å respectively. If we doped BTO with Mg^{2+} and Zr^{4+} whose ionic radii are both 0.72 Å Mg occupies A site and Zr occupies B site of BTO. The pure BMZT single phased tertagoganl structure when the Mg content is <1.5% at-% (9) and Zr is 0.1%, if Zr content is <0.42% at-% (10) the sample is changes into the cubic structure. By doping with Zr the diffraction angles are shifted towards the lower angle side indicating the increase in lattice parameters due to the incorporation of smaller content of Zr in place of Ba. In **Figure 1** BMZT 10 sample possesses the tetragonal structure and BMZT 20 and BMZT 40 samples posses the cubic structure.

3.2. Microstructural Analysis

Figure 2 shows The SEM micrographs BMZT 10, BMZT 20 and BMZT 40 samples. It is found that the average grain size of samples are ~1.00, ~1.10 and ~1.66 μm increased as the Zr content increases from 10% to 40%. This increase is in agreement with our XRD pattern. Moreover the surface observation shows a good density of grains with some porosity.

3.3. Dielectric Properties

3.3.1. Temperature Dependence Dielectric Properties

Figure 3 shows the temperature dependence of the dielectric constant and loss of Zr doped BMZT samples measured at 1 MHz. The figure shows, the value of dielectric constant increases gradually to a maximum value (ε_m) with increase in temperature up to transition temperature and then decreases indicating a phase transition. It is also found that the Curie temperature T$_c$ of BMZT samples with Zr dopant of (0.10, 0.20, 0.40) corresponding to the maximum dielectric constant is 373, 323 and 180 respectively. The results indicates that the curie temperature of BMZT decreased may be due to Ti ions replaced by Zr ions and Zr ionic radius is little more, it can increase the grain size and exactly not joining the Zr atoms in Ti sites, due to the Zr ions conducts the little current then the dielectric constant and curie temperature both may be decreased. According to **Figures 3(a)-(c)** the peak value of the dielectric constant of BMZT samples with the Zr dopant of (0.10, 0.20, 0.40) is 1406, 1040 and 563 respectively. The result indicates that the peak value of dielectric constant for low doped sample is the maximum and the peak value decreases with Zr content.

(a)

(b)

(c)

Figure 2. SEM micrograph of (a) BMZT 10, (b) BMZT 20, (c) BMZT 40 samples.

In **Figure 3(c)** shows that the dielectric loss initially increases with temperature reaches maximum. Further increase in temperature loss is decreased but for BMZT samples of (0.10, 0.20) it is at lower temperature little bit high value of loss due to the presence of all types of polarisation and may be due to the contribution of finite resistivity of the materials. Further increase in temperature loss decreases minimum and further increasing temperature loss also increased.

3.3.2. Frequency Dependence Dielectric Properties
As shown in **Figure 4(a)** it is found that the dielectric constant of BMZT (0.10, 0.20, 0.40) decreased rapidly at low frequencies. At very high frequencies dielectric constant is very low and it maintains constant value. It may be due to there must be defects with opposite charges (dipoles) to preserve charge neutrality. Theses dipoles could be oriented to align the direction of the applied electric field. When the frequency increases, the dipoles do not catch up with the change of the electric field to complete polarisation so that the dielectric constant decreases.

In the **Figure 4(b)** the dielectric losses were a combined result of electrical conduction and orientational polarisation of the matter. The energy losses, which occur in dielectrics due to dc conductivity and dipole relaxation. The loss factor of a dielectric material is a useful indicator of the energy loss as heat.

4. Conclusion

Perovskite types $(Ba_{0.9}Mg_{1.0})(Zr_xTi_{1−x})O_3$ (with x = 0.10, 0.20, 0.40) ceramics have prepared through solid state reaction route. The XRD study at room temperature suggests that the composition of BMZT 10 has single phase

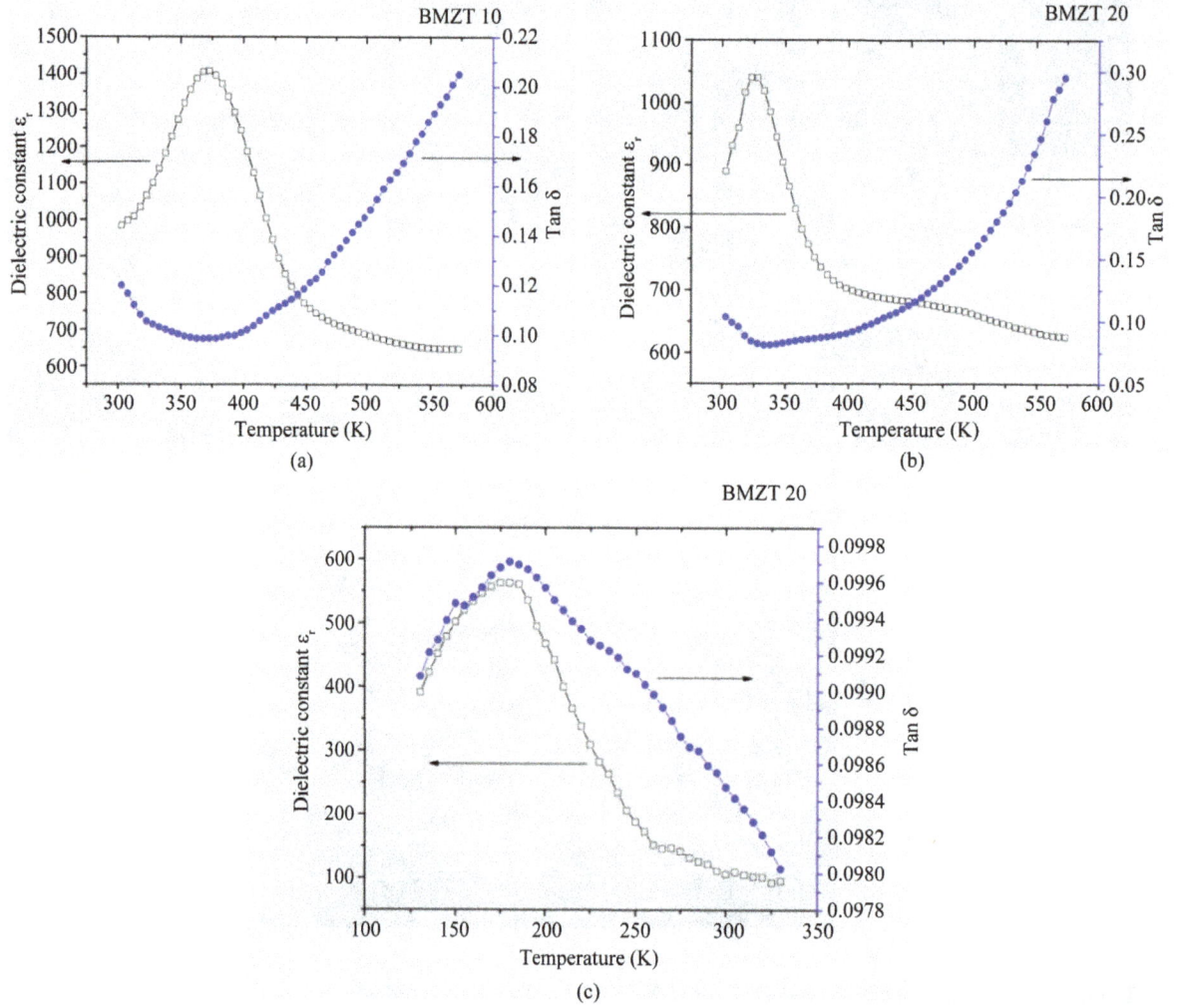

Figure 3. Temperature dependence of Dielectric constant and Dielectric loss of (a) BMZT 10, (b) BMZT 20, (c) BMZT 40 samples.

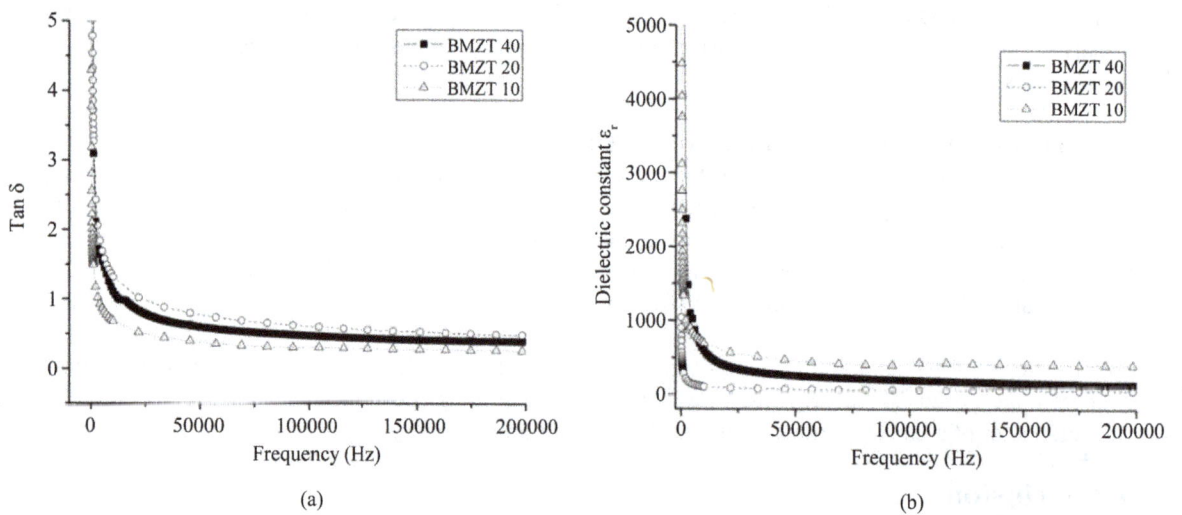

Figure 4. Frequency dependence of (a) Dielectric constant, (b) Dielectric loss of BMZT 10, BMZT 20 and BMZT 40 samples.

tetragonal and BMZT 20, BMZT 40 cubic symmetry with space group pm-3m. The dielectric study reveals that the materials undergo BMZT 20, BMZT 40 a diffuse type ferroelectric phase transition. The transition temperature decreased with Zr content and the maximum dielectric constant also decreased with Zr content.

References

[1]   Takenaka, T. and Nagata, H. (2005) Current Status and Prospects of Lead-Free Piezoelectric Ceramics. *Journal of the European Ceramic Society*, **25**, 2693-2700. http://dx.doi.org/10.1016/j.jeurceramsoc.2005.03.125

[2]   Saito, Y., Takao, H., Tani, T., *et al.* (2004) Lead-Free Piezoceramics. *Nature*, **432**, 84-87. http://dx.doi.org/10.1038/nature03028

[3]   Dixit, A., Majumder, S.B., Katiyar, R.S. and Bhalla, A.S. (2003) Relaxor Behavior in Sol-Gel-Derived $BaZr_{(0.40)}Ti_{(0.60)}O_3$ Thin Films. *Applied Physics Letters*, **82**, 2679. http://dx.doi.org/10.1063/1.1568166

[4]   Dobal, P.S., Katiyar, R.S. and Raman, J. (2002) Studies on Ferroelectric Perovskites and Bi-Layered Compounds Using Micro-Raman Spectroscopy. *Journal of Raman Spectroscopy*, **33**, 405-423. http://dx.doi.org/10.1002/jrs.876

[5]   Dixit, A., Majumder, S.B., Savvinov, A., Katiyar, R.S., Guo, R. and Bhalla, A.S. (2002) Investigations on the Sol-Gel-Derived Barium Zirconium Titanate Thin Films. *Materials Letters*, **56**, 933-940. http://dx.doi.org/10.1016/S0167-577X(02)00640-7

[6]   Paik, D.S., Park, S.E., Wada, S., Liu, S.F. and Shrout, T.R. (1999) E-Field Induced Phase Transition in <001>-Oriented Rhombohedral $0.92Pb(Zn_{1/3}Nb_{2/3})O_3$-$0.08PbTiO_3$ Crystals. *Journal of Applied Physics*, **85**, 1080. http://dx.doi.org/10.1063/1.369252

[7]   Yu, Z., Ang, C., Guo, R. and Bhalla, A.S. (2002) Dielectric Properties and High Tunability of $Ba(Ti_{0.7}Zr_{0.3})O_3$ Ceramics under dc Electric Field. *Applied Physics Letters*, **81**, 1285. http://dx.doi.org/10.1063/1.1498496

[8]   Dash, S.K., Kant, S., Dalai, B., Swain, M.D. and Swain, B.B. (2014) Characterization and Dielectric Properties of Barium Zirconium Titanate Prepared by Solid State Reaction and High Energy Ball Milling Processes. *Journal of Applied Physics*, **88**, 129-135.

[9]   Badapanda, T., Cavalcante, L.S., da Luz Jr., G.E., Batista, N.C., Anwar, S. and Longo, E. (2013) Effect of Yttrium Doping in Barium Zirconium Titanate Ceramics: A Structural, Impedance, and Modulus Spectroscopy Study. *Metallurgical and Materials Transactions A*, **44**, 4296.

[10]   Chen, T., Zhang, T., Zhou, J.F., Zhang, J.W., Liu, Y.H. and Wang, G.C. (2012) Piezoelectric Properties of $[(K_{1−x}Na_x)0.95Li_{0.05}]0.985Ca_{0.015}(Nb_{0.95}Sb_{0.05})0.985Ti_{0.015}O_3$ Lead-Free Ceramics. *Indian Journal of Physics*, **86**, 443-446. http://dx.doi.org/10.1007/s12648-012-0087-1

[11]   Mitic, V.V., Nikolic, Z.S., Pavlovic, V.B., Paunovic, V., Miljkovic, M., Jordovic, B. and Zivkovic, L. (2010) Influence of Rare-Earth Dopants on Barium Titanate Ceramics Microstructure and Corresponding Electrical Properties. *Journal of the American Ceramic Society*, **93**, 132-137. http://dx.doi.org/10.1111/j.1551-2916.2009.03309.x

[12]   Jung, W.S., Kim, J.H., Kim, H.T. and Yoon, D.H. (2010) Effect of Temperature Schedule on the Particle Size of Barium Titanate during Solid-State Reaction. *Materials Letters*, **64**, 170-172. http://dx.doi.org/10.1016/j.matlet.2009.10.035

[13]   Cao, W.Q., Xiong, J.W. and Sun, J.P. (2007) Dielectric Behavior of Nb-Doped $Ba(Zr_xTi_{1−x})O_3$. *Materials Chemistry and Physics*, **106**, 338-342. http://dx.doi.org/10.1016/j.matchemphys.2007.06.017

[14]   Singh, S.V., Thakur, A.N., Singh, O.P., Kumar, S.C. and Ahmad, A. (2009) Dielectric Properties of $PbSrWO_4$ and $PbBaWO_4$ Compounds. *Indian Journal of Physics*, **83**, 375-381. http://dx.doi.org/10.1007/s12648-009-0125-9

[15]   Chen, T., Wang, H.L., Zhang, T., Zhou, J.F., Zhang, J.W., Liu, Y.H. and Wang, G.C. (2013) Piezoelectric Properties of La and Nb Co-Modified $Bi_4Ti_3O_{12}$ High-Temperature Ceramics. *Indian Journal of Physics*, **87**, 629-631. http://dx.doi.org/10.1007/s12648-013-0278-4

[16]   Parkash, O., Kumar, D., Dwivedi, R.K., Srivastava, K.K., Singh, P. and Singh, S. (2007) Effect of Simultaneous Substitution of La and Mn on Dielectric Behavior of Barium Titanate Ceramic. *Journal of Materials Science*, **42**, 5490-5496. http://dx.doi.org/10.1007/s10853-006-0985-8

[17]   Langhammer, H.T., Müller, T., Böttcher, R. and Abicht, H.P. (2008) Structural and Optical Properties of Chromium-Doped Hexagonal Barium Titanate Ceramics. *Journal of Physics: Condensed Matter*, **20**, Article ID: 085206. http://dx.doi.org/10.1088/0953-8984/20/8/085206

[18]   Lu, D.Y., Toda, M. and Sugano, M. (2006) High-Permittivity Double Rare-Earth-Doped Barium Titanate Ceramics with Diffuse Phase Transition. *Journal of the American Ceramic Society*, **89**, 3112-3123. http://dx.doi.org/10.1111/j.1551-2916.2006.00893.x

[19]   Chen, Z.W. and Chu, J.Q. (2008) Piezoelectric and Dielectric Properties of $Bi_{0.5}(Na_{0.84}K_{0.16})_{0.5}TiO_3$-$Ba(Zr_{0.04}Ti_{0.96})O_3$ Lead Free Piezoelectric Ceramics. *Advances in Applied Ceramics*, **107**, 222-226.

http://dx.doi.org/10.1179/174367608X263403

[20] Fu, C.L., Cai, W., Chen, H.W., Feng, S.C., Pan, F.S. and Yang, C.R. (2008) Voltage Tunable $Ba_{0.6}Sr_{0.4}TiO_3$ Thin Films and Coplanar Phase Shifters. *Thin Solid Films*, **516**, 5258-5261. http://dx.doi.org/10.1016/j.tsf.2007.07.059

[21] Cai, W., Fu, C.L., Gao, J.C. and Chen, H.Q. (2009) Effects of Grain Size on Domain Structure and Ferroelectric Properties of Barium Zirconate Titanate Ceramics. *Journal of Alloys and Compounds*, **480**, 870-873. http://dx.doi.org/10.1016/j.jallcom.2009.02.049

[22] Du, F.T., Yu, P.F., Cui, B., Cheng, H.O. and Chang, Z.G. (2009) Preparation and Characterization of Monodisperse Ag Nanoparticles Doped Barium Titanate Ceramics. *Journal of Alloys and Compounds*, **478**, 620-623. http://dx.doi.org/10.1016/j.jallcom.2008.11.099

[23] Yuan, Y., Zhang, S.R., Zhou, X.H. and Tang, B. (2009) Effects of Nb_2O_5 Doping on the Microstructure and the Dielectric Temperature Characteristics of Barium Titanate Ceramics. *Journal of Materials Science*, **44**, 3751-3757. http://dx.doi.org/10.1007/s10853-009-3502-z

[24] Xiao, S.X. and Yan, X.P. (2009) Preparation and Characterization of Si-Doped Barium Titanate Nanopowders and Ceramics. *Microelectronic Engineering*, **86**, 387-391. http://dx.doi.org/10.1016/j.mee.2008.11.042

[25] Rath, M.K., Pradhan, G.K., Pandey, B., Verma, H.C., Roul, B.K. and Anand, S. (2008) Synthesis, Characterization and Dielectric Properties of Europium-Doped Barium Titanate Nanopowders. *Materials Letters*, **62**, 2136-2139. http://dx.doi.org/10.1016/j.matlet.2007.11.033

[26] Gulwade, D. and Gopalan, P. (2008) Diffuse Phase Transition in La and Ga Doped Barium Titanate. *Solid State Communications*, **146**, 340-344. http://dx.doi.org/10.1016/j.ssc.2008.02.018

[27] Unruan, M., Sareein, T., Tangsritrakul, J., Prasetpalichatr, S., Ngamjarurojana, A., Anata, S. and Yimnirun, R. (2008) Changes in Dielectric and Ferroelectric Properties of Fe^{3+}/Nb^{5+} Hybrid-Doped Barium Titanate Ceramics under Compressive Stress. *Journal of Applied Physics*, **104**, Article ID: 124102. http://dx.doi.org/10.1063/1.3042228

[28] Yaseen, H., Baltianski, S. and Tsur, Y. (2006) Effect of Incorporating Method of Niobium on the Properties of Doped Barium Titanate Ceramics. *Journal of the American Ceramic Society*, **89**, 1584-1589. http://dx.doi.org/10.1111/j.1551-2916.2006.00966.x

[29] Cha, S.H. and Han, Y.H. (2006) Effects of Mn Doping on Dielectric Properties of Mg-Doped $BaTiO_3$. *Journal of Applied Physics*, **100**, Article ID: 104102. http://dx.doi.org/10.1063/1.2386924

[30] Shen, Z.J., Chen, W.P., Qi, J.Q., Wang, Y., Chan, H.L.W., Chen, Y. and Jiang, X.P. (2009) Dielectric Properties of Barium Titanate Ceramics Modified by SiO_2 and by $BaO-SiO_2$. *Physica B: Condensed Matter*, **404**, 2374-2376. http://dx.doi.org/10.1016/j.physb.2009.04.039

[31] Kirianov, A., Hagiwara, T., Kishi, H. and Ohsato, H. (2002) Effect of Ho/Mg Ratio on Formation of Core-Shell Structure in $BaTiO_3$ and on Dielectric Properties of $BaTiO_3$ Ceramics. *Japanese Journal of Applied Physics*, **41**, 6934-6937. http://dx.doi.org/10.1143/JJAP.41.6934

[32] Wang, S., Zhang, S.R., Zhou, X.H., Li, B. and Chen, Z. (2005) Effect of Sintering Atmospheres on the Microstructure and Dielectric Properties of Yb/Mg Co-Doped $BaTiO_3$ Ceramics. *Materials Letters*, **59**, 2457-2460. http://dx.doi.org/10.1016/j.matlet.2005.03.016

[33] Rout, S.K., Sinha, E. and Panigrahi, S. (2007) Dielectric Properties and Diffuse Phase Transition in $Ba_{1-x}Mg_xTi_{0.6}Zr_{0.4}O_3$ Solid Solutions. *Materials Chemistry and Physics*, **101**, 428-432. http://dx.doi.org/10.1016/j.matchemphys.2006.08.002

[34] Henning, D., Schnell, A. and Simon, G. (1982) Diffuse Ferroelectric Phase Transitions in $Ba(Ti_{1-y}Zr_y)O_3$ Ceramics. *Journal of the American Ceramic Society*, **65**, 539-544. http://dx.doi.org/10.1111/j.1151-2916.1982.tb10778.x

[35] Yu, Z., Guo, R. and Bhalla, A.S. (2000) Dielectric Behavior of $Ba(Ti_{1-x}Zr_x)O_3$ Single Crystals. *Journal of Applied Physics*, **88**, 410. http://dx.doi.org/10.1063/1.373674

7

Temperature-Frequency Characteristics of Dielectric Properties of Compositions LDPE + $x$vol%Bi$_2$Te$_3$

E. M. Gojayev[1], A. Y. Ismailova[2], S. I. Mammadova[1], G. S. Djafarova[1]

[1]Azerbaijan Technical University, Baku, Azerbaijan
[2]Ganja State University, Ganja, Azerbaijan
Email: geldar-04@mail.ru

Abstract

In the paper, the results of investigations of temperature and frequency dependences of dielectric permeability and dielectric loss of compositions LDPE + $x$vol%Bi$_2$Te$_3$ are stated. The investigations were carried out at frequency $10 - 10^5$ Hz and temperature 20°C - 150°C intervals, respectively. It was revealed that increase of percentage of the filler Bi$_2$Te$_3$ in the matrix, reduces to increase of dielectric permeability and dielectric loss of composites LDPE + $x$vol%Bi$_2$Te$_3$ in connection with the change reducing to Maxwell-Wagner's volume polarization and emergence of comparative strong inner field in semiconductor clusters.

Keywords

Composites LDPE + $x$vol%Bi$_2$Te$_3$, Dielectric Permeability, Dielectric Loss, Semiconductor Clusters

1. Introduction

Creation of composite materials is one of the basic directions in development of new prospective materials. Filling of polymers reduces to changes in characteristics of supramolecular structuration and in density of packaging as solid high-dispersive fillers may serve as builders of nucleus of crystals or their imperfections [1]. The fillers have considerable influence on mobility of different kinetic units of a polymer and on spectrum of its relaxation time. The filler particles play the role of a structuration, and the boundary layer of a polymer with filler has a special saturation structure. These are trapping cites with different values of energy of activation where the electrons are stabilized and as a result, electroactive properties of polymers are improved. It should be noted that depending on the nature, size, form and distribution character of filler, the obtained polymeric composition may be electro conducting, antistatic or dielectric [2]-[4].

Recently, in place of a filler semi-conductor compounds are frequently used, and the materials of scientific-practical interest have already been obtained [5]. It was revealed that with use of a filler of threefold compounds as $A^{III}B^{III}C_2^{VI}$ based on polyethylene, one can obtain a new class of electret materials with record time of life [6].

In the present paper, we give the results of investigations of dielectric properties of composite materials based on lower density polyethylene (LDPE) filled with semiconductor compound Bi_2Te_3.

2. Experimental Technique

Composition samples were obtained by mechanical mixing of powder Bi_2Te_3 with powder of LDPE in a porcelain motor. The mixing is continued up to receiving homogeneous mixture. The mixture some time is maintained at melting temperature of polymer under pressure 5 MPa. At the same temperature, by pressing the homogeneous mixture, the pressure slowly increases to 15 MPa. At this pressure, the sample is maintained within 5 minutes, and then is cooled in water. Herewith the sizes of the samples are: the thickness is about 80 - 120 mkm, diameter of the obtained samples 35 mm. In order to provide reliable electric contact between the samples and electrodes made of the stainless steel, the electrodes made of a thin aluminum foil of 7 mkm in thickness pressed on both working faces of the samples, are used.

Dielectric permeability and dielectric loss of LDPE + $x$vol%Bi_2Te_3 were measured in the range of 296 - 520 K at linear growth of temperature with velocity 2.5 degree/min and in frequency range 1 by the technique described in the paper [6].

The reliable electric contact of electrodes made of stainless steel of diameter 15 mm was provided by using pressed electrodes made of aluminum foil of 7 mkm in thickness. The value of electric capacitance (C) and tangent of dielectric loss angle $(\mathrm{tg}\delta)$ of the investigated sandwich structures were determined by means of the device of the brand E7−20. The samples were located into a measuring cell with pressing electrodes that in its turn was located into the heating system. Measuring of capacity and $\mathrm{tg}\delta$ was carried out in freshly prepared samples, and the quantity of the modulus of the complex of dielectric permeability (ε) were determined by the known formula

$$\varepsilon = \frac{Cd}{\varepsilon_0 S}$$

where, C is the measured electrical capacitance of the sample, F; electrical constant $\varepsilon_0 = 8.85 \times 10^{-12}$ F/m; d—thickness of the sample, m; S—area of the sample. In experiments, frequency of the given electric field changed from 10^2 to 10^6 Hz, and temperature range 20°C - 140°C. Measuring voltage amplitude is 1 B. The temperature was determined by means of a standard thermocouple copper-constantan.

3. Experimental Results and Discussions

Temperature dependences of dielectric permeability and dielectric losses of composite materials LDPE + $x$vol%Bi_2Te_3 at temperature range 20°C - 160°C were studied. The results of investigations are given in **Figures 1-4**. As it follows from **Figure 1** for the composite 5 vol%Bi_2Te_3 in the investigated temperature range in $\varepsilon(T)$ dependence at 50°C and 130°C weak maxima are observed, and on the whole ε decreases due to temperature growth.

For the composite 10 vol%Bi_2Te_3 in all temperature range, the dielectric permeability remains constant. With increasing the content of Bi_2Te_3 to 15 vol%, on temperature dependence of dielectric permeability weak maxima at 70°C and 140°C, minimum for 100°C were revealed. Temperature dependences of dielectric loss of composites LDPE + $x$vol%Bi_2Te_3 were also studied. The investigations were carried out in temperature interval 20°C - 150°C. The results are given in **Figure 3**. As is seen from **Figure 4** for a composite with the filler LDPE + 5 vol%Bi_2Te_3 on the curve of $\mathrm{tg}\delta(T)$ dependence for 80,102 and 116°C the weakly expressed maxima are observed. On the whole, in the studied temperature range, $\mathrm{tg}\delta$ increases due to temperature growth. With increasing the volumetric content of the filler to LDPE + 10vol%Bi_2Te_3 the curve smoothies out and maxima disappear. However, also for this composite $\mathrm{tg}\delta$ increases according to temperature growth. For the composite LDPE + 15vol%Bi_2Te_3 the $\mathrm{tg}\delta(T)$ dependence becomes linear and change of $\mathrm{tg}\delta$ with temperature is slight.

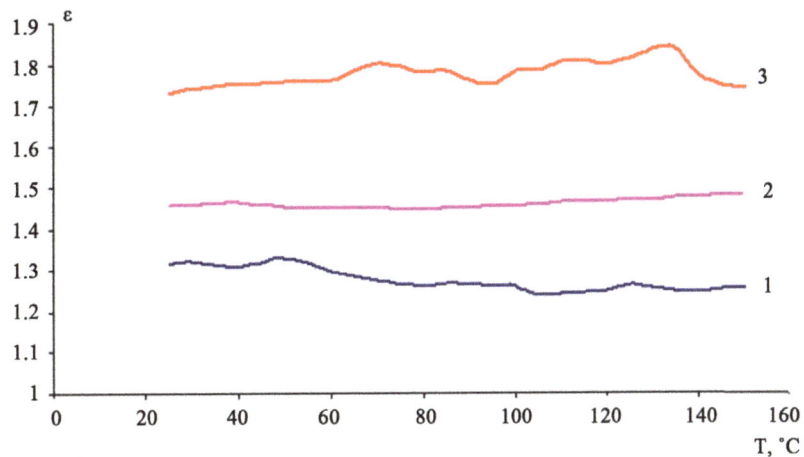

Figure 1. Temperature dependences of dielectric permeability of composite materials LDPE + $x$vol%Bi$_2$Te$_3$, 1 - 5, 2 - 10; 3 - 15.

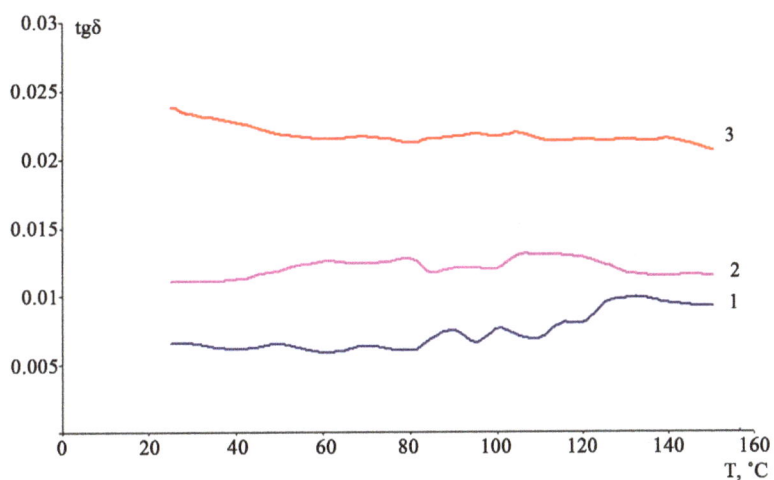

Figure 2. Temperature dependences of dielectric loss of composite materials LDPE + $x$vol%Bi$_2$Te$_3$, 1 - 5, 2 - 10; 3 - 15.

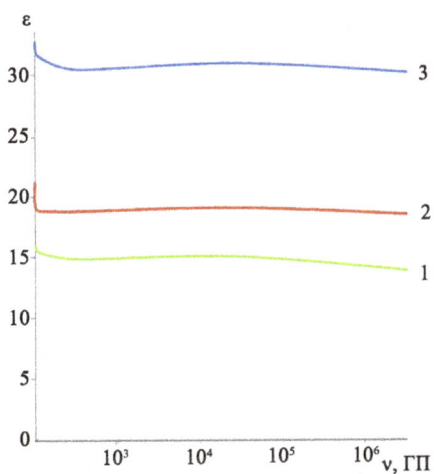

Figure 3. Frequency dependence of dielectric permeability of composite materials LDPE + $x$vol%Bi$_2$Te$_3$, 1 - 5, 2 - 10; 3 - 15.

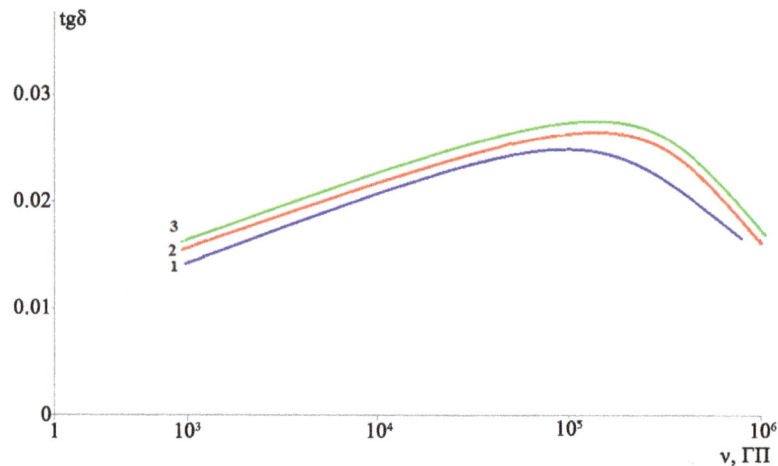

Figure 4. Frequency dependence of dielectric loss of composite materials LDPE + $x$vol%Bi$_2$Te$_3$, 1 - 5, 2 - 10; 3 - 15.

The results of investigation of frequency dependence of dielectric permeability of composites LDPE + $x$vol%Bi$_2$Te$_3$ are given in **Figure 3**. As it follows from **Figure 3**, in the frequency range 0 - 10^3 Hz, the dielectric permeability slightly decreases, and with further increase of frequency to 10^6 Hz remains practically constant. With increasing volumetric content of the filled, ε increases. Dispersions in $\varepsilon(v)$ dependence were not found.

The results of investigations of frequency dependence of dielectric loss of composites LDPE + $x$vol%Bi$_2$Te$_3$ are reduced in **Figure 4**. As it follows from **Figure 4**, in all the investigated composites in frequency range 10^3 - 10^5 Hz the typical is that dielectric loss increases, while in the range 10^5 - 10^6 Hz decreases. Dispersions in tg$\delta(v)$ dependence were not found.

Thus, the analysis of the obtained results shows that with increasing the volumetric content of the filler Bi$_2$Te$_3$ dielectric permeability and dielectric loss increase. Apparently this is connected with the fact that increase of the volumetric content of the Bi$_2$Te$_3$ filler reduces to growth of the number of particles Bi$_2$Te$_3$ per cross sections of the composite, and this is equivalent to the part of Bi$_2$Te$_3$ in the general thickness of sample. The clusters closed with each other in the sample's thickness may be considered as pure resistance included between electrodes. Since Bi$_2$Te$_3$ has high conductivity compared with LDPE, we can assume that the composite's resistance will be especially determined between the particles Bi$_2$Te$_3$ on the boundaries of clusters (the clusters are surrounded with PE layers with small ε). Accumulation and redistribution of free electric charges that distort initial inner electric field, occurs in alternating current. At lower frequencies, the inner electric fields are distributed according to conductivities. Consequently, change of dielectric parameters due to increase of frequency and including temperature may be explained by emergence of a comparatively strong inner field in semi-conductor clusters.

4. Conclusion

By researches of temperature and frequency characteristics of dielectric permeability and dielectric loss of composites PELD + $x$vol%Bi$_2$Te$_3$, it was revealed that with a variation of the volume maintenance of a filler, temperature and frequency, it is possible to receive composites with the demanded dielectric parameters.

References

[1]   Ushakov, N.M., Ulzutueva, N. and Kosobudsky, I.D. (2008) Termo dielectric Properties of Polymer Composite Na-nomaterials Based on Copper and Copper-Oxide Matrix Density Polyethylene. *TFF*, **78**, 65-69.

[2]   Alyoshin, A.N. and Alexandrova, W.L. (2008) Switching and Memory Effects Caused by Hopping Mechanism of Charge Transport in Composite Films Based on Conducting Polymers and Inorganic Nanoparticles. *Physics State Solids*, **50**, 1895-1900.

[3]   Deadlock, A.B. and Garmashov, S.I. (2011) Dielectric Loss in Statistical Mixtures. *FTT*, **53**, 1129-1132.

[4]   Sokolov, E.M., Babenko, S.D. and Morivsky, A.P. (2010) The Dielectric Properties of the Composites Modified with

Carbon Nano Struktur Microwave. *JSF*, **80**, 83-87.

[5]   Gojayev, E.M., Maharramov, V.M., Osmanov, S.S. and Allahyarov, E.M. (2007) The Charge State of the Compositions Based on Polyethylene with a Semiconductor Filler TlInSe2. *Electronic Processing of Materials*, **2**, 84-88.

[6]   Gojayev, E.M., Ahmadova, Kh.R., Safarova, S.I., Djafarova, G.S. and Mextiyeva, Sh.M. (2015) Effect of Aluminum Nano-Particles on Microrelief and Dielectric Properties of PE + TlInSe$_2$ Composite Materials. *Open Journal of Inorganic Non-Metallic Materials*, **5**, 11-19. http://dx.doi.org/10.4236/ojinm.2015.51002

8

Effect of NaCl Doping on Growth, Characterization, Optical and Dielectric Properties of Potassium Hydrogen Phthalate (KHP) Crystals

R. K. Raju[1], S. M. Dharamaprakash[2], H. S. Jayanna[1]*

[1]Department of Physics, Kuvempu University, Shankaraghatta, India
[2]Department of Physics, Mangalore University, Mangalore, India
Email: *jayanna60@gmail.com

Abstract

Single crystals of sodium chloride doped Potassium Hydrogen Phthalate (KHP) were grown from aqueous solution by slow evaporation method at room temperature. The powder X-ray diffraction analysis was carried out and lattice cell parameters estimated. FTIR studies confirm the presence of functional groups and slight distortion of groups due to doping of sodium metal. UV-visible and photoluminescence spectral studies revealed to understand the optical properties. The NLO property of grown crystals has been confirmed by Kurtz powder technique. Dielectric studies of samples showed that dielectric constant decreased slowly with increasing frequency and attains saturation at higher frequencies.

Keywords

Crystal Growth, Second Harmonic Generation, Lattice Parameters, Dielectric Constant, Photoluminescence

1. Introduction

Second order nonlinear optical (NLO) materials have recently attracted much attention due to their potential applications in emerging optoelectronic technologies [1]. Materials with large optical nonlinearities with stable physical and thermal efficiencies required for many of these applications. Non-linear optical (NLO) materials

---

*Corresponding author.

have attracted and gained enormous demand due to their wide applications in the recent technologies like optoelectronics, optical communication and data storage systems [2] [3]. Potassium hydrogen phthalate (KHP) crystal, with chemical formula K ($C_6H_4COOH \cdot COO$), is well known material for its application in the production of crystal analyzers for long-wave X-ray spectrometers [4]-[7]. Phthalate single crystals are piezoelectrics with high coefficients of acousto-optical interaction [8] [9]. KHP crystallizes in orthorhombic structure with space group $Pca2_1$ [10] [11]. It has platelet morphology with cleavage along (010) plane. This feature allows one to use them for data processing and intra laser modulation in various acousto optical devices [12]-[14]. It is important to search for new NLO material, which possesses large NLO coefficient, shorter cutoff wavelength, transparency in the UV region and higher laser damage threshold [15] [16]. Influence of alkali metal has strong effect on the material properties like morphology, optical and thermoluminescene [17] [18].

2. Experimental

2.1. Crystal Growth

Potassium Hydrogen Phthalate (KHP) analytical reagent (AR) grade was purified by repeated recrystallisation using double distilled water as solvent. The crystals were grown by slow evaporation solution growth technique (SEST) at room temperature. A saturated solution was prepared (12 g/100ml from literature) under slightly acidic conditions pH is 4.8. Sodium Chloride (AR grade) of different concentration (1M%, 3M% and 5M%), were prepared and added 5 ml each separately to the supersaturated solution of KHP. Pure and doped solution was stirred for 7 - 8 hr using magnetic stirrer for homogeneous mixing. After homogeneous mixing solutions were transferred to clean Petri dish covered with polythene cover and perorations were made on polythene covers for proper evaporation of the solvent. The whole setup was kept in dust free area and closely monitored. Small crystals appeared in the beginning about 4 - 5 days for both pure and doped KHP, due to slow evaporation and grew larger in considerable time of about 15 - 20 days. At higher concentration of dopant, the adsorption film blocks the growth surface and inhibits the growth process [19].

Good quality optically transparent large size crystals were selected for carrying out the measurements. Photographs of grown doped and undoped crystals as shown in **Figure 1**.

2.2. Characterisation

In order to confirm the material of the crystal powder X-ray diffraction (PXRD) in the 2θ range 5° to 70° with Cu K_α ($\lambda = 1.5418$ Å). The TG-DTA studies were carried out using TA-Instruments model NETZSCH TG 209 F1 with heat range 10°C/min. The FT-IR spectrum was recorded using Bruker-alpha for all crystals grown in the range of 400 cm^{-1} to 4000 cm^{-1}. Ocean Optics UV-Vis-NIR spectrometer was used to study optical transparency

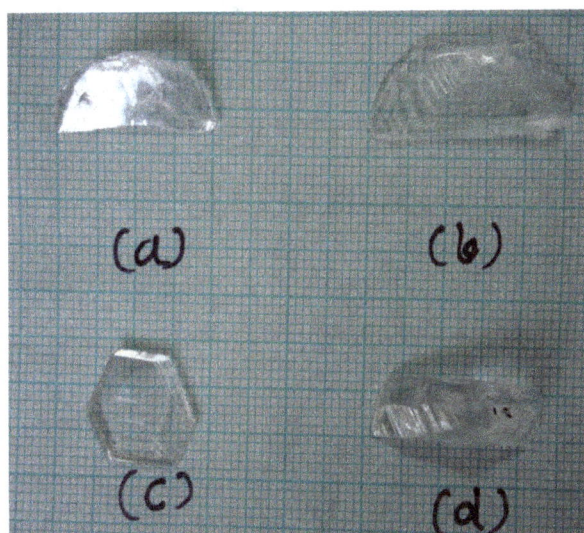

Figure 1. Photograph of grown crystals: (a) K0: Pure; (b) K1: 1M%; (c) K2: 3M% and (d) K3: 5M% doped KHP.

of the crystals between 200 nm to 900 nm. The SGH test was carried by the Kurtz powder method. AN Nd:YAG laser with 1.064 μm was made fall on samples packed in micro-capillary tube. SHG generated by randomly oriented microcrystals were detected by photomultiplier tube after filtration of incident radiation of 1.064 μm. The frequency doubling was confirmed by Green colour of the output radiation. The SHG were carried out for all grown crystals. The Photoluminescence studies were carried using Horiba Jobin YVON LabRam equipment. The dielectric study was carried out for pure and doped samples using impendence analysis Interface LCR meter model PSM 1735 N4L at room temperature.

3. Results and Discussions

3.1. X-Ray Diffraction Analysis

The pure and NaCl doped crystals were subjected to powder X-ray diffraction. Using the JCPDS with X'pert high Score plus software data were analyzed. It showed that doped KHP crystallizes in orthorhombic system with space group $Pca2_1$. The calculated lattice parameters were in agreement with reported values [20]. The Ionic radius of K atom is slightly large compare to that of Na atom [21]. Hence it is reasonable to agree that the dopant can enter the crystalline matrix without much distortion. Very small minor changes in peak intensity due to lattice strains as result of doping but the basic crystalline structure remains the same. The obtained cell parameters are a = 9.653 Å, b = 13.461 Å, c = 6.431 Å & $\alpha = \beta = \gamma = 90°$ for pure KHP and a = 9.577 Å, b = 13.246 Å, c = 6.463 Å; a = 9.628 Å, b = 13.555 Å, c = 6.518 Å; a = 9.844 Å, b = 13.389 Å, c = 6.340 Å and $\alpha = \beta = \gamma = 90°$ respectively for 1M%, 3M% & 5M% of NaCl doped KHP crystals. The XRD spectra shown in **Figure 2**.

3.2. FT-IR Spectral Studies

The FT-IR spectral analysis was carried out in the region 4000 - 400 cm^{-1}. The spectrum is shown in **Figure 3**, the functional groups present were identified and a small shift is observed as result of NaCl doping. It is due to the lattice strain developed. In The spectra, that characteristic OH stretching peaks occur at 3441 cm^{-1}, 3443 cm^{-1}, 3452 cm^{-1} for pure (0M%) and doped (3M% & 5M%) respectively shows shifting of vibrational absorptions. This could be due to lattice strain because of doping of Na$^+$ ion into crystal lattice [20]. The some of the stretching frequencies are given below (**Table 1**).

3.3. Thermal Studies

The TG-DTA curves of pure and doped crystals are shown in **Figure 4**. The analyses were carried out for pure

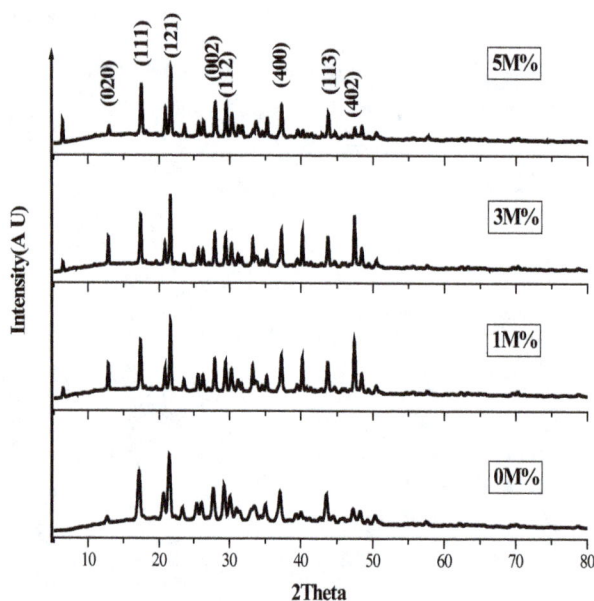

Figure 2. XRD pattern of pure and doped crystals.

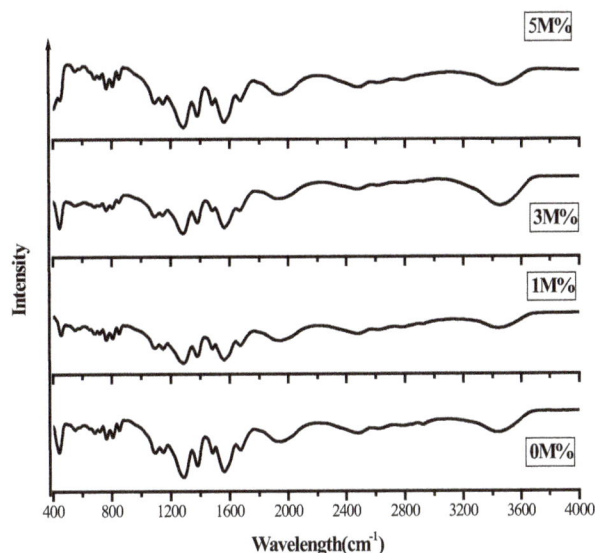

Figure 3. FTIR spectra of pure and doped crystals.

Table 1. FT-IR frequencies assignments of pure and doped crystals.

Functional groups	Pure KHP cm^{-1}	KHP + 1M% cm^{-1}	KHP + 3M% cm^{-1}	KHP + 5M% cm^{-1}
C=O symmetrical stretching	1566	1566	1565	1560
O=C stretching	1482	1484	1484	1484
C-C=O stretching	1085	1085	1085	1092
C-H plane bend	870	859	859	854
O-H stretching	3441	3441	3443	3452
C=C asymmetric stretching	1945	1942	1940	1955
C=O symmetrical stretching	441	454	444	447
C=C stretching	1381	1382	1381	1379

and doped crystals between 30°C to 1000°C in Nitrogen atmosphere with heating rate of 10°C/min. From the TG curve, it is evident that the pure KHP is stable up to 303°C [22] [23]. In doped crystals are 308°C, 308°C & 309°C respectively for 1M%, 3M% & 5M%, no decomposition. This shows that, there is a slight increase in melting point due to effect of doping ensures the stability. The TG-DTA curves for doped and undoped samples shows two stage weight loss patterns. The first major weight loss occurred between the temperatures 269°C and 303°C with 40%, this weight loss. The second stage weight loss noticed between the temperatures 303°C and 494°C experiences a weight loss of about 60% with complete decomposition of Phthalic acid. The TG curves shows that pure and doped KHP crystals are stable upto 303°C and can be designed for device application in this temperature range.

3.4. UV-Visible Spectral Studies

The grown crystals of pure and urea doped KHP were subjected to optical absorption studies. The samples scanned in the wavelength range 200 - 900 nm. The high percentage of transmission in the entire visible region is observed for all samples and an important property for NLO applications. It is observed that, no significant change in the cut-off wavelength for doped crystals. The UV-Vis spectra are shown in **Figure 5** which shows these crystals can be used for optical device fabrication.

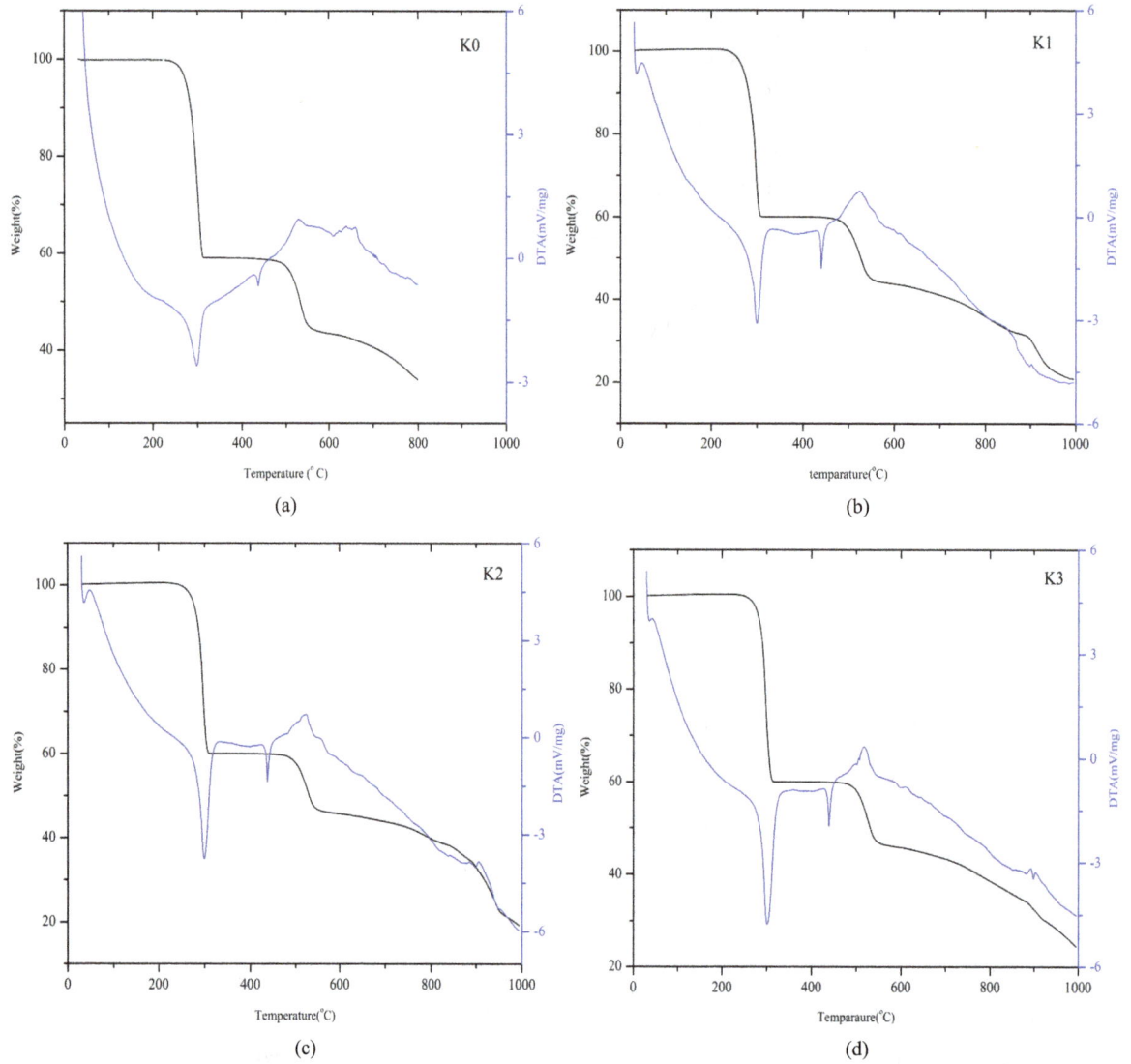

Figure 4. (a) TG-DTA of K0; (b) TG-DTA of K1; (c) TG-DTA of K2; (d) TG-DTA of K3.

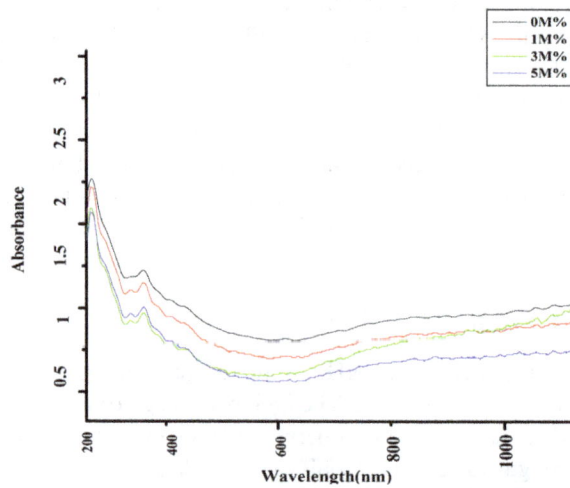

Figure 5. UV-visible spectra.

3.5. Photoluminescence Studies

The PL spectrum of pure and doped crystals when excited with laser of wavelength 325 nm are shown in **Figure 6**. Broad peaks from green to violet emissions for pure and doped KHP crystals observed. There is no significant change in peak but shift from 416 nm to 406 nm is observed with increase in dopent concentration, which shows the enhancement fluorescence which act as ligand. The intermolecular energy transfer is also facilitated which may enhance fluorescence [24] [25]. A peak at 493 nm and 498 nm corresponds to 3M% & 5M% doping of alkali metal.

3.6. NLO Studies

Second harmonic generation (SHG) test was performed on these crystals by Kurtz powder technique [26] with input radiation of 5.0 mJ/pulse. The Nd:YAG laser of 1064nm radiation was used as optical source and directed on powder sample filled in microcapillary tubes. The frequency doubling was confirmed with green radiation emission. The output SHG intensities for pure and doped samples give relative NLO efficiencies of the measured samples are tabulated below (**Table 2**) with Urea as reference sample. It is observed that enhancement of SHG efficiency with dopant and found that SGH efficiency is concentration dependent. Many materials have been identified with higher molecular non-linearity, the attainment of SHG effects requires favorable alignments of the molecule within the crystal structure which can be achieved facilitating nonlinearity in the presence of solvent [27]. The SGH can be enhanced by attaining the molecular alignment through inclusion complexation [28]. It is reported that enhancement in crystalline perfection could lead to the increase in NLO efficiency [29].

3.7. Dielectric Studies

The Dielectric measurements were done on pure and NaCl doped KHP crystals using Impendence analysis Interface LCR meter Model PSM 1735 N4L. Dielectric permittivity measurements were carried out for silver pasted pure and doped samples for electrical contact at room temperature for different frequencies. The dielectric constant have been calculated using the equation $\epsilon_r = Cd/A\epsilon_0$ where d is the thickness of the sample, A is the area of the sample. Measurements are made in the frequency in the range 1000 Hz to 35 MHz. The variations of dielectric constant with frequency for pure and doped samples are shown in the **Figure 7**. It is observed that, at low frequencies values of ϵ_r were maximum and decreased with increasing frequency and attains saturation at higher frequencies. The high value of dielectric constant of the crystal at low frequency is due to space charge polarization [30] [31]. According to the Miller rule, the lower value of dielectric constant at higher

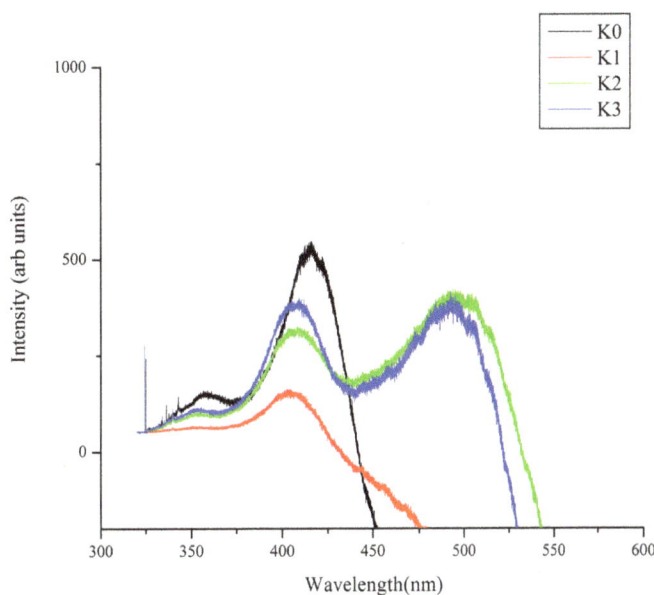

Figure 6. Photoluminiscience spectra.

Table 2. The SGH output.

Sample	SGH output I_{2w}/(mV)
Urea (ref)	320
Pure KHP	92
KHP + 1M%	110
KHP + 3M%	94

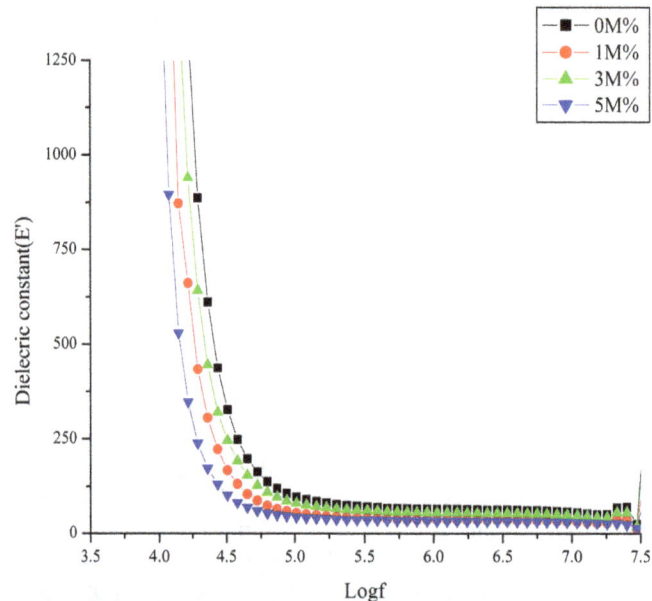

Figure 7. Dielectric studies.

frequencies is a suitable parameter for enhancement of SHG coefficient [32]. The variation of dielectric constant is due to the incorporation of dopant inside the KHP crystal lattice and also lower value for higher concentration of dopant with frequency suggests that the crystal possess enhanced optical quality with lesser defects and this property plays important role for the optoelectronic devices [33].

4. Conclusion

Single crystals of pure and NaCl doped Potassium hydrogen phthalate (KHP) have been grown by slow evaporation technique at room temperature. The powder X-ray diffraction studies confirmed the incorporation of sodium ions into the crystal lattice of KHP and also calculated lattice parameters. The FT-IR spectral analysis confirms presence of functional groups of KHP. TG-DTA analysis showed that the crystals were stable upto melting point without any decomposition. The UV-visible spectra show that pure & doped crystals have good optical transmittance in the entire visible region which is desirable property for opto-electronics. The powder SHG test confirms the NLO property of pure and doped KHP crystals. The dielectric studies of all samples show that dielectric constant decreases slowly with increasing frequency and attains saturation at higher frequencies.

References

[1]   Marcy, H.O., Warren, L.F., Webb, M.S., Ebbers, C.A., Velsko, S.P., Kennedy, G.C. and Catella, G.C. (1992) Second-Harmonic Generation in Zinc Tris(Thiourea) Sulfate. *Applied Optics*, **31**, 5051-5060. http://dx.doi.org/10.1364/AO.31.005051

[2]   Jones, L., Paschen, K.W. and Nicholson, J.B. (1963) Performance of Curved Crystals in the Range 3 to 12A. *Applied Optics*, **2**, 955-961.

[3]   Miyashita, Y.O., Murakami, A., Aoki, K. and Yamaguchi, S. (1991) *Proceedings of the SPIE—The International Society for Optical Engineering*, **1503**, 463.

[4]   Sudhahar, S., Krishnakumar, M., Jayaramakrsishanan, V., Muralidharan, R. and Mohankumar, R. (2014) Effect of Sm+ Rare Earth Ion on the Structural, Thermal, Mechanical and Optical properties of Potassium Hydrogen Phthalate Single Crystals. *Journal of Materials Science and Technology*, **30**, 13-18. http://dx.doi.org/10.1016/j.jmst.2013.08.017

[5]   George, J. and Premachanran, S.K. (1981) Dislocation and Indentation Studies on Potassium Acid Phthalate Crystals. *Journal of Physics D: Applied Physics*, **14**, 1277-1281. http://dx.doi.org/10.1088/0022-3727/14/7/015

[6]   Khant, M.D.S. and Narasimhamurthy, T.S. (1982) Elasto-Optic Studies on Potassium Acid Phthalate Single Crystal. *Journal of Materials Science Letters*, **1**, 268-270.

[7]   Comoretto, D., Rossi, L. and Borghesi, A. (1997) Optical Properties of Potassium Acid Phthalate. *Journal of Materials Science Letters*, **12**, 1262-1267.

[8]   Bhat, S.G. and Dharmaprakash, S.M. (1998) A New Metal-Organic Crystal: Bismuth Thiourea Chloride. *Materials Research Bulletin*, **33**, 833. http://dx.doi.org/10.1016/S0025-5408(98)00049-X

[9]   Meenakshisundaram, S., Parthiban, S., Sarathi, N., Kalavathy, R. and Bhagavannarayana, G. (2006) Effect of Organic Dopants on ZTS Single Crystals. *Journal of Crystal Growth*, **293**, 376-381. http://dx.doi.org/10.1016/j.jcrysgro.2006.04.094

[10]   Belyaev, L.M., Belikova, G.S., Gil'varg, A.B., *et al.* (1969) *Soviet Physics-Crystallography*, **14**, 544-549.

[11]   Belikova, G.S., Belyaev, L.M., Goloveï, M.P., *et al.* (1974) *Soviet Physics Crystallography*, **19**, 351-355.

[12]   Okaya, Y. (1965) The Crystal Structure of Potassium Acid Phthalate, $KC_6H_4COOH.COO$. *Acta Crystallographica*, **19**, 879-882. http://dx.doi.org/10.1107/S0365110X65004590

[13]   Enculescu, M. (2010) Morphological and Optical Properties of Doped Potassium Hydrogen Phthalate Crystals. *Physica B: Condensed Matter*, **405**, 3722-3727. http://dx.doi.org/10.1016/j.physb.2010.05.074

[14]   Timpanaro, S., Sassella, A., Borghesi, A.Z., Porzio, W., Fontaine, P. and Goldmann, M. (2001) Crystal Structure of Epitaxial Quaterthiophene Thin Films Grown on Potassium Acid Phthalate. *Advanced Materials*, **3**, 127-130. http://dx.doi.org/10.1002/1521-4095(200101)13:2<127::AID-ADMA127>3.0.CO;2-Y

[15]   Stemmler, B.L. and Legrand, P. (1995) Measurement of the X-Ray Spectrometric Properties of Cesium Hydro Phthalate (CsAP) Crystal with the Synchrotron Radiation. *Review of Scientific Instruments*, **66**, 1601. http://dx.doi.org/10.1063/1.1145920

[16]   Benedict, J.B., Wallace, P.M., Reid, P.J. and Jang, S.-H. (2003) Up-Conversion Luminescence in Dye-Doped Crystals of Potassium Hydrogen Phthalate. *Advanced Materials*, **15**, 1068-1070. http://dx.doi.org/10.1002/adma.200303715

[17]   Prasad, P.N. and Williams, D.J. (1991) Introduction to Nonlinear Effects in Molecules and Polymers. Wiley, Berlin.

[18]   Zhang, H.W., Batra, A.K. and Lal, R.B. (1994) Growth of Large Methyl-(2,4-Dinitrophenyl)-Aminopropanoate: 2-Methyl-4-Nitroaniline Crystals for Nonlinear Optical Applications. *Journal of Crystal Growth*, **137**, 141-144. http://dx.doi.org/10.1016/0022-0248(94)91262-9

[19]   Kuznetsov, V.A., Okhrimenko, J.M. and Rak, M. (1998) Growth Promoting Effect of Organic Impurities on Growth Kinetics of KAP and KDP Crystals. *Journal of Crystal Growth*, **193**, 164-173. http://dx.doi.org/10.1016/S0022-0248(98)00489-8

[20]   Ramsamy, G., Parthiban, S., Meenakshisundaram, S.P. and Mojumdar, S.C. (2010) Influence of Alkali Metal Sodium Doping on the Properties of Potassium Hydrogen Phthalate (KHP) Crystals. *Journal of Thermal Analysis and Calorimetry*, **100**, 861-865. http://dx.doi.org/10.1007/s10973-010-0678-z

[21]   Shannon, R.D. (1976) Revised Effective Ionic Radii and Systematic Studies of Interatomic Distances in Halides and Chalcogenides. *Acta Crystallographica Section A*, **A32**, 751-767. http://dx.doi.org/10.1107/S0567739476001551

[22]   Vasudevan, G., Anbusrinivasan, P., Madhurambal, G. and Mojumdar, S.C. (2009) Thermal Analysis, Effect of Dopants, Spectral Characterisation and Growth Aspects of KAP Crystals. *Journal of Thermal Analysis and Calorimetry*, **96**, 99-102. http://dx.doi.org/10.1007/s10973-008-9880-7

[23]   Meenakshisundaram, S.P., Parthiban, S., Madhurambal, G. and Mojumdar, S.C. (2008) Effect of Chelating Agent (1,10-Phenanthroline) on Potassium Hydrogen Phthalate Crystals. *Journal of Thermal Analysis and Calorimetry*, **94**, 21-25. http://dx.doi.org/10.1007/s10973-008-9182-0

[24]   Muthu, K., Bhagavannarayana, G., Chandrasekaran, C., Parthiban, S., Meenakshisundaram, S.P., and Mojumdar, S.C. (2010) Os(VIII) Doping Effects on the Properties and Crystalline Perfection of Potassium Hydrogen Phthalate (KHP) Crystals. *Journal of Thermal Analysis and Calorimetry*, **100**, 793-799. http://dx.doi.org/10.1007/s10973-010-0759-z

[25]   Gayathri, K., Krishnan, P., Rajkumar, P.R. and Anbalagan, G. (2014) Growth, Optical, Thermal and Mechanical Characterization of an Organic Crystal: Brucinium 5-Sulfosalicylate Trihydrate. *Bulletin of Materials Science*, **37**, 1589-1595. http://dx.doi.org/10.1007/s12034-014-0721-y

[26] Kurtz, S.K. and Perry, J.J. (1968) A Powder Technique for the Evaluation of Nonlinear Optical Materials. *Journal of Applied Physics*, **39**, 3798. http://dx.doi.org/10.1063/1.1656857

[27] Meenakshisundaram, S., Parthiban, S., Bhagavannarayana, G., Madhurambal, G. and Mojumdar, S.C. (2006) Influence of Organic Solvent on Tristhioureazinc(II)Sulphate Crystals. *Journal of Thermal Analysis and Calorimetry*, **96**, 125-129. http://dx.doi.org/10.1007/s10973-008-9884-3

[28] Rak, M., Eremin, N.N., Eremina, T.A., Kuznetsov, V.A., Okhrimenko, T.M. and Furmanova, N.G. (2005) On the Mechanism of Impurity Influence on Growth Kinetics and Surface Morphology of KDP Crystals—I: Defect Centres Formed by Bivalent and Trivalent Impurity Ions Incorporated in KDP Structure—Theoretical Study. *Journal of Crystal Growth*, **273**, 577-585. http://dx.doi.org/10.1016/j.jcrysgro.2004.09.067

[29] Wang, Y. and Eaton, D.F. (1985) Optically Non-Linear Organic Molecules Cyclodextrin Inclusion Complexes. *Chemical Physics Letters*, **120**, 441-444. http://dx.doi.org/10.1016/0009-2614(85)85637-2

[30] Narasimha, B., Choudhary, R.N. and Roa, K.V. (1988) Dielectric Properties of LaPO$_4$ Ceramics. *Journal of Materials Science*, **23**, 1416-1418. http://dx.doi.org/10.1007/BF01154610

[31] Roa, K.V. and Samakula, C. (1965) Dielectric Properties of Cobalt Oxide, Nickel Oxide, and Their Mixed Crystals. *Journal of Applied Physics*, **36**, 2031. http://dx.doi.org/10.1063/1.1714397

[32] von Hundelshausen, U. (1971) Electrooptic Effect and Dielectric Properties of Cadmium-Mercury-Thiocyanate Crystals. *Physics Letters A*, **34**, 405-406. http://dx.doi.org/10.1016/0375-9601(71)90939-x

[33] Balarew, C. and Duhlew, R. (1984) Application of the Hard and Soft Acids and Bases Concept to Explain Ligand Coordination in Double Salt Structures. *Journal of Solid State Chemistry*, **55**, 1-6. http://dx.doi.org/10.1016/0022-4596(84)90240-8

Radiation Forces on a Dielectric Sphere Produced by Finite Olver-Gaussian Beams

Salima Hennani, Lahcen Ez-zariy*, Abdelmajid Belafhal*

Laboratory of Nuclear, Atomic and Molecular Physics Department of Physics, Faculty of Sciences, Chouaïb Doukkali University, El Jadida, Morocco
Email: *ezzariy@gmail.com, *belafhal@gmail.com

Abstract

In this work, we use the analytical expression of the propagation of Finite Olver-Gaussian beams (FOGBs) through a paraxial ABCD optical system to study the action of radiation forces produced by highly focused FOGBs on a Rayleigh dielectric sphere. Our numerical results show that the FOGBs can be employed to trap and manipulate particles with the refractive index larger than that of the ambient. The radiation force distribution has been studied under different beam widths. The trapping stability under different conditions is also analyzed.

Keywords

Finite Olver-Gaussian Beams, Radiation Forces, Scattering Force, Gradient Force, Rayleigh Dielectric Sphere

1. Introduction

In recent years, the accelerating finite Airy beam has been introduced within the framework of laser optics field by Siviloglou *et al.* [1] [2], on the basis of the result published by Berry and Balzas [3], in the context of the quantum mechanics, whose have introduced the Airy wave packet function as a solution of the Schrödinger equation in 1979. This new laser beams family exhibits many important characteristics, which permit them good candidates in several applications such as manipulating, trapping and transport of particles. It has an intensity

---
*Corresponding author.

profile that tends to be accelerated transversely during propagation. This leads to its experimental realization where most of the interesting properties were observed directly in many configurations [2], and the study of its ballistic dynamics shows that these waves follow parabolic trajectories similar to these of projectiles moving under the action of uniform gravitational field [4] [5]. This can absolutely help in handling micron particles. This topic was discussed in the literature first by Ashkin [6] and demonstrated how to manipulate three-dimensional trapping of a dielectric particle by using a highly focused laser beam thanks to their wide applications in manipulating various particles such as neutral atoms [7], molecules [8], micron-sized dielectric particles [9], DNA molecules [10] and living biological cells [11]-[13]. The optical traps or tweezers have also attracted attention in many literature works [14]-[18]. The radiation forces explained by scattering and gradient forces are generated by the exchange of momentum and energy between photons and particles.

As known, the first conventional optical trap is constructed by the fundamental Gaussian beams because it has a Gaussian peak in the cross profile and it is suitable for trapping the particle with the index of refraction higher than that of ambient [7]. However, many researchers have demonstrated that other beams are also useful in trapping particles such as doughnut laser beams including Bessel beams which are more available in the trapping and manipulation of process [19]-[21]. The trapping characteristics of different beams, such as Laguerre-Gaussian [22], hollow-Gaussian [23], Bessel-Gaussian [24], cylindrical vector [25], Gaussian Schell model [26] and flat-topped-Gaussian beams [27], have been studied.

The "non-diffracting" Airy beam has also attracted attention for its use in trapping particles because of its potential applications in several domains of the same context, such as: guidance plasma [28], acceleration of electrons in vacuum [29], production of three optical bullets dimensions [30] and optical micromanipulation [31] [32]. Opposed to other laser beams, the Airy family beam can transport microparticles along curved self-healed paths, and remove particles or cells from a section of a sample chamber [33]. Its novelty is that the trapping potential landscape tends to freely self-bend during propagation, and also its bend degree can be controlled, and the direction of acceleration can be switched by a nonlinear optical method [33], and the micromanipulation by Airy beam reported to date is related to Mie particles [31] [32] whose radii are larger than the wavelength. All these tunable properties make the Airy beam as a versatile and powerful tool for many optical manipulations. Our work done in this review is based on the above cited theoretical studies dealing the radiation force on spherical particles [8]-[10] [33]-[37]. Of these, we cite the theoretical analysis of the radiation pressure force of laser light on a dielectric sphere in the Rayleigh regime developed by Harada and Asakura [33], the method proposed by Rohrbach and Stelzer [34], to calculate the forces trapping dielectric particles and the investigation made by Zemánek et al. [35] about the rule of Gaussian beams in optical trapping of Rayleigh particles. Yet, Cheng et al. [36] have developed an analysis of optical trapping and propulsion of Rayleigh particles using the ordinary Airy beam. And in Ref. [37], Svoboda and Block have discussed some biological applications of optical forces.

In this paper, we consider a theoretical description of the radiation forces produced by the Finite Olver-Gaussian beam in Rayleigh scattering regime when the radius of particles is much smaller than the wavelength. Note that the new beams family, called "Finite Olver-Gaussian beams", is introduced within the optic field and their characteristics and propagation properties in aligned and misaligned optical systems are studied and examined for the first time by our research group [37]-[39].

2. Fields of Finite Olver-Gaussian Beams through an ABCD Optical System

The electrical field of FOG Bs at the input plane ($z = 0$) is defined by Belafhal et al. [38] as

$$U_1\left(x_1, z = 0\right) = O_n\left(\frac{x_1}{\omega_0}\right)\exp\left\{a_0\,\frac{x_1}{\omega_0}\right\}\exp\left\{-b_0\,\frac{x_1^2}{\omega_0^2}\right\}, \tag{1}$$

where n is the order of FOGBs. For $n = 0$, we will treat the ordinary finite Airy beam with a_0 is the truncation coefficient and b_0 is a coefficient to be equal to 1 or 0, depend that the incident beam is modulated by a Gaussian envelope or not, respectively. Due to the special properties of Airy Beam, it can absolutely be used to guide and trap neutral particles.

Under the paraxial approximation, the electrical field of FOGBs beam passing via any ABCD optical system can be expressed as [39].

$$U_2(x_2,z) = \left(\frac{1}{2\pi}\right)\sqrt{\frac{ik}{2b}}\sqrt{\frac{1}{\frac{ikA}{2B}+\frac{b_0}{\omega_0^2}}}\exp\left\{-\frac{ikD}{2B}x_2^2\right\}\exp\left\{\frac{\left(\frac{a_0}{\omega_0^2}+\frac{ik}{B}x_2\right)^2}{4\left(\frac{ikA}{2B}+\frac{b_0}{\omega_0^2}\right)}\right\}$$

$$\times\exp\left\{\frac{1}{96\omega_0^6\left(\frac{ikA}{2B}+\frac{b_0}{\omega_0^2}\right)^3}\right\}\exp\left\{\frac{\left(\frac{a_0}{\omega_0^2}+\frac{ik}{B}x_2\right)}{8\omega_0^3\left(\frac{ikA}{2B}+\frac{b_0}{\omega_0^2}\right)^2}\right\}O_n\left(\frac{\left(\frac{a_0}{\omega_0^2}+\frac{ik}{B}x_2\right)}{2\omega_0\left(\frac{ikA}{2B}+\frac{b_0}{\omega_0^2}\right)}+\frac{1}{16\omega_0^4\left(\frac{ikA}{2B}+\frac{b_0}{\omega_0^2}\right)^2}\right),$$

(2)

where $k = 2\pi/\lambda$ is the wave number and λ being the wavelength. A, B, C and D are the transfer matrix elements of the paraxial optical system. z is the distance between the input and the output planes. Consider now the FOGBs propagates through a free space described by the following transfer matrix

$$\begin{pmatrix} A & B \\ C & D \end{pmatrix} = \begin{pmatrix} 1 & z \\ 0 & 1 \end{pmatrix}.$$

(3)

Substituting Equation (3) in Equation (2), one obtains the intensity distribution of the FOGBs through a free space.

3. Theory of the Radiation on a Dielectric Sphere

Assume that the particle is sufficiently small compared to the wavelength of the laser beam *i.e.* $a \ll \lambda$, where a is the radius of the particle and λ is the wavelength of the incident beam. So, in the Rayleigh scattering the particle could be treated as a point dipole. Two types of radiation forces can be examined: the scattering force and the gradient one. The electric field is polarized in the *x*-direction. The center of the particle is located at the position (x, z). The physical quantities of the field vectors of the electromagnetic wave are real functions of time and space given by [33]

$$\begin{cases} E(r,t) = \mathrm{Re}\left[E(r)\mathrm{e}^{i\omega t}\right], \\ H(r,t) = \mathrm{Re}\left[H(r)\mathrm{e}^{i\omega t}\right], \end{cases}$$

(4)

with ω is the temporal angular frequency of the laser beam. The instantaneous energy flux crossing an area per time unit in the beam propagation direction corresponding to the propagation vector is given by [33]

$$S(r,t) = E(r,t)\times H(r,t)$$
$$= \frac{1}{2}\mathrm{Re}\left[E(r)\times H(r)\mathrm{e}^{2i\omega t}\right]+\frac{1}{2}\mathrm{Re}\left[E(r)\times H(r)\right].$$

(5)

The measurable physical quantity which evaluates the radiation force, of the light, is the beam intensity at the position $r(x,y,z)$ which is defined as

$$I(r)e_z = \langle S(r,t)\rangle = \frac{1}{2}\mathrm{Re}\left[E(r)\times H^*(r)\right] = \frac{n_2\varepsilon_0 c}{2}|E(r)|^2\,e_z.$$

(6)

In the last equation, n_2 is the refractive index of a surrounding medium, $c = 1/\sqrt{\varepsilon_0\mu_0}$ is the speed of the light in the vacuum. ε_0 and μ_0 are the dielectric constant and the magnetic permeability in the vacuum, respectively, and e_z is a unity vector along the wave vector.

The radiation pressure force exerted on the particle in the Rayleigh regime is described by two components actions on the dipole. One of these forces is the scattering force. As known, the electric field oscillates harmonically in time, and the induced point dipole follows with synchronization the electric field which gives that the particle acts as an oscillating electric dipole which radiates secondary or scattered waves in all directions. The scattering force is given and discussed in [37]

$$F_{scat}(r) = I(r)e_z = \frac{n_2}{c}C_{pr}I(r)e_z = \frac{n_2}{c}C_{pr}|E(r)|^2 e_z, \quad (7)$$

where C_{pr} is the cross section force of the radiation pressure of the particle in the Rayleigh approximation and is given by [37]

$$C_{pr} = \frac{8}{3}\pi(ka)^4 a^2 \left(\frac{m^2-1}{m^2+2}\right)^2, \quad (8)$$

with $m = n_1/n_2$ is the relative refractive index of the particle, n_2 is the refractive index of the ambient, n_1 is the refractive index of the particle and a is the radius of the particle.

The other component is a gradient force due to the Lorentz force acting on the dipole induced by the electromagnetic field [37]. The instantaneous gradient force is defined by [37]

$$F_{grad}(r,t) = 4\pi n_2^2 \varepsilon_0 a^3 \left(\frac{m^2-1}{m^2+2}\right)\frac{1}{2}\nabla E^2(r,t), \quad (9)$$

where $\nabla E^2 = 2(E\nabla)E + 2E \times (\nabla \times E)$ As a result of the Maxwell equations, one have $\nabla \times E = 0$, which gives the following expression

$$F_{grad}(r) = \langle F_{grad}(r,t)\rangle = 4\pi n_2^2 \varepsilon_0 a^3 \left(\frac{m^2-1}{m^2+2}\right)\frac{1}{2}\nabla \langle E^2(r,t)\rangle.$$

$$= \pi n_2^2 \varepsilon_0 a^3 \left(\frac{m^2-1}{m^2+2}\right)\frac{1}{2}\nabla |E(r,t)|^2 \quad (10)$$

with $\frac{1}{2}\nabla \langle E^2(r,t)\rangle = |E(r,t)|^2$

4. Numerical Simulations and Results Discussions

By the use of the theory of the radiation on a dielectric sphere, exposed in the previous section, we will prove numerically that the considered beams family drags particles into their intensity peaks. The parameters chosen in the calculations are: the radius of the particle is $a = 60\,\mu m$, the refractive index are $n_1 = 1.332$, $n_2 = 1.59$ and the wavelength is $\lambda = 632.8\,nm$ and the spot size in the medium is $\omega_0 = 5\,nm$. The coefficients a_0 and b_0 take the value 0.1. In our numerical calculations, we simulate the radiation force produced by the incident focused FOGBs considered in one-dimension. We choose the peak intensity of the input beam as $I_0 = 3.10^{11}\,W/m^2$ with the above numerical parameters of λ and a. As well known, for a stable optical trapping, when a particle moves away, the radiation forces will pull the particle it back to the equilibrium position. Apparently, there exist two types of radiation forces: the gradient and the scattering forces. The both could be used to pulls, manipulates and traps the particles. In our case, we assume that the radius of the particle is much smaller than the wavelength of the laser beam, so the Rayleigh approximation theory is applied. Under this approximation, the radiation forces include the scattering and the gradient ones which are regarded as the key for trapping the particle are responsible to pull particles towards the center and the scattering force tends to push the small Rayleigh particle out along the direction of the beam propagation and destabilize the optical trap. So, it leads to particle departure from the equilibrium position. While for the large Rayleigh distance, there is a competition between the axial gradient force and the scattering force. This may determine whether the particle can be trapped or not in the axial direction.

Figures 1-4 show the transverse gradient force along x-axis for the various orders $n = 0$, 1 and 2 at different axial positions z and with different values of the beam size ω_0. The gradient force alternates between the positive and negative directions. The positive gradient force means the direction of transverse gradient force is in +x-direction and the negative gradient force means in the negative x-direction. So it depends upon the particle's position relative to nearby position of the optical intensity peak. These suggest that the Rayleigh particle in the Finite Olver-Gaussian beam will be pulled into nearby intensity peak and transported along the propagation direction.

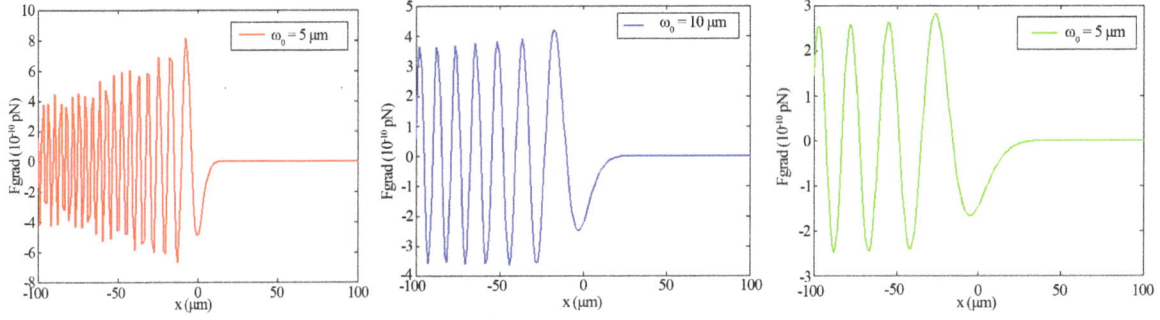

Figure 1. Transverse gradient force of finite zeroth-order Olver-Gaussian beam versus x for different values of ω_0 and for z = 500 µm.

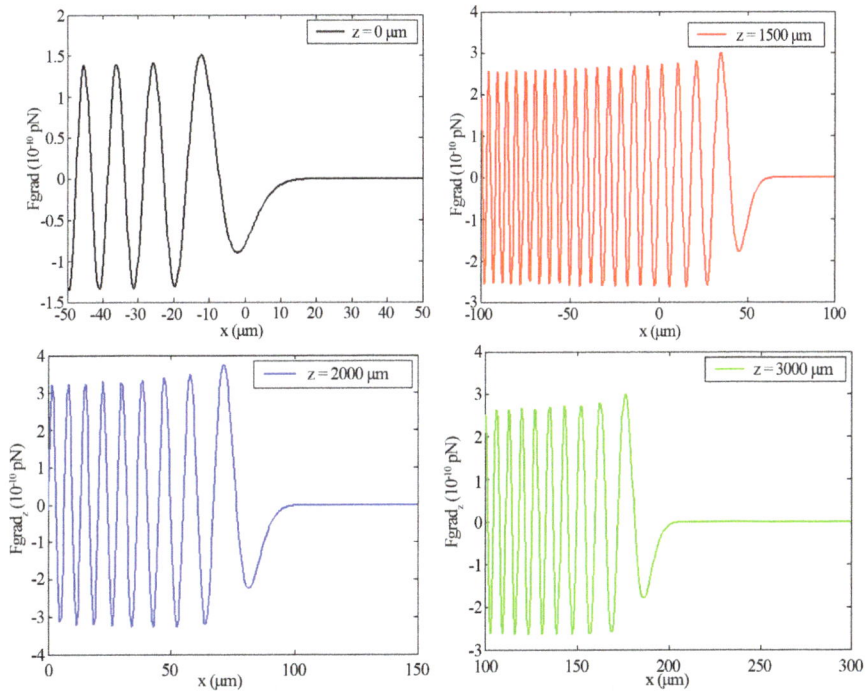

Figure 2. Transverse gradient force of finite zeroth-order of Olver-Gaussian beam versus x for different propagation distances z and ω_0 = 5 µm.

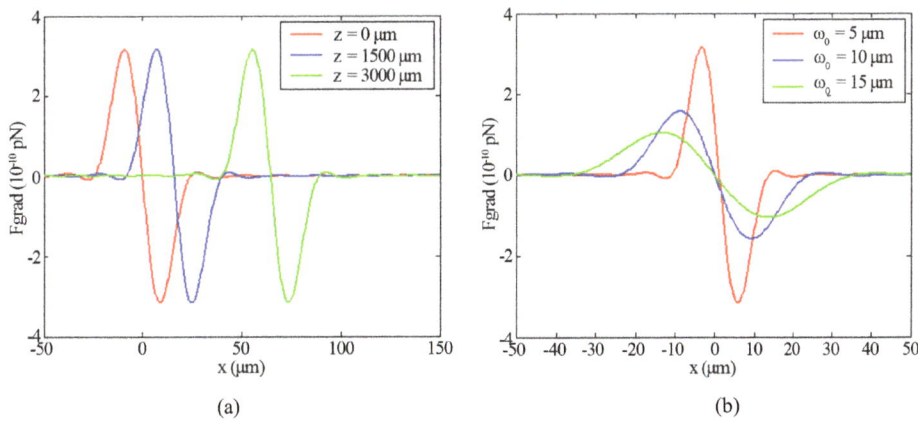

(a)   (b)

Figure 3. Transverse gradient force of Finite Olver-Gaussian beam of second order (n = 2) versus x for different: (a) Propagation distances z, (b) Beam width size.

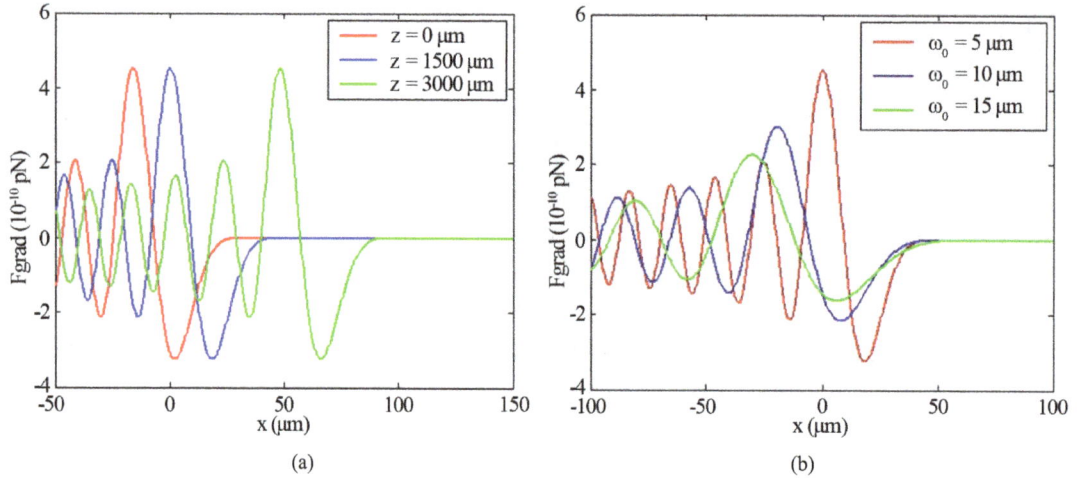

Figure 4. Scattering force of Finite Olver-Gaussian beam of first order (n = 1) versus x for different: (a) Propagation distance z ($\omega_0 = 5$ μm), (b) Beam witdth size ω_0 (z = 500 μm).

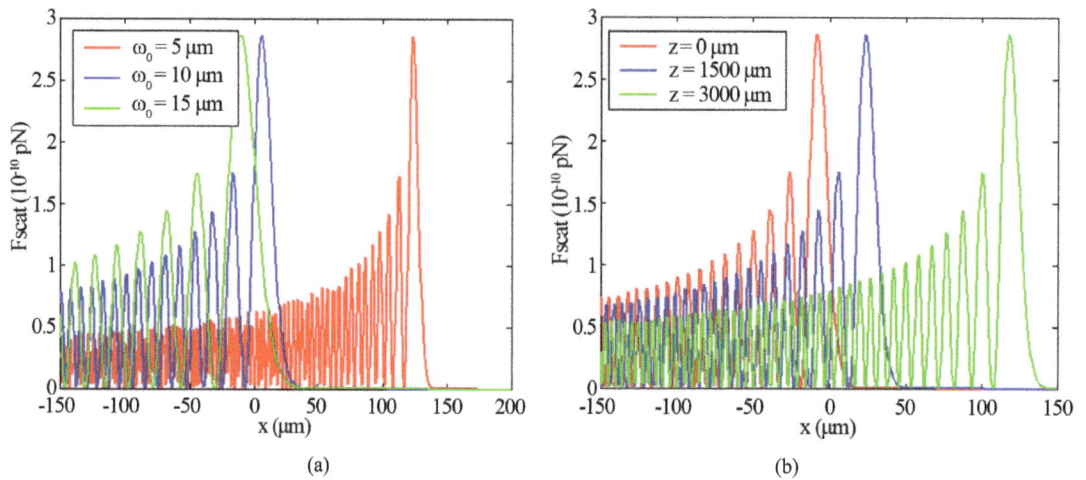

Figure 5. Transverse gradient force of Finite Olver-Gaussian beam of first order (n = 1) versus x for different: (a) Propagation distances z ($\omega_0 = 5$ μm), (b) Beam witdth size ω_0 (z = 500 μm).

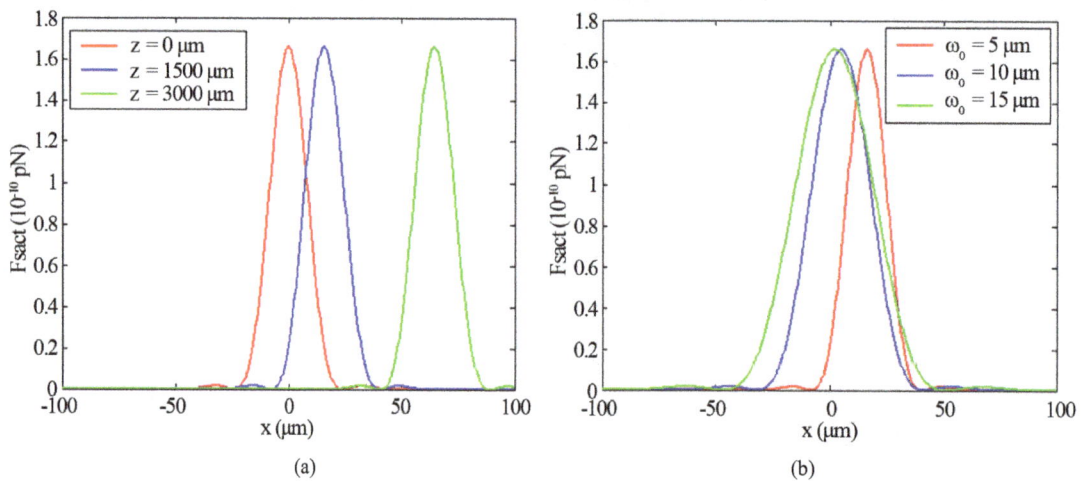

Figure 6. Scattering force of finite zeroth-order Olver-Gaussian beam versus x for different: (a) Propagation distances z ($\omega_0 = 5$ μm), (b) Beam witdth size ω_0 (z = 500 μm).

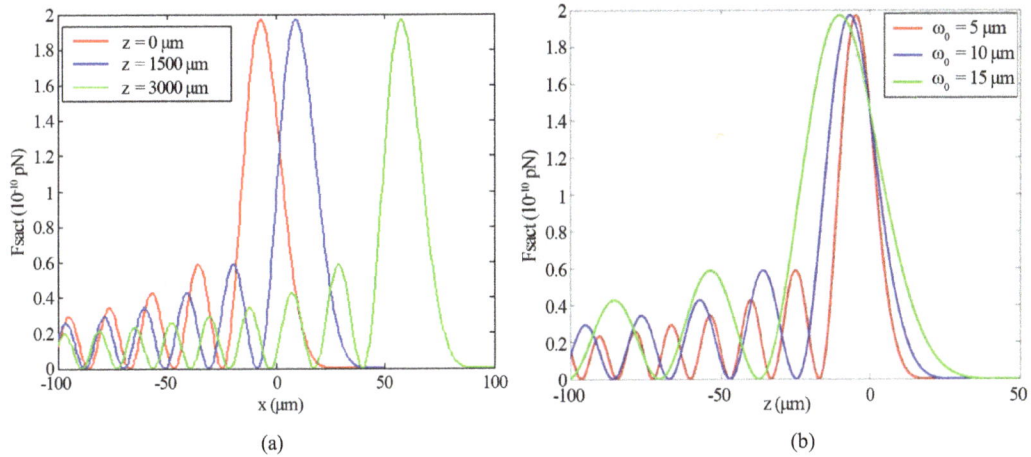

Figure 7. Scattering force of Finite Olver-Gaussian beam of second order (n = 2) versus x for different: (a) propaga- tion distances z (ω_0 = 5 μm), (b) Beam witdth size ω_0 (z = 500 μm).

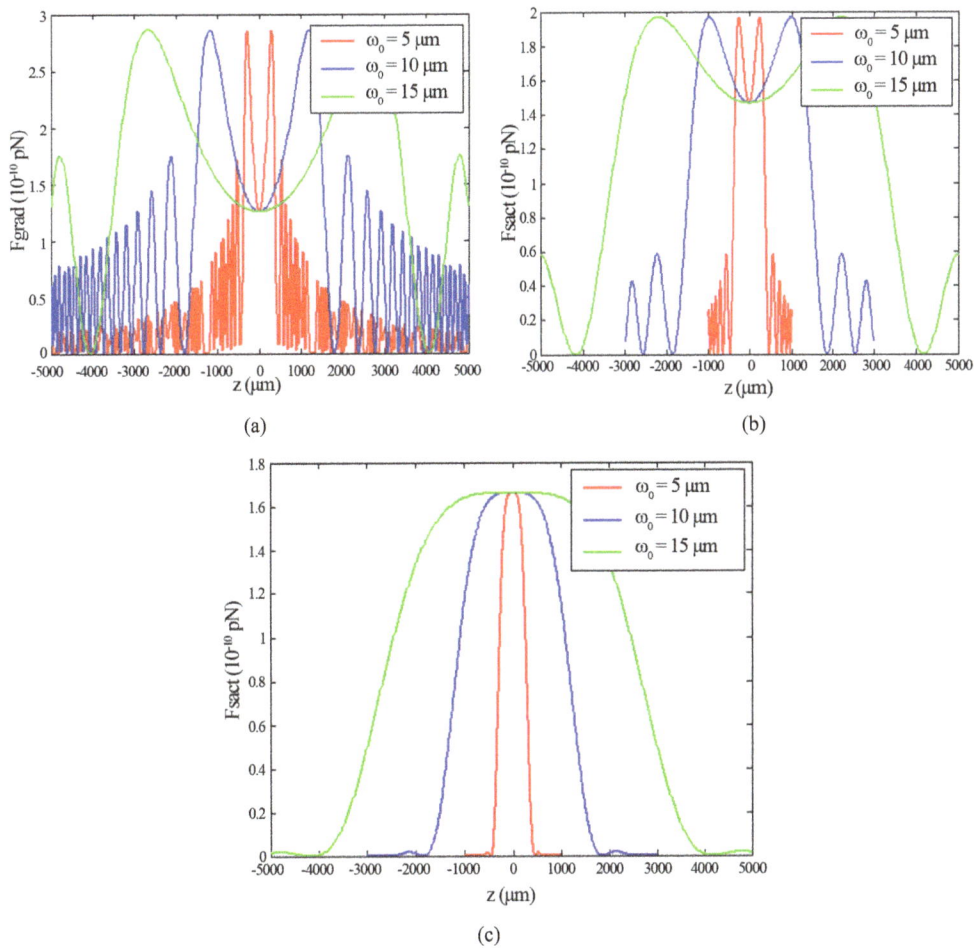

Figure 8. Example of a scattering force of Finite Olver-Gaussian beam at varying z planes for different beam width size ω_0 (z = 500 μm). for: (a): zeroth order (n = 0), (b): first order (n = 1) and (c): second order (n = 2).

Illustrations of **Figures 1-4** show that as far as the propagation distance z increases as much as the number of oscillations of the gradient force increases and the zeros of the central lobe and sidelobes move towards the positive x. However, the maxima of the lobes decrease insignificantly with the increase of z. Yet, the gradient force reaches less of zeros and the lobes maximums decrease with the increase of the waist size of the incident Gaussian envelope. The decrease of the maximums lobes with the waist is very significant. In **Figures 5-7**, we plot the scattering force of the various orders n of the Finite Olver-Gaussian beam versus x for different propagation distances z and different beam width sizes ω_0.

As the same as in **Figures 1-4** concern the gradient force, from the plots of **Figure 5** and **Figure 7** about the scattering force, it is pointed that the same remarks are repeated in this case. **Figure 8**, shows the scattering force of the Finite Olver-Gaussian beam for different orders n along z-axis and with different values of beam size ω_0. Our results obtained for the ordinary Airy Beams are compared with those presented in Ref. [36]. Our analysis shows that the radiation forces of focused finite Olver-Gaussian beams may be used to trap and manipulate the Rayleigh dielectric particles and in order to stability trap particles. The axial gradient force must be greatly larger than the scattering force $\left|F_{grad}\right|/\left|F_{scat}\right| \geq 1$. This report is called the stability criterion.

5. Conclusion

We have investigated the radiation force exerted on a dielectric particle in the Rayleigh scattering regime illuminated by FOGBs with different beams order n. Several numerical simulations are carried out in the work, in order to study and analyze the effects of some parameters on the radiation gradient and transverse forces, for instance, the parameters of the laser beam illuminate the dielectric sphere and the propagation distance. The numerical results show that the Finite Olver-Gaussian beams could drag particles by the peaks of intensity and could transport and manipulate them in the inter-lobes. Finally, we conclude that Finite Olver-Gaussian beams may be used in many practical applications of nanotechnology and biotechnology.

References

[1] Siviloglou, G.A. and Christodoulides, D.N. (2007) Accelerating Finite Energy Airy Beams. *Optics Letters*, **32**, 979-981. http://dx.doi.org/10.1364/OL.32.000979

[2] Siviloglou, G.A., Broky, J., Dogariu, A. and Christodoulides, D.N. (2007) Observation of Accelerating Airy Beams. *Physical Review Letters*, **99**, 213901. http://dx.doi.org/10.1103/PhysRevLett.99.213901

[3] Berry, M.V. and Balazs, N.L. (1979) Nonspreading Wave Packets. *American Journal of Physics*, **4**, 264-267. http://dx.doi.org/10.1119/1.11855

[4] Siviloglou, G.A., Broky, J., Dogariu, A. and Christodoulides, D.N. (2008) Ballistic Dynamics of Airy Beams. *Optics Letters*, **33**, 207-209. http://dx.doi.org/10.1364/OL.33.000207

[5] Besieris, I.M. and Shaarawi, A.M. (2007) A Note on an Accelerating Finite Energy Airy Beam. *Optics Letters*, **32**, 2447-2449. http://dx.doi.org/10.1364/OL.32.002447

[6] Ashkin, A. (1970) Atomic-Beam Deflection by Resonance-Radiation Pressure. *Physical Review Letters*, **25**, 1321. http://dx.doi.org/10.1103/PhysRevLett.25.1321

[7] Yavuz, D.D., Kulatunga, P.B., Urban, E., Johnson, T.A., Proite, N., Henage, T., Saffman, M., *et al.* (2006) Fast Ground State Manipulation of Neutral Atoms in Microscopic Optical Traps. *Physical Review Letters*, **96**, 063001. http://dx.doi.org/10.1103/PhysRevLett.96.063001

[8] Calander, N. and Willander, M. (2002) Optical Trapping of Single Fluorescent Molecules at the Detection Spots of Nanoprobes. *Physical Review Letters*, **89**, 143603. http://dx.doi.org/10.1103/PhysRevLett.89.143603

[9] Taguchi, K., Ueno, H. and Ikeda, M. (1997) Rotational Manipulation of a Yeast Cell Using Optical Fibres. *Electronics Letters*, **33**, 1249-1250. http://dx.doi.org/10.1049/el:19970827

[10] Day, C. (2006) Optical Trap Resolves the Stepwise Transfer of Genetic Information from DNA to RNA. *Physics Today*, **59**, 26-27. http://dx.doi.org/10.1063/1.2180165

[11] Mao, F.L., Xing, Q.R., Wang, K., Lang, L.Y., Wang, Z., Chai, L. and Wang, Q.Y. (2005) Optical Trapping of Red Blood Cells and Two-Photon Excitation-Based Photodynamic Study Using a Femtosecond Laser. *Optics Communications*, **256**, 358-363. http://dx.doi.org/10.1016/j.optcom.2005.06.076

[12] Chang, Y.R., Hsu, L. and Chi, S. (2005) Optical Trapping of a Spherically Symmetric Rayleigh Sphere: A Model for Optical Tweezers upon Cells. *Optics Communications*, **246**, 97-105. http://dx.doi.org/10.1016/j.optcom.2004.10.066

[13] Xie, C.G. and Li, Y.Q. (2003) Confocal Micro-Raman Spectroscopy of Single Biological Cells Using Optical Trapping and Shifted Excitation Difference Techniques. *Journal of Applied Physics*, **93**, 2982-2986. http://dx.doi.org/10.1063/1.1542654

[14] Neuman, K.C. and Block, S.M. (2004) Optical Trapping. *Review of Scientific Instruments*, **75**, 2787-2809. http://dx.doi.org/10.1063/1.1785844

[15] Furst, E.M. (2003) Interactions, Structure, and Microscopic Response: Complex Fluid Rheology Using Laser Tweezers. *Soft Materials*, **1**, 167-185. http://dx.doi.org/10.1081/SMTS-120022462

[16] Ashkin, A. (1992) Forces of a Single-Beam Gradient Laser Trap on a Dielectric Sphere in the Ray Optics Regime. *Biophysical Journal*, **61**, 569-582. http://dx.doi.org/10.1016/S0006-3495(92)81860-X

[17] Ashkin, A. (2000) History of Optical Trapping and Manipulation of Small-Neutral Particle, Atoms, and Molecules. *IEEE Journal of Selected Topics in Quantum Electronics*, **6**, 841-856. http://dx.doi.org/10.1109/2944.902132

[18] Grzegorczyk, T.M., Kemp, B.A. and Kong, J.A. (2006) Stable Optical Trapping Based on Optical Binding Forces. *Physical Review Letters*, **96**, 113903. http://dx.doi.org/10.1103/PhysRevLett.96.113903

[19] O'Neil, A.T. and Padgett, M.J. (2000) Three-Dimensional Optical Confinement of Micron-Sized Metal Particles and the Decoupling of the Spin and Orbital Angular Momentum within an Optical Spanner. *Optics Communications*, **185**, 139-143. http://dx.doi.org/10.1016/S0030-4018(00)00989-5

[20] Kuga, T., Torii, Y., Shiokawa, N., Hirano, T., Shimizu, Y. and Sasada, H. (1997) Novel Optical Trap of Atoms with a Doughnut Beam. *Physical Review Letters*, **78**, 4713. http://dx.doi.org/10.1103/PhysRevLett.78.4713

[21] Gahagan, K.T. and Swartzlander, G.A. (1996) Optical Vortex Trapping of Particles. *Optics Letters*, **21**, 827-829. http://dx.doi.org/10.1364/OL.21.000827

[22] Dienerowitz, M., Mazilu, M., Reece, P.J., Krauss, T.F. and Dholakia, K. (2008) Optical Vortex Trap for Resonant Confinement of Metal Nanoparticles. *Optics Express*, **16**, 4991-4999. http://dx.doi.org/10.1364/OE.16.004991

[23] Zhao, C.L., Wang, L.G. and Lu, X.H. (2007) Radiation Forces on a Dielectric Sphere Produced by Highly Focused Hollow Gaussian Beams. *Physics Letters A*, **363**, 502-506. http://dx.doi.org/10.1016/j.physleta.2006.11.028

[24] Zhao, C., Wang, L. and Lu, X. (2008) Radiation Forces of Highly Focused Bessel-Gaussian Beams on a Dielectric Sphere. *Optik-International Journal for Light and Electron Optics*, **119**, 477-480. http://dx.doi.org/10.1016/j.ijleo.2006.11.013

[25] Zhan, Q.W. (2003) Radiation Forces on a Dielectric Sphere Produced by Highly Focused Cylindrical Vector Beams. *Journal of Optics A*: *Pure and Applied Optics*, **5**, 229. http://dx.doi.org/10.1088/1464-4258/5/3/314

[26] Wang, L.G. and Zhao, C.L. (2007) Dynamic Radiation Force of a Pulsed Gaussian Beam Acting on Rayleigh Dielectric Sphere. *Optics Express*, **15**, 10615-10621. http://dx.doi.org/10.1364/OE.15.010615

[27] Zhao, C., Cai, Y., Lu, X. and Eyyuboğlu, H.T. (2009) Radiation Force of Coherent and Partially Coherent Flat-Topped Beams on a Rayleigh Particle. *Optics Express*, **17**, 1753-1765. http://dx.doi.org/10.1364/OE.17.001753

[28] Polynkin, P., Kolesik, M., Moloney, J.V., Siviloglou, G.A. and Christodoulides, D.N. (2009) Curved Plasma Channel Generation Using Ultraintense Airy Beams. *Science*, **324**, 229-232. http://dx.doi.org/10.1126/science.1169544

[29] Li, J.X., Zang, W.P. and Tian, J.G. (2010) Vacuum Laser-Driven Acceleration by Airy Beams. *Optics Express*, **18**, 7300-7306. http://dx.doi.org/10.1364/OE.18.007300

[30] Chong, A., Renninger, W.H., Christodoulides, D.N. and Wise, F.W. (2010) Airy-Bessel Wave Packets as Versatile Linear Light Bullets. *Nature Photonics*, **4**, 103-106. http://dx.doi.org/10.1038/nphoton.2009.264

[31] Baumgartl, J., Mazilu, M. and Dholakia, K. (2008) Optically Mediated Particle Clearing Using Airy Wavepackets. *Nature Photonics*, **2**, 675-678. http://dx.doi.org/10.1038/nphoton.2008.201

[32] Baumgartl, J., Hannappel, G.M., Stevenson, D.J., Day, D., Gu, M. and Dholakia, K. (2009) Optical Redistribution of Microparticles and Cells between Microwells. *Lab on a Chip*, **9**, 1334-1336. http://dx.doi.org/10.1039/b901322a

[33] Harada, Y. and Asakura, T. (1996) Radiation Forces on a Dielectric Sphere in the Rayleigh Scattering Regime. *Optics Communications*, **124**, 529-541. http://dx.doi.org/10.1016/0030-4018(95)00753-9

[34] Rohrbach, A. and Stelzer, E.H. (2001) Optical Trapping of Dielectric Particles in Arbitrary Fields. *Journal of Optics A*: *Pure and Applied Optics*, **18**, 839-853. http://dx.doi.org/10.1364/JOSAA.18.000839

[35] Zemánek, P., Jonáš, A., Šrámek, L. and Liška, M. (1998) Optical Trapping of Rayleigh Particles Using a Gaussian Standing Wave. *Optics Communications*, **151**, 273-285. http://dx.doi.org/10.1016/S0030-4018(98)00093-5

[36] Cheng, H., Zang, W., Zhou, W. and Tian, J. (2010) Analysis of Optical Trapping and Propulsion of Rayleigh Particles Using Airy Beam. *Optics Express*, **18**, 20384-20394. http://dx.doi.org/10.1364/OE.18.020384

[37] Svoboda, K. and Block, S.M. (1994) Biological Applications of Optical Forces. *Annual Review of Biophysics and Biomolecular Structure*, **23**, 247-285. http://dx.doi.org/10.1146/annurev.bb.23.060194.001335

[38] Belafhal, A., Ez-Zariy, L., Hennani, S. and Nebdi, H. (2015) Theoretical Introduction and Generation Method of a Novel Nondiffracting Waves: Olver Beams. *Optics and Photonics Journal*, **5**, 234-246.

http://dx.doi.org/10.4236/opj.2015.57023

[39] Hennani, S., Ez-zariy, L. and Belafhal, A. (2015) Propagation Properties of Finite Olver-Gaussian Beams Passing through a Paraxial ABCD Optical System. *Optics and Photonics Journal*, **5**, 273-294.
http://dx.doi.org/10.4236/opj.2015.59026

10

Limiting the Migration of Bisphenol A from Polycarbonate Using Dielectric Barrier Discharge

Emad A. Soliman[1], Ahmed Samir[2*], Ali M. A. Hassan[3], Mohamed S. Mohy-Eldin[1], Gamal Abd El-Naim[1]

[1]Department of Polymer Materials Research, Advanced Technology and New Materials Research Institute, SRTA-City, New Bourg El-Arab City, Alexandria, Egypt
[2]Center of Plasma Technology, Al-Azhar University, Nasr City, Cairo, Egypt
[3]Department of Chemistry, Faculty of Science, Al-Azhar University, Nasr City, Cairo, Egypt
Email: [*]ahmed_samir_aly@yahoo.com

Abstract

Dielectric barrier discharge is used as a cheap technique for surface treatment of polycarbonate. The discharge system is working in open air at atmospheric pressure. The treatments are carried out at low discharge powers (1.5 and 2 W) for treatment time (2.5 - 15 min). The treated samples show decrease in the contact angle and increase in the crystallinity, thermal stability and surface roughness. The effect of ozone on the increase in the oxygen containing functional groups is discussed. The treatment process shows effective limitation of the migration of bisphenol A from the surface of polycarbonate due to the cross linking. Zero migration of bisphenole A is recorded as the sample is treated for 7.5 min. The treatment process is found to be very efficient with very low cost.

Keywords

Dielectric Barrier Discharge, Polycarbonate, Bisphenole A, Surface Treatment

1. Introduction

Bisphenol A, (2, 2-bis (4-hydroxyphenyl) propane, (BPA)) is a chemical used primarily as a monomer in the production of polycarbonate plastic (PC). Polycarbonate (PC) is widely used, especially in developing countries,

---

[*]Corresponding author.

in food contact materials such as infant feeding bottles, tableware (plates, mugs, jugs, and beakers), microwave ovenware, food containers, water bottles, milk and beverage bottles, processing equipment and water pipes. BPA can migrate into food from food containers made of polycarbonate plastic. The human BPA exposure from different sources has been proved [1] [2]. Later BPA has been discovered to be a carcinogen and cause many other health problems [3]-[7]. There are many recent studies that confirm the migration of BPA, especially from baby bottles made of polycarbonate, with dangerous rates affecting the health of infants [8]-[12].

Plasma has been introduced as an effective technique for treatment of polycarbonate surface. Different properties of the surface of polycarbonate can be changed by plasma treatment [13]-[17]. One of the important advantages of plasma treatment is that: it changes the surface properties of the polycarbonate without altering the bulk properties. The economical impact prevents the wide spread of plasma treatments of surfaces in commercial applications. Usually low pressure plasma systems are very expensive where a vacuum system is needed. Also the power supplies raise the price of plasma system, especially when using RF or microwave power supplies. In addition, the running cost of the treatment process including gases and electric power consumption is another charge that makes the plasma treatment relatively an expensive treatment technique.

In the present work a dielectric barrier discharge (DBD) system, working in open air at atmospheric pressure, is introduced as a cheap source of plasma that overcomes the economical disadvantages of plasma treatment systems. For the first time, DBD has been used in the treatment of polycarbonate surface for the sake of limiting the migration of BPA. The effect of DBD treatment on the surface properties of polycarbonate samples has been studied.

2. Experimental Setup

The experimental arrangement of DBD plasma reactor used in the present treatment is shown in **Figure 1**. The DBD cell consisted of two electrodes of stainless steel discs, and each one has a diameter of 20 cm and thickness of 2 mm. The lower and upper electrodes were fixed to a perspex base of 40 cm diameter and 2 cm thickness. A dielectric material made of glass has a thickness of 1.2 mm was pasted. The upper and lower perspex discs were collected to each other via rubber O-ring. The gap distance between the dielectric glass and the lower electrode was 3 mm. The cell was fed by air via the gas inlet where the air filled the gap space and then exhausted through the gas outlet. Before any treatments the gas is left to flow in the cell for about 5 minutes to sweep any impurities in the reactor. The gas pressure was kept at atmospheric pressure and the gas flow rate is 0.1 L/min. The two electrodes were connected to a high voltage AC power supply of 50 Hz frequency and a variable voltage of 0 - 10 kV. A limiting resistance of 250 kΩ was used to limit the discharge current. The applied voltage was measured via a resistive potential divider (PD) 500:1. The discharge current was measured by measuring the potential drop across a resistance of 1 kΩ. The accumulated charge on the electrodes was measured by measuring the potential drop across a capacitor of 14 μF. The waveforms of the voltage, current and charge was measured by a digital storage oscilloscope (Model HM-407). The concentration of ozone formed inside the DBD system was measured using ozone analyzer (model AFX H1). The exhaust gas of the DBD cell was collected by the analyzer

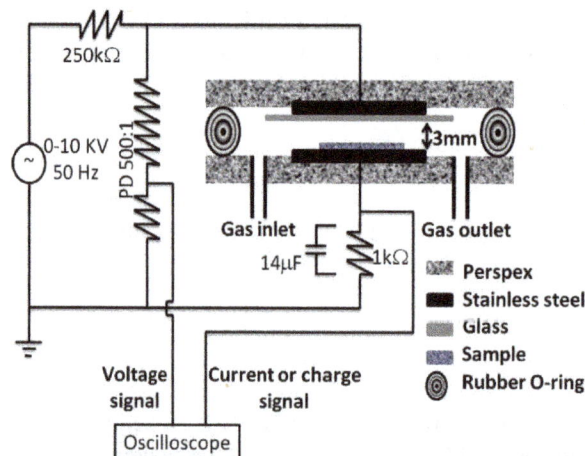

Figure 1. Schematic diagram of DBD plasma reactor.

and measured with accuracy less than 0.1 g/m^3.

PC samples were taken from baby bottles which were collected from the local market in Egypt. The samples were cut into 3×2 cm^2. The treated samples were fixed at the lower electrode where the upper surface of the sample was exposed to the plasma reactive species. For double face treatments, the samples were retreated on the other surface at the same treatment conditions. The samples were treated in the DBD system at different treatment times (2.5 - 15 min) and different discharge currents (1 and 1.5 mA).

The properties of the samples were examined using different techniques. X-Ray Diffraction (XRD) patterns of the samples were recorded by Shimadzu-XRD-7000 Diffractometer—Japan, operating at room temperature with Cu (Kα) radiations of wavelength ($\lambda = 1.5406$ Å), generated at 30 Kv - 30 mA. The 2θ range for all samples was in the range from 4° to 80° with scan speed 0.2°/s.

Thermo-Gravimetric Analysis (TGA) (Model TGA-50H, Shimadzu—Japan) was used to study the changes in thermal stability of the treated samples. The measurements were carried out with a heating rate of 10°C/min under flow of N2 gas where the temperature were elevated up to 600°C.

The contact angle measurements were carried out using the sessile drop method. A contact angle goniometer (ramé-hatr Model 500) using an optical subsystem was used to capture the profile of a pure drop of water on the surface of the sample and then the contact angle was measured.

Scanning Electron Microscope (SEM) was used to study the morphology of the surface of the samples. The samples were coated with gold by the use of Carbon Sputter Coater SPI, Module control, SPI supplies—USA. After coating the samples were placed in the cavity of JEOL JSM-6390 LA, Scanning Electron Microscope (SEM), with a power supply of 30 kV was used as an electron source, and the magnification was X = 7500.

Gas Chromatography/Mass spectrometry GC/MS (Model GC-2010 with GC/MS-QP 2010 Shimadzu—Japan) was used in the present work to measure the migration of BPA from the polycarbonate samples. The samples were put in a beaker with the stimulant (Methy-tert-butyl ether, from Fisher Sci., UK, HPLC Grade), then shacked at a water bath shaker for one hour, then collected at a glass vials to the chromatographic analysis. GC-MS is frequently used for the determination of BPA concentration because of its high accuracy [18]-[21].

3. Results and Discussion

Figure 2 shows the voltage and current waveforms of the DBD. The onset voltage of the discharge in the present conditions has been found to be 3.5 kV. As the discharge is set in the gap space the discharge current flow as filamentary current, see **Figure 2**. The appearance of the filamentary current in DBD is attributed to the fact that: in the case of DBD, plasma is formed in narrow channels (with diameter of few tens to few hundreds of nm) with short duration time (few tens of ns), these channels are called micro-discharge filaments and the discharge in this case is called filamentary discharge [22] [23]. The filamentary discharge is formed according to the following mechanism. As the applied voltage between the two electrodes exceeds the onset voltage, the free electrons in the gap space are accelerated by the electric field to energies that equal or exceed the ionization energy of the gas, and hence create an avalanche in which the number of electrons doubles with each generation of ionizing collision. The high mobility of the electrons compared to the ions allows the electron swarm to move across the gap in nanoseconds. The electrons leave behind the slower ions, and various excited and active species that may undergo further chemical and physical reactions with the treated samples in the discharge gap. When the electron swarm reaches the opposite electrode, the electrons spread out over the dielectric layer, counteracting the positive charge on the instantaneous anode. This factor combined with the cloud of slower ions left behind reduces the electric field in the vicinity of the filament and terminates any farther ionization along the original track in time scales of tens of nanoseconds [24]. So, micro-discharge filaments are generated individually in the discharge gap. The phase difference between current and voltage, shown in **Figure 2**, indicates to the capacitive reactance of the DBD cell.

According to the design of present treatment system, the running cost of the treatment process depends only on the consumed power in the discharge cell. To evaluate the consumed power in the DBD cell, it is wrong to express it as the product of current and voltage because of the capacitive reactance of the cell and the filamentary behavior of the current. In the present work, the consumed power has been measured using Lissajous method [25] where the voltage difference between the two electrodes has been measured as a function of the charge on the electrodes (Lissajous curve). **Figure 3** shows both the discharge current and the consumed power in the

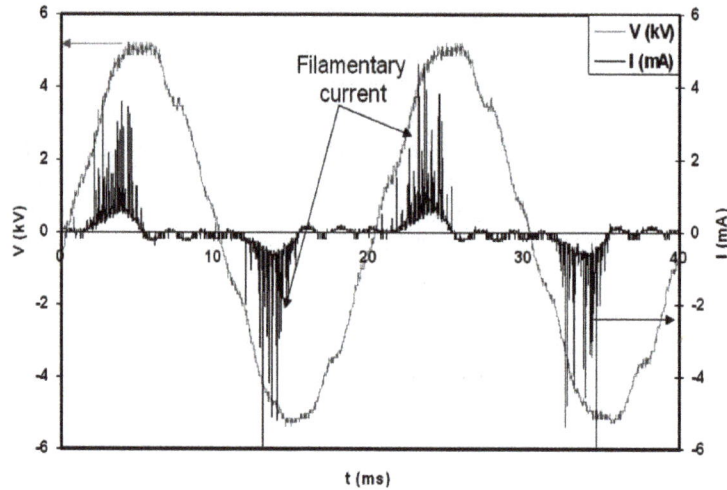

Figure 2. Voltage and current waveforms of DBD.

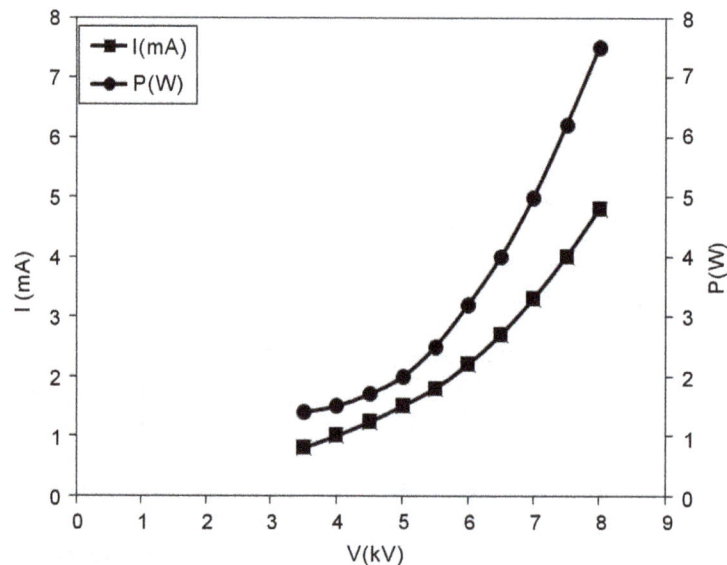

Figure 3. Discharge current and consumed power in the DBD cell as a function of the discharge voltage.

DBD cell as a function of the discharge voltage. The consumed power in the DBD cell has been found to be very low, even at high voltage (around 8 kV) the consumed power is less than 8 W. In the present work the treatment processes has been carried out at two applied voltages 4 and 5 kV which correspond to discharge currents 1 and 1.5 mA and consumed powers 1.5 and 2 W respectively. The gas temperature under these conditions varies from 30°C to 40°C. At higher discharge voltages the gas temperature is elevated to unwanted values which decrease the efficiency of the treatment process.

Figure 4 shows the XRD spectra of the untreated and treated PC samples. The two spectra are characterized by the appearance of halos extending in 2θ range from 15° to 18°. The ratio of diffraction peak areas was used for the analysis of crystal structure. The profile of the halos shows that the polymer is a partly crystalline polymer with a dominant amorphous phase. The peak has been found to be little more intense for the plasma treated sample which indicates to a little increase in crystallinity. There are no remarkable changes in the shape or position of the diffraction peaks after plasma treatment.

The decomposition behavior and thermal stability of the untreated and plasma treated PC samples have been studied using TGA. The thermograms of the untreated and the treated samples are shown in **Figure 5**. The

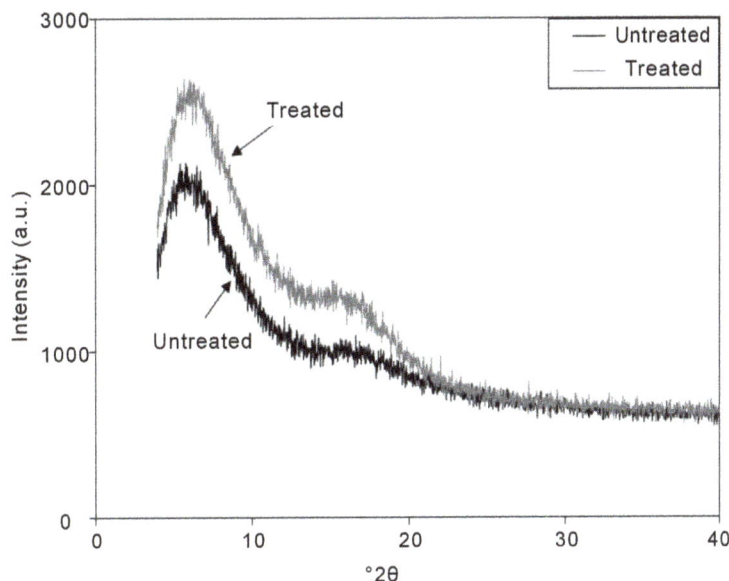

Figure 4. XRD spectra of the untreated and treated (1.5 W & 7.5 min) PC samples.

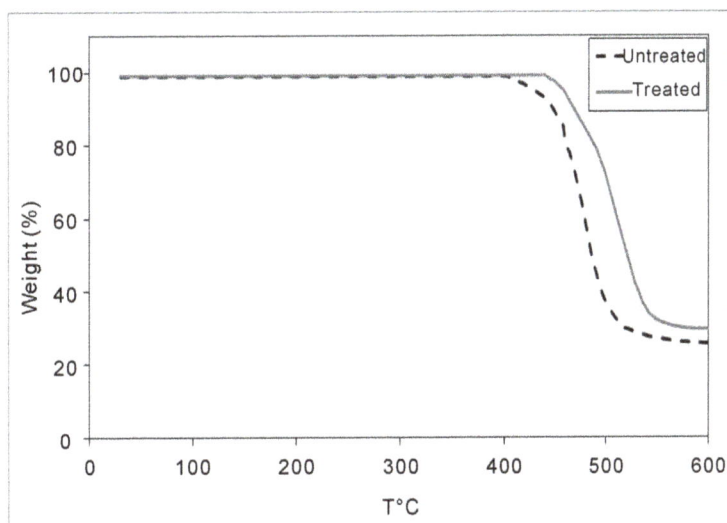

Figure 5. TGA thermograms of the untreated and the treated (1.5 W & 7.5 min) PC samples.

weight loss of the untreated PC sample starts at 400°C, while it starts at 440°C for the plasma treated sample. The weight loss reaches a maximum at 570°C with a value of around 76% for the untreated sample and reaches a maximum at 590°C with a value of around 70% for the plasma treated samples. These results show that the thermal stability of the PC samples has been increased due to the plasma treatment.

The weight loss of PC sample in TGA is attributed to the decomposition (thermal and oxidative) of carbonate link in between the monomers of BPA with vaporization and elimination of volatile products [26]. The increase in the resistance of PC by plasma treatment reveals that the cross-linking dominants or formation of more organized structure seems to be happening as due to the plasma treatment.

Figure 6 shows the contact angle between a drop of distilled water (around 2 μL in volume) and the surface of PC samples (untreated and treated at different treatment times and powers). The measurements of the contact angle have been carried out 7 days after the treatment by DBD. It can be noticed that the contact angle decreases

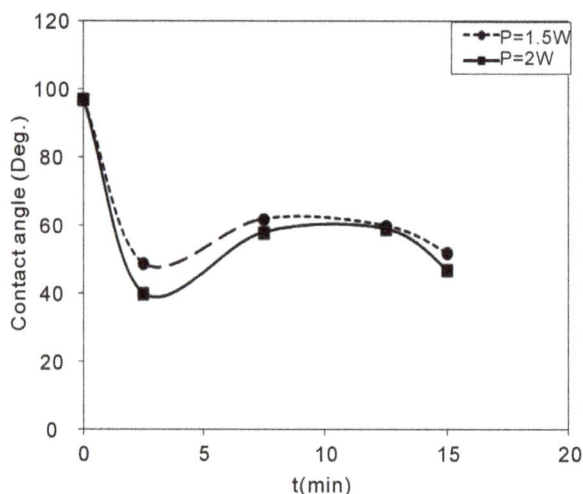

Figure 6. Contact angle as a function of the treatment time at two different discharge powers.

to around its half value by plasma treatment showing a saturation behavior with both the treatment time (from 2.5 to 15 min) and the discharge power (1.5 and 2 W). The decrease in contact angle can be attributed to increase in surface roughness and incorporation of hydrophilic functional groups which increases the surface energy and hence decreases the contact angle.

The increase in surface roughness has been proved by the SEM examination of untreated and treated samples, shown in **Figure 7**. Such increase in the surface roughness of the treated sample can be referred to the etching process of PC surface by plasma active species (electrons, ions and UV). In the beginning, the etching process is carried out by removal of low-molecular contaminates such as additive and absorbed species. After that the etching process starts to ablate the polymer chain itself. These etching processes are due to the physical removal of molecules by the impact of energetic plasma species and by the breaking up of bonds and chain scission [27].

The incorporation of hydrophilic functional groups is mainly related to the increase in the concentration of oxygen containing functional groups at the PC surface [28]. In low pressure plasma treatments, the increase in the concentration of oxygen containing functional groups was referred to the oxidation of the polymer surface by atomic oxygen generated by plasma dissociation of oxygen molecules [29]. In the present work we are attributing the oxidation process of PC surface not only to atomic oxygen but also to ozone molecules. Inside the micro-discharge filaments the energetic electrons react with the oxygen molecules forming excited atomic oxygen [30]:

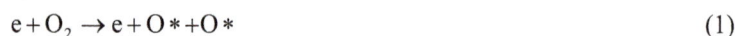

$$e + O_2 \rightarrow e + O* + O* \tag{1}$$

At atmospheric pressure there is a high probability of the reaction between the excited atomic oxygen and the oxygen molecules to form ozone molecules [31]:

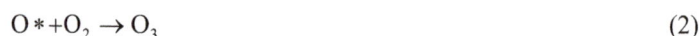

$$O* + O_2 \rightarrow O_3 \tag{2}$$

The formation of ozone molecules has been studied as a function of the discharge power by measuring the ozone concentration in the exhaust gas of the DBD cell using ozone analyzer (model AFX H1). This relation is shown in **Figure 8**. The increase of the ozone concentration with the discharge power can be attributed to the increase in the electron density and electron energies which increase the reactions of ozone formation (reaction 1 and 2). The saturation effect at discharge powers higher than 4 W can be referred to the elevation of the gas temperature which increases the dissociation rate of ozone molecules. At discharge powers (1.5 and 2 W) valuable concentrations of ozone (2 and 3 g/m^3 respectively) have been measured. The oxidation potential of ozone (2.1 V) is comparable to that of atomic oxygen (2.4 V) [32]. These results support our claim about the important role of ozone in the oxidation process of PC surface.

Figure 9 shows the density of BPA migrated from untreated and treated PC samples at different discharge powers and treatment time. The migration density of BPA from the untreated PC sample is around 5.5 ppm. The migration decreases for both discharge powers with treatment time up to 7.5 min and increases again at treat-

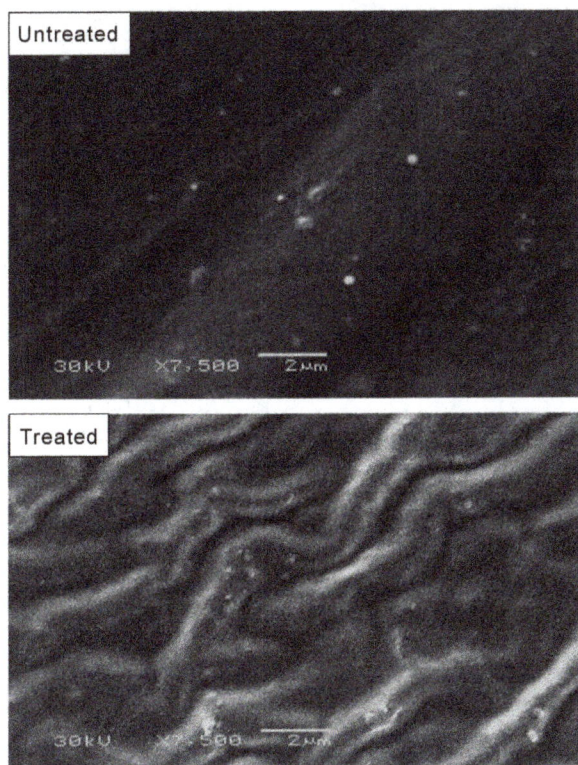

Figure 7. SEM micrographs of the untreated and treated (1.5 W & 7.5 min) PC samples.

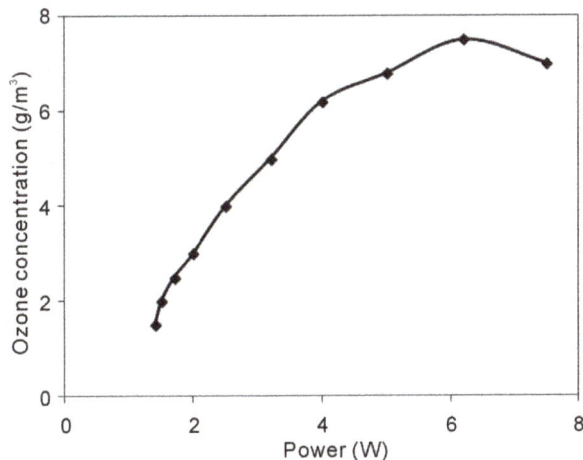

Figure 8. Ozone concentration in the exhaust gas of the DBD cell as a function of the discharge power.

ment time of 10 min. The limitation of BPA migration by plasma treatment up to 7.5 min can be referred to the effect of surface treatment of PC samples by plasma species. Plasma species break chemical bonds leaving free radicals at the PC surface. Ozone molecules and atomic oxygen react with the free radicals to produce new functional groups. The new fictional groups increase the cross-linking at the PC surface and hence increase the resistance to the dissociation of BPA monomers. The increase in the resistance to the dissociation by the treatment process has been indicated in TGA measurements. The cross linking at the surface of PC can explain the increase in the crystallinity of the treated samples showed by XRD pattern.

The increase in BPA migration at 10 min can be explained by the effect of etching process. At long treatment

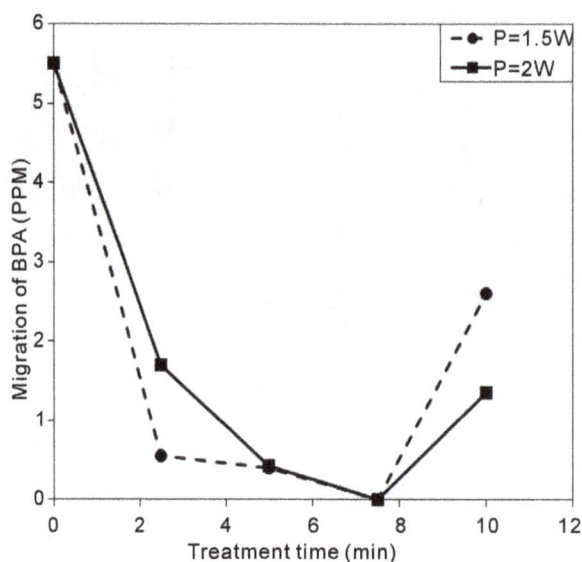

Figure 9. The migration of BPA from PC samples as a function of the treatment time at different consumed powers.

Table 1. Migration of BPA and the cost of the treatment process expressed as the surface density of consumed energy at different treatment conditions.

Power (W)	Treatment Time (min)	Migration of BPA (ppm)	% Limitation	(Cost) Surface density of consumed energy (kWh/m^2)
1.5	2.5	0.55	90	0.003
1.5	5	0.4	92	0.006
1.5	7.5	0	100	0.009
1.5	10	2.6	53	0.012
2	2.5	1.7	70	0.004
2	5	0.4	92	0.008
2	7.5	0	100	0.012
2	10	1.35	75	0.016
	Untreated	5.5	0	0

time (10 min) the temperature of a thin layer at the surface of the sample is elevated due to the impact of the plasma species on the sample surface and due to the elevation of the gas temperature. Such elevation of the surface temperature of the sample increases the etching rate whereas the etching of the treated layer is faster than the treatment of new layer. So, keeping the sample in low temperature is very important for efficient treatments.

As discussed above the most important advantage of the present system is the low cost of the treatment process. The cost of the treatment process here depends only on the consumed electric energy. The surface density of consumed energy, *i.e.*, the consumed electric energy per unit area of the treated PC samples, has been calculated and shown in **Table 1**. It can be notice that: at optimum treatment conditions, discharge power 1.5 W and treatment time of 7.5 min, the surface density of the consumed energy is only about 0.009 kWh/m^2 which is very low cost compared with the cost of the syntheses of PC. According to the present work, DBD is recommended as an economical treatment technique of PC instead of using more expensive materials or more expensive treatment techniques.

4. Conclusion

Dielectric barrier discharge has been found to be a cheap and effective technique for surface treatment of polycarbonate. The system has shown effective treatment of PC surface even at low consumed power and short treatment time. Examinations of the untreated and treated samples have shown decrease in the contact angle and increase in the crystallinity, thermal stability, and surface roughness. Effect of etching processes on the surface roughness has been indicated. The effect of ozone on the increase in the oxygen containing functional groups has been discussed. Effect of the treatment process on the migration of BPA from the surface of PC has been studied. Optimum limitation of the migration of BPA has been found at treatment time of 7.5 min where there is no any migration of BPA recorded. At longer time of treatment the migration of BPA increases again where the ablation of the treated layers due to high etching rate becomes more effective than the treatment process. The treatment process has been found to be very efficient in the limitation of the migration of BPA with very low cost.

References

[1]   Vandenberg, L.N., Hauser, R., Marcus, M., Olea, N. and Welshons, W.V. (2007) Human Exposure to Bisphenol A (BPA). *Reproductive Toxicology*, **24**, 139-177. http://dx.doi.org/10.1016/j.reprotox.2007.07.010

[2]   Biles, J.E., McNeal, T.P., Begley, T.H. and Hollifield, H.C. (1997) Determination of Bisphenol-A in Reusable Polycarbonate Food-Contact Plastics and Migration to Food-Simulating Liquids. *Journal of Agricultural and Food Chemistry*, **43**, 3541-3544. http://dx.doi.org/10.1021/jf970072i

[3]   Swan, S.H. (2000) Intrauterine Exposure to Diethylstilbestrol: Long-Term Effects in Humans. *APMIS*, **108**, 793-804. http://dx.doi.org/10.1111/j.1600-0463.2000.tb00001.x

[4]   Klip, H., Verloop, van Gool, J.D., Koster, M.E., Burger C.W. and van Leeuwin, F.E. (2002) Hypospadias in Sons of Women Exposed to Diethylstilbestrol in Utero: A Cohort Study. *The Lancet*, **359**, 1102-1107. http://dx.doi.org/10.1016/S0140-6736(02)08152-7

[5]   Troisi, R., Hatch, E.E., Titus-Ernstoff, L., Hyer, M., Palmer, J.R., Robboy, S.J., Strohsnitter, W.C., Kaufman, R., Herbst, A.L. and Hoover, R.N. (2007) Cancer Risk in Women Prenatally Exposed to Diethylstilbestro. *International Journal of Cancer*, **121**, 356-360. http://dx.doi.org/10.1002/ijc.22631

[6]   Li, D., Zhou, Z., Qing, D., He, Y., Wu, T., Miao, M., Wang, J., Weng, X., Ferber, J.R., Herrinton, L.J., Zhu, Q., Gao, E., Checkoway, H. and Yuan, W. (2010) Occupational Exposure to Bisphenol-A (BPA) and the Risk of Self-Reported Male Sexual Dysfunction. *Human Reproduction*, **25**, 519-527. http://dx.doi.org/10.1093/humrep/dep381

[7]   Kubwabo, C., Kosarac, I., Stewart, B., Gauthier, B.R., Lalonde, K. and Lalonde, P.J. (2009) Migration of Bisphenol A from Plastic Baby Bottles, Baby Bottle Liners and Reusable Polycarbonate Drinking Bottles. *Food Additives & Contaminants. Part A, Chemistry, Analysis, Control, Exposure & Risk Assessment*, **26**, 928-937. http://dx.doi.org/10.1080/02652030802706725

[8]   Cao, X.-L. and Corriveau, J. (2008) Migration of Bisphenol A from Polycarbonate Baby and Water Bottles into Water under Severe Conditions. *Journal of Agricultural and Food Chemistry*, **56**, 6378-6381. http://dx.doi.org/10.1021/jf800870b

[9]   De Coensel, N., David, F. and Sandra, P. (2009) Study on the Migration of Bisphenol-A from Baby Bottles by Stir Bar Sorptive Extraction-Thermal Desorption-Capillary GC-MS. *Journal of Separation Science*, **32**, 3829-3836. http://dx.doi.org/10.1002/jssc.200900349

[10]   Bredey, C., Fjeldalz, P., Skjevraky, I. and Herikstady, H. (2003) Increased Migration Levels of Bisphenol A from Polycarbonate Baby Bottles after Dishwashing, Boiling and Brushing. *Food Additives and Contaminants*, **20**, 684-689. http://dx.doi.org/10.1080/0265203031000119061

[11]   Nam, S.-H., Seo, Y.-M. and Kim, M.-G. (2010) Bisphenol A Migration from Polycarbonate Baby Bottle with Repeated Use. *Chemosphere*, **79**, 949-952. http://dx.doi.org/10.1016/j.chemosphere.2010.02.049

[12]   Palmer, J.R., Wise, L.A., Hatch, E.E., Troisi, R., Titus-Ernstoff, L., Strohsnitter, W., Kaufman, R., Herbst, A.L., Noller, K.L., Hyer, M. and Hoover, R.N. (2006) Prenatal Diethylstilbestrol Exposure and Risk of Breast Cancer. *Cancer Epidemiology, Biomarkers Prevention*, **15**, 1509. http://dx.doi.org/10.1158/1055-9965.EPI-06-0109

[13]   Hofrichter, A., Bulkin P. and Drevillon, B. (2002) Plasma Treatment of Polycarbonate for Improved Adhesion. *Journal of Vacuum Science & Technology A: Vacuum, Surfaces, and Films*, **20**, 245. http://dx.doi.org/10.1116/1.1430425

[14]   Kitova, S., Minchev, M. and Danev, G. (2005) RF Plasma Treatment of Polycarbonate Substrates. *Journal of Optoelectronics and Advanced Materials*, **7**, 2607-2612.

[15]   Subedi, D.P., Madhup, D.K., Adhikari, K. and Joshi, U.M. (2008) Plasma Treatment at Low Pressure for the Enhancement of Wettability of Polycarbonate. *Indian Journal of Pure & Applied Physics*, **46**, 540-544.

[16]   Qureshi, A., Shah, S., Pelagade, S., Singh, N.L., Mukherjee, S., Tripathi, A., Despande, U.P. and Shripathi, T. (2010) Surface Modification of Polycarbonate by Plasma Treatment. Journal of Physics: Conference Series, **208**, Article ID: 012108.

[17] Vijayalakshmi, K.A., Mekala, M., Yoganand, C.P. and Navaneetha Pandiyaraj, K. (2011) Studies on Modification of Surface Properties in Polycarbonate (PC) Film Induced by DC Glow Discharge Plasma. *International Journal of Polymer Science*, **2011**, Article ID: 426057. http://dx.doi.org/10.1155/2011/426057

[18] Biles, J.E., McNeal, T.P. and Begley, T.H. (1997) Determination of Bisphenol-A in Reusable Polycarbonate Food-Contact Plastics and Migration to Food-Simulating Liquids. *Journal of Agricultural and Food Chemistry*, **45**, 3541. http://dx.doi.org/10.1021/jf970072i

[19] Gonzalez-Casado, A., Navas, N., Del Olmo, M. and Vilchez, J.L. (1998) Determination of Bisphenol A in Water by Micro Liquid-Liquid Extraction Followed by Silylation and Gas Chromatography-Mass Spectrometry Analysis. *Journal of Chromatographic Science*, **36**, 565-569.

[20] Casajuana, N. and Lacorte, S. (2003) Presence and Release of Phthalic Esters and Other Endocrine Disrupting Compounds in Drinking Water. *Chromatographia*, 57, 649-655. http://dx.doi.org/10.1007/BF02491744

[21] Liu, R., Zhou, J.L. and Wilding, A. (2004) Simultaneous Determination of Endocrine Disrupting Phenolic Compounds and Steroids in Water by Solid-Phase Extraction-Gas Chromatography-Mass Spectrometry. *Journal of Chromatography A*, **1022**, 179-189. http://dx.doi.org/10.1016/j.chroma.2003.09.035

[22] Tay, W.H., Yap, S.L. and Wong, C.S. (2014) Electrical Characteristics and Modeling of a Filamentary Dielectric Barrier Discharge in Atmospheric Air. Sains Malaysiana, **43**, 583

[23] El-Zeer, D.M., Salem, A.A., Rashed, U.M., Abd Elbaset, T.A. and Ghalab, S. (2014) A Comparative Study between the Filamentary and Glow Modes of DBD Plasma in the Treatment of Wool Fibers. *International Journal of Engineering Research and Applications*, **4**, 401-410.

[24] Konelschatz, U., Eliasson, B. and Egli, W. (1997) Dielectric-Barrier Discharges. Principle and Applications. *Journal de Physique III d'octobre*, **7**, C4-47

[25] Kostov, K.G., Honda, R. Y., Alves, L.M.S. and Kayama, M.E. (2009) Characteristics of Dielectric Barrier Discharge Reactor for Material Treatment. *Brazilian Journal of Physics*, **39**, 322. http://dx.doi.org/10.1590/S0103-97332009000300015

[26] Šíra, M., Trunec, D., St'ahel, P., Buršíková, V. and Navrátil, Z. (2008) Surface Modification of Polycarbonate in Homogeneous Atmospheric Pressure Discharge. *Journal of Physics D: Applied Physics*, **41**, Article ID: 015205. http://dx.doi.org/10.1088/0022-3727/41/1/015205

[27] Kokkoris, G., Vourdas, N. and Gogolides, E. (2008) Plasma Etching and Roughening of Thin Polymeric Films: A Fast, Accurate, in Situ Method of Surface Roughness Measurement. *Plasma Processes and Polymers*, **5**, 825-833. http://dx.doi.org/10.1002/ppap.200800071

[28] Vargo, T.G., Gardella, J.A. and Salvati, L. (1989) Multitechnique Surface Spectroscopic Studies of Plasma Modified Polymers III. H_2O and O_2/H_2O Plasma Modified Poly(Methyl Methacrylate)s. *Journal of Polymer Science Part A: Polymer Chemistry*, **27**, 1267-1286. http://dx.doi.org/10.1002/pola.1989.080270413

[29] Yun, Y.I., Kim, K.S., Uhm, S.-J., Kkatua, B.B., Cho, K., Kim, J.K. and Park, C.E. (2004) Aging Behavior of Oxygen Plasma-Treated Polypropylene with Different Crystallinities. *Journal of Adhesion Science and Technology*, **18**, 1279-1297. http://dx.doi.org/10.1163/1568561041588200

[30] Garamoon, A.A., Elakshar, F.F., Nossair, A.M. and Kotp, E.F. (2002) Experimental Study of Ozone Synthesis. Plasma Sources Science and Technology, **11**, 254. http://dx.doi.org/10.1088/0963-0252/11/3/305

[31] Garamoon, A.A., Elakshar, F.F. and Elsawah, M. (2009) Optimizations of Ozone Generator at Low Resonance Frequency. *The European Physical Journal Applied Physics*, **48**, 21002. http://dx.doi.org/10.1051/epjap/2009144

[32] Eliasson, B., Hirth, M. and Kogelschatz, U. (1987) Ozone Synthesis from Oxygen in Dielectric Barrier Discharges. *Journal of Physics D: Applied Physics*, **20**, 1421. http://dx.doi.org/10.1088/0022-3727/20/11/010

11

Dependence of the Structure, Optical Phonon Modes and Dielectric Properties on Pressure in Wurtzite GaN and AlN

Huanyou Wang[1*], Yaqi Chen[1], Yalan Li[1,2], Xiangyan He[1]

[1]Department of Physics and Electronic Information Engineering, Xiangnan University, Chenzhou, China
[2]College of Physical Science and Technology, Huazhong Normal University, Wuhan, China
Email: *whycs@163.com

Abstract

The density functional perturbation theory (DFPT) is employed to study the structure, optical phonon modes and dielectric properties for wurtzite GaN and AlN under hydrostatic pressure. In order to calculate accurately the Born effective charges and high frequency dielectric tensors, we utilize two sum rules to monitor this calculation. The calculated optical phonon frequencies and longitudinal-transverse splitting show an increasing with pressure, whereas the Born effective charges and high frequency dielectric tensors are found to decrease with pressure. In particular, we analysed the reason for discrepancy between this calculation and previous experimental determination of pressure dependence of the LO-TO splitting in AlN. The different pressure behavior of the structural and lattice-dynamical properties of GaN and AlN is discussed in terms of the strengths of the covalent bonds and crystal anisotropy. Our results regarding dielectric Grüneisen parameter are predictions and may serve as a reference.

Keywords

GaN AlN, Lattice Dynamics, Dielectric, Pressure

1. Introduction

The group-III nitrides GaN and AlN are currently being actively investigated in view of their promising potential for short-wavelength electroluminescence devices and high-temperature, high-power, and high-frequency electronics [1]-[5]. An important motivation for high-pressure investigations stems from the fact that group-III-nitride layers are commonly subjected to large built-in strain since they are often grown on different substrates

---

*Corresponding author.

having considerable lattice mismatch.

The understanding the effect of pressure on the vibrational properties is quite important. Its knowledge allows one to correlate macroscopic thermodynamic parameters with properties on the atomic scale. The neutron scattering, electron energy loss spectroscopy, IR absorption, Raman spectroscopy, and diamond anvil cell etc. experimentally have been used to study phonons and related properties. Perlin et al. [6] and Kuball et al. [7] performed the high-pressure Raman studies for AlN, estimating the pressure coefficients of Raman-active modes. They reported that under pressure, the LO-TO (E₁) splitting slightly decreased and the LO-TO (A₁) splitting increases. Afterwards, Goñi et al. [8] compared the pressure dependence of the Raman-active modes in GaN and AlN with ab initio calculations and found a small but increasing LO-TO (E₁) splitting under pressure. However, a decrease of the LO-TO splitting for both A₁ and E₁ modes in AlN was estimated in recent Raman measurements of Yakovenko et al. [9] and Francisco et al. [10]. Meanwhile, Francisco et al. studied the optical phonon modes and pressure dependences of AlN by means of ab initio lattice dynamical calculations, but being a increase of the LO-TO splitting for both A₁ and E₁ modes. Perlin et al. [11] compared pressure dependence of the A_1 (TO), A_1 (LO) phonon modes and effective transverse charge of wurtzite GaN by Raman scattering and means of tight-binding formalism. In contrast with the extensive range of experimental studies on the pressure effect on the phonon dispersion of semiconductors, theoretical works on the topic are relatively sparse. The pressure dependence of the LO-TO splitting in GaN and AlN is an issue of controversy. In a polar lattice, the splitting of the optical phonon modes is determined by two parameters, Born effective charge of the lattice ions and the screening of the Coulomb interaction, which depends on the electronic part of the dielectric constant in the phonon frequency regime.

In this work, we study the pressure effect on phonon and relevant properties for GaN up to 50 *GPa* and AlN up to 20 *GPa* by DFPT computations. Firstly, we calculate and analyse the evolution with pressure of the unit cell shape (i.e., *c/a* ratio) and unit cell geometry (i.e., internal parameter *u*) of GaN and AlN. In the following section, it will study pressure dependence of zone-center optical phonon modes and the LO-TO splitting of both the A₁ and E₁ modes. Next, it will calculate and discuss the pressure dependence of Born effective charge tensors Z^* and the high frequency dielectric tensor ε_∞. Finally, it will predict dielectric Grüneisen paramete γ^ε.

2. Theory and Computational Details

2.1. Theory

The interatomic force constants (IFC's) describing the atomic interactions in a crystalline solid are defined in real space as [12]

$$C_{k\alpha,k'\beta}(a,b) = \frac{\partial^2 E}{\partial \tau_{k\alpha}^a \partial \tau_{k'\beta}^b} \tag{1}$$

Here, $\tau_{k\alpha}^a$ is the displacement vector of the kth atom in the ath primitive unit cell (with translation vector R_a) along the α axis. E is the Born-Oppenheimer (BO) total energy surface of the system (electrons plus clamped ions).

The vibration frequencies $\left(\omega_{j,q}\right)$ and polarization vectors $\left[e_k\left(q|j\right)\right]$ of the phonon modes with wave vector q are determined by solving the eigenvalue matrix equation

$$\sum_{k\beta}\tilde{D}_{k\alpha,k'\beta}(\bar{q})e_{k'\beta}(\bar{q}|j) = \omega_{j,\bar{q}}^2 e_{k,\alpha}(\bar{q}|j) \tag{2}$$

where $\tilde{D}_{k\alpha,k'\beta}(q)$ is the dynamical matrix, which is related to the Fourier transform of the IFC's.

The dielectric constant mainly is influenced by two factors which are the electron and phonon,

$$\varepsilon_{\alpha\beta}(\omega) = \varepsilon_{\alpha\beta}^\infty + \frac{4\pi}{\Omega_0}\sum_{kk'}\sum_{\alpha'\beta'} Z_{k,\alpha\alpha'}^*\left[\tilde{C}(q=0) - M\omega^2\right]_{k\alpha',k'\beta'}^{-1} Z_{k',\beta\beta'}^*, \tag{3}$$

where $\varepsilon_{\alpha\beta}^\infty = \delta_{\alpha\beta} - \frac{4\pi}{\Omega_0}2E_{el}^{\varepsilon_\alpha\varepsilon_\beta}$ is electronic contribution to the dielectric constant; $E_{el}^{\varepsilon_\alpha\varepsilon_\beta}$ is the second derivative of the total electronic energy with respect to a perturbing electric field along directions α and β; Ω_0 and M are the unit cell volume and mass; to other variables, see Ref. [12].

The Born effective charge is defined as the variation of the force on a given atom under the application of an electric field

$$Z^*_{k,\beta\alpha} = \Omega_0 \frac{\partial P_{mac,\beta}}{\partial \tau_{K\alpha}(q=0)} = \frac{\partial F_{k,\alpha}}{\partial \varepsilon_\beta} \qquad (4)$$

where $P_{mac,\beta}$ is the macroscopic electric polarization induced by the screened electric field. In order to calculate accurately the Born effective charges and dielectric constant, we utilize two sum rules to monitor calculation. The first is the acoustic-sum rule: the dynamical matrix at the zone center should admit the homogenous translations of the solid

$$\sum_{k'} \tilde{C}_{k\alpha,k'\beta}(q=0) = 0 . \qquad (5)$$

The second sum rule guarantees that the charge neutrality is also fulfilled at the level of the Born effective charges. For every direction α and β, one must have

$$\sum_k Z^*_{k,\alpha\beta} = 0 . \qquad (6)$$

By the above sum rules, we can monitor whether the calculation is well converged with respect to numerical parameters, like the number of plane waves, the sampling of BZ, and the number of points of the exchange-correlation grid.

2.2. Computational Details

We use a first-principles pseudopotential method base on the density functional perturbation theory with wave function represented in a plane-wave basis set. This work is performed employing the ABINT package [13]. A review of the method (and of the algorithm used for the convergence of electronic density and atomic positions) can be found in Ref. [14]. The effect of the approximation to the exchange-correlation (XC) energy is considered. The pseudopotential for Ga, Al and N atoms are generated according to scheme of Troullier and Martin [15]. Brillouin-zone integrations were performed using $12 \times 12 \times 8$ k-point mesh, and phonon frequencies were computed on a $6 \times 6 \times 4$ q-point mesh. Plane-wave basis sets with a cutoff of 40 Hartree were used. These calculating parameters are chosen to guarantee the total energy error in 0.1 mHartree.

3. Results and Discussion

In **Figure 1**, we show the evolution with pressure of the unit cell shape (*i.e.*, *c/a* ratio) and unit cell geometry (*i.e.*, internal parameter u) of the wurtzite structure GaN and AlN.

The calculations were performed in two steps. In the first step, we calculate the total energy of the bulk wurtzite crystal as a function of the unit cell volume. Then, using the definition of pressure, $P = -\partial E_{tot}/\partial V$, one can find the unit cell volume corresponding to the certain value of the external pressure P. In this step, for a given unit cell volume, *c/a* and *u* are optimized. As can be seen from these **Figure 1(a)** and **Figure 1(b)**, one can correlate the magnitude of changes in *c/a* and *u* under the hydrostatic pressure with deviations of the nitride structures from "ideal" wurtzite. The weaker dependences of *c/a* and *u* on the hydrostatic pressure are obtained for GaN, which possesses the smaller deviation of *c/a* from the ideal value 1.633. We find linear pressure coefficients of $\left(\partial^c_a/\partial p\right)_{p=0} = -5.45 \times 10^{-5}$ and $\left(\partial u/\partial p\right)_{p=0} = 1.81 \times 10^{-6}$ for GaN. In the case of AlN, the situation is completely different. The values *c/a* (*u*) are remarkably smaller (larger) than the ideal values and decrease (increase) with rising hydrostatic pressure, with a slope of $\left(\partial^c_a/\partial p\right)_{p=0} = -6.86 \times 10^{-4}$ and $\left(\partial u/\partial p\right)_{p=0} = 1.27$ $\times 10^{-4}$. Our results for *c/a* are in reasonable agreement with experimental studies of the lattice constants under pressure [16] [17].

For most stable wurtzite-type structures *c/a* ratio and the *u* parameter are strongly correlated; If *c/a* decreases, then *u* increases in such a way that the inequivalent bond lengths $R^{(1)}$ (along the *c* direction with bond length $R^{(1)} = uc$) and $R^{(2)}$ are nearly equal, however, the tetrahedral angles are distorted. The bond lengths $R^{(1)}$ and $R^{(2)}$

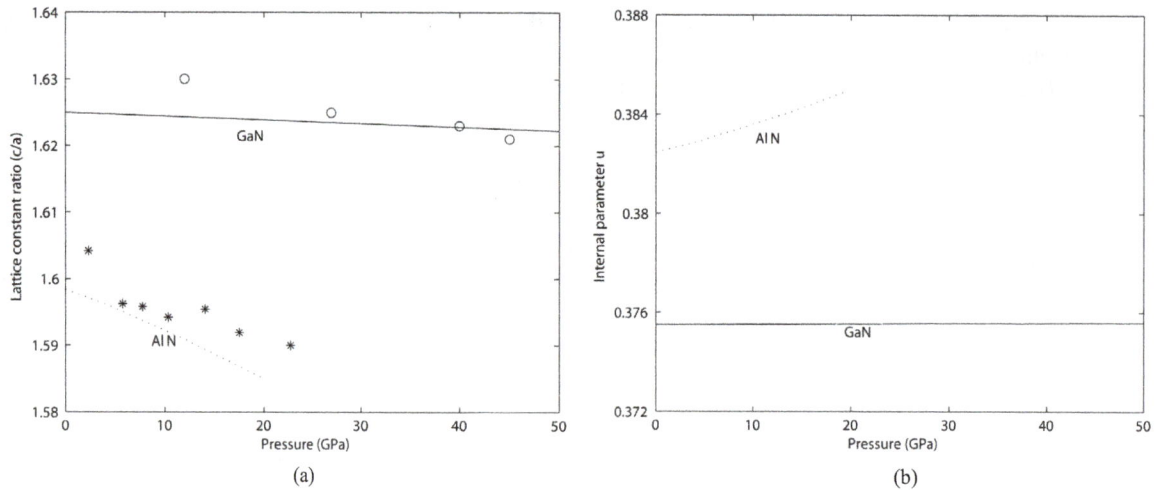

Figure 1. Structural parameters of wurtzite GaN and AlN under hydrostatic pressure. Asterisk and open circles are experimental results from Ref. [16] for GaN and Ref. [17] for AlN.

(in the hexagonal plane and is threefold degenerate with bond length $R^{(2)} = \sqrt{1/3 + \left(1/2 - u\right)^2 \left(c/a\right)^2}$) would be equal if $u = a^2/\left(3c^2\right) + 1/4$. The so-estimated value of the internal parameter u of GaN (0.3754) nearly agrees with the calculated one (0.3745). In the case of AlN there is, however, a larger deviation between the estimated value (0.3784) and the calculated one (0.3803). This finding can be attributed to the stronger covalent bonding of AlN, which preserves the ideal tetrahedral bond angles.

The pressure dependence of zone-center optical phonon modes and the LO-TO splitting of both the A_1 and E_1 modes are plotted in **Figure 2** (GaN) and **Figure 3** (AlN). **Table 1** summarizes fitting pressure coefficients of the A_1 and E_1. In general, the agreement between our calculated values at zero pressure and other theoretical values [8] is reasonably good. The calculated frequencies are slightly above other experimental data, which is likely due to underestimation of the lattice parameters as is usual in DFT-LDA calculations. Also, the calculated pressure coefficients are typically smaller than other experimental values. Also interesting is the difference between the pressure dependence of the LO-TO splittings in AlN. According to our calculations both the LO-TO splitting of both the A_1 and E_1 modes are almost constant or even increase slightly with increasing pressure. An increase of the LO-TO splitting for the A1 mode was found Manjón et al. [18] and Goñi et al. [8] and attributed to the decrease of the refractive index with pressure in AlN as suggested by ab initio calculations [19]. The experimental discrepancies could be due to differences in sample preparation and pressure environment that may affect the pressure response of the Raman modes in AlN [9]. Additionally, the weak intensity of the TO and LO modes reported by Kuball et al. [7] might have resulted in an inaccurate determination of the pressure coefficients and LO-TO splittings.

Taking the angular dispersion of the TO modes, $\left[\omega_{TO}\left(E_1\right) - \omega_{TO}\left(A_1\right)\right]/\omega_{TO}\left(E_1\right)$ as a measure of the crystal anisotropy, we find 0.081 for AlN and 0.045 for GaN. AlN is thus more anisotropic than GaN, which is coincident with former discuss.

The LO-TO splitting is a function of Born effective charges Z^* and infrared dielectric constant ε_∞. For modes of the same symmetry with atomic displacements along direction α ($\alpha = x, z$) one finds [20]

$$\omega_{LO}^2\left(\alpha\right) - \omega_{TO}^2\left(\alpha\right) = \frac{2e^2 \left(Z^*\right)_{\alpha\alpha}^2}{\varepsilon_0 \left(\varepsilon_\infty\right)_{\alpha\alpha} V \mu} \tag{7}$$

where ε_0 is the vacuum permittivity, μ is the reduced mass of an anion-cation pair, V is the available volume per pair, and ω is the angular mode frequency given in hertz. The change of the Born effective charge under compression can be determined from the measured frequencies of the optical phonons using Equation (7). Born effective charge tensors determine, with the high frequency dielectric tensor ε_∞, the strength of Coulomb interaction which is responsible for the splitting between the transverse (TO) and the longitudinal (LO) optical

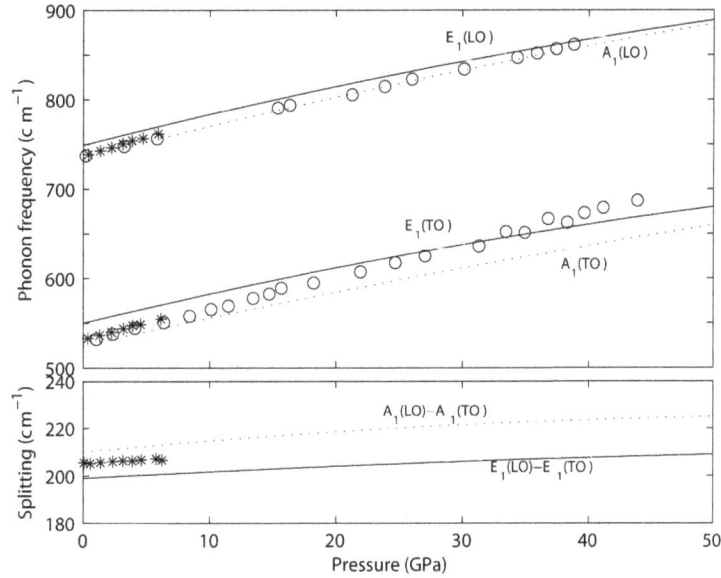

Figure 2. The zone-center optical frequencies for the wurtzite GaN as functions of pressure (upper panel). Pressure dependence of the LO-TO splitting for both the A_1 and E_1 modes in GaN (lower panel). Open circles and asterisk are taken from experimental results for A_1 from Ref. [11] and Ref. [8] for GaN respectively.

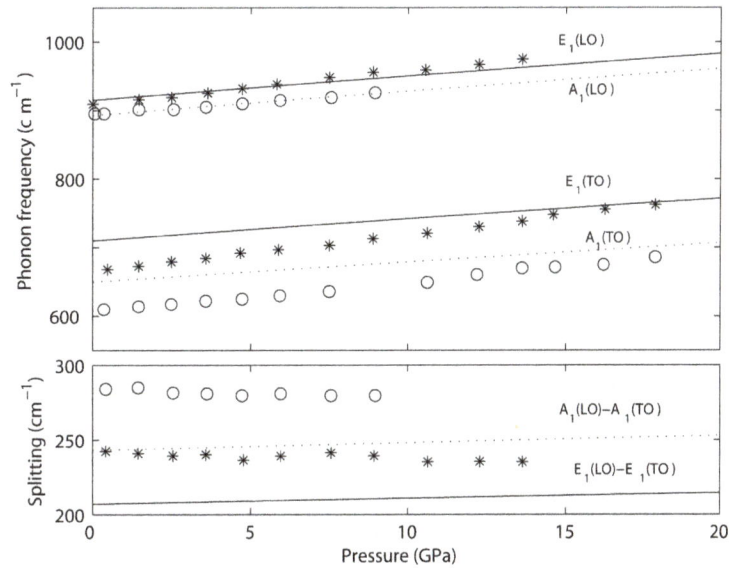

Figure 3. The zone-center optical frequencies for the wurtzite AlN as functions of pressure (upper panel). Pressure dependence of the LO-TO splitting for both the A_1 and E_1 modes in AlN (lower panel). Open circles and asterisk are taken from experimental results for A_1 and E_1 from Ref. [10] for AlN respectively.

modes. It is a measure of the change in electronic polarization due to ionic displacements. For atom k, $Z^*_{k,\beta\alpha}$ quantifies to linear order the polarization per unit cell, created along the direction β when the atoms of sublattice k are displaced along the direction α, under the condition of zero electric field.

Although there are four atoms in the unit cell of the wurtzite structure the nonsymmorphic space group C^4_{6v} with a screw along the c axis enforces that only two of them are independent. Furthermore, because of the

Table 1. Fitting parameters used for the pressure dependence of the phonon frequencies in GaN and AlN . For comparison theoretical and experimental results in other literatures are also shown.

		$A_1(TO)$	$E_1(TO)$	$A_1(LO)$	$E_1(LO)$
GaN	Calc. [a]	3.10	3.08	3.42	3.432
	Expt. [b]	3.9	3.94	4.4	
	Expt. [c]	3.55		3.2	
	Calc. [d]	3.1	3.3	3.5	3.6
AlN	Calc. [a]	3.05	2.96	3.48	3.50
	Expt. [e]	4.08	5.07	4.00	
	Expt. [f]	4.35	5.33	3.70	4.77
	Expt. [g]	4.05	4.52	4.00	3.60
	Calc. [h]	3.00	3.80	3.50	4.00

[a]This work, [b]Ref. [8], [c]Ref. [11], [d]Ref. [8], [e]Ref. [7], [f]Ref. [10], [g]Ref. [9], [h]Ref. [8].

acoustic sum rule: $\sum_k Z^*_{k,\alpha\beta} = 0$, Only two independent components $Z^*_{//}$ and $Z^*_{\perp}$ of Born effective charge tensor are existent. Contrary to the effective charges, the form of the dielectric tensor is determined by the symmetry of the crystal and is expected to be diagonal for the wurtzite structure. The dielectric tensors ε_∞ should have two independent components $\varepsilon^{//}_\infty$ and $\varepsilon^{\perp}_\infty$ along and perpendicular to the c axis, respectively.

The calculated Born effective charge tensors at zero pressure agree well with the experimental data obtained from first-order Raman-scattering experiments [21]. To the best of our knowledge, no other experimental data of the Born effective charge tensors for GaN and AlN exist. The calculated Born effective charges $Z^*_{//}$ and $Z^*_{\perp}$ decrease with increasing pressure. The pressure-induced reduction of the dynamical ion charges indicates a charge redistribution from the nitrogen atoms to the gallium or aluminum atoms in comparison with the pressure-free situation.

Our results concerning $\varepsilon^{\perp}_\infty$ and $\varepsilon^{//}_\infty$ at zero pressure for GaN are found to be 6.18 and 6.26, and for AlN 4.41 and 4.70 respectively. Comparing our calculated data with those computed by the orthogonalized linear combination of atomic orbitals (OLCAO) (Ref. [22]), and Pseudopotential calculations [19] [23], the agreement is good. However, Our values are bigger than those through full-potential linear-muffin-tin-orbital (LMTO) method [24]. The authors in Ref. [24] neglect the influence of local-field effects on the dielectric tensor, which, as reported, reduces the value of the dielectric constants by about 10% - 15% [25].

The average value $\varepsilon(\infty) = (1/3) Tr\varepsilon_\infty$ have also been calculated and found to be 4.5 for AlN and 6.2 for wurtzite GaN. These values agree to within 10% with the experimental some obtained by infrared reflectivity [26] and Raman-scattering [21]. However, one should note that the experimental data available for the stable structures of GaN and AlN are scarce and may suffer from the relatively low quality of the crystal samples. In addition, the screening tends to overestimate average dielectric tensor in theoretical calculations performed within the LDA approximation.

The pressure dependence of $\varepsilon^{\perp}_\infty$ and $\varepsilon^{//}_\infty$ for GaN and AlN is shown in **Figure 4** and **Figure 5**. Note that as pressure rises, $\varepsilon^{\perp}_\infty$ and $\varepsilon^{//}_\infty$ decrease monotonically, which is similar to what has been found for the most tetrahedrally coordinated semiconductors. Similar to that mode Grüneisen parameter, one can define a dielectric Grüneisen parameter

$$\gamma^\varepsilon = -\frac{d\ln\varepsilon}{d\ln V} \tag{8}$$

to characterize the pressure dependence of the dielectric constants. For GaN and AlN, we have found that the perpendicular and parallel components of γ^ε are both negative with $\gamma^{\varepsilon^{//}_\infty} = -0.397$, $\gamma^{\varepsilon^{\perp}_\infty} = -0.381$ for GaN, and $\gamma^{\varepsilon^{//}_\infty} = -0.582$, $\gamma^{\varepsilon^{\perp}_\infty} = -0.473$ for AlN.

4. Conclusion

In summary, first-principles calculations in the framework of the DFPT are carried out to study the pressure

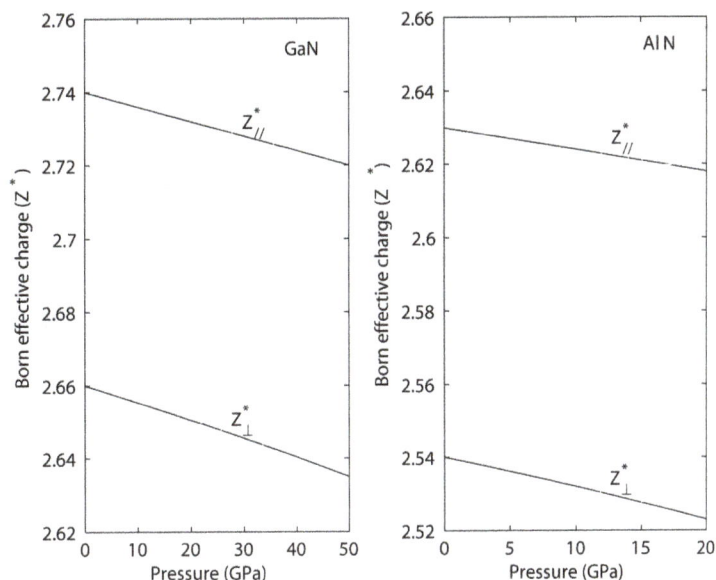

Figure 4. Born effective charge versus pressure for wurtzite GaN and AlN.

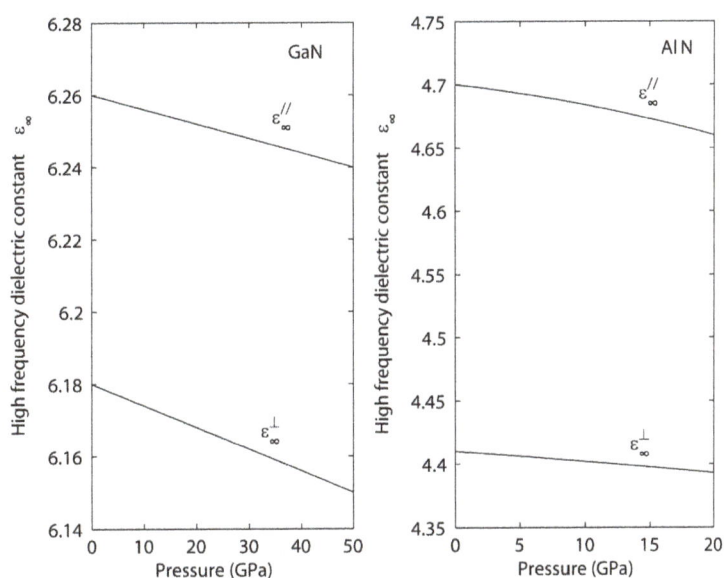

Figure 5. High-frequency dielectric constant versus pressure for wurtzite GaN and AlN.

dependences of structure, phonon and dielectric properties for wurtzite GaN and AlN. Our results show that pressure dependence of the wurtzite parameters c/a are reasonably well, as measured by high pressure X-Ray diffraction studies. The calculated pressure dependence of optical phonon frequencies, Born effective charges, dielectric constants is agreement with other theoretical data available. However, the calculated pressure coefficients of optical phonon frequencies are smaller than other experimental values, and pressure dependence of the LO-TO splittings in AlN between calculated and experimental data is contrary. The pressure dependence of $Z_\perp^*$, $Z_{//}^*$, $\varepsilon_\infty^\perp$ and $\varepsilon_\infty^{//}$ for wurtzite GaN and AlN is decreased monotonically as pressure rises, which is similar to most tetrahedrally coordinated semiconductors.

References

[1]   Schubert, E.F. and Kim, J.K. (2005) Solid-State Light Sources Getting Smart. *Science*, **308**, 1274-1278.

http://dx.doi.org/10.1126/science.1108712

[2]   Kuykendall, T., Ulrich, P., Aloni, S. and Yang, P. (2007) Complete Composition Tunability of InGaN Nanowires Using a Combinatorial Approach. *Nature Materials*, **6**, 951-956. http://dx.doi.org/10.1038/nmat2037

[3]   Simon, J., Protasenko, V., Lian, C., Xing, H. and Jena, D. (2010) Polarization-Induced Hole Doping in Wide—Band-Gap Uniaxial Semiconductor Heterostructures. *Science*, **327**, 60-64. http://dx.doi.org/10.1126/science.1183226

[4]   Dong, L., Yadav, S.K., Ramprasad, R. and Alpay, S.P. (2010) Band Gap Tuning in GaN through Equibiaxial In-Plane Strains. *Applied Physics Letters*, **96**, Article ID: 202106. http://dx.doi.org/10.1063/1.3431290

[5]   Wang, H.Y., Xu, H., Huang, T.T. and Deng, C.S. (2008) Thermodynamics of Wurtzite GaN from First-Principle Calculation. *The European Physical Journal B*, **62**, 39-43. http://dx.doi.org/10.1140/epjb/e2008-00121-2

[6]   Perlin, P., Polian, A. and Suski, T. (1993) Raman-Scattering Studies of Aluminum Nitride at High Pressure. *Physical Review B*, **47**, 2874-2879. http://dx.doi.org/10.1103/PhysRevB.47.2874

[7]   Kuball, M., Hayes, J.M. and Prins, A.D. (2001) Raman Scattering Studies on Single-Crystalline Bulk AlN under High Pressures. *Applied Physics Letters*, **78**, 724-726. http://dx.doi.org/10.1063/1.1344567

[8]   Goñi, A.R., Siegle, H., Syassen, K., Thomsen, C. and Wagner, J.M. (2001) Effect of Pressure on Optical Phonon Modes and Transverse Effective Charges in GaN and AlN. *Physical Review B*, **64**, Article ID: 035205. http://dx.doi.org/10.1103/PhysRevB.64.035205

[9]   Yakovenko, E.V., Gauthier, M. and Polian, A. (2004) High-Pressure Behavior of the Bond-Bending Mode of AlN. *JETP*, **98**, 981-985. http://dx.doi.org/10.1134/1.1767565

[10]   Manjón, F.J., Errandonea, D., Romero, A.H., Garro, N., Serrano, J. and Kuball, M. (2008) Lattice Dynamics of Wurtzite and Rocksalt AlN under High Pressure: Effect of Compression on the Crystal Anisotropy of Wurtzite-Type Semiconductors. *Physical Review B*, **77**, Article ID: 205204. http://dx.doi.org/10.1103/physrevb.77.205204

[11]   Perlin, P., Suski, J. and Ager, W. (1999) Transverse Effective Charge and Its Pressure Dependence in GaN Single Crystals. *Physical Review B*, **60**, 1480-1483. http://dx.doi.org/10.1103/PhysRevB.60.1480

[12]   Gonze, X. and Lee, C. (1997) Dynamical Matrices, Born Effective Charges, Dielectric Permittivity Tensors, and Interatomic Force Constants from Density-Functional Perturbation Theory. *Physical Review B*, **55**, 10355-10368. http://dx.doi.org/10.1103/PhysRevB.55.10355

[13]   Gonze, X., Beuken, J.M. and Caracas, R. (2002) First-Principles Computation of Material Properties: The ABINIT Software Project. *Computational Materials Science*, **25**, 478-492. http://dx.doi.org/10.1016/S0927-0256(02)00325-7

[14]   Payne, M.C., Teter, M.P., Allan, D.C., Arias, T.A. and Jonannopoulos, J.D. (1992) Iterative Minimization Techniques for *Ab initio* Total-Energy Calculations: Molecular Dynamics and Conjugate Gradients. *Reviews of Modern Physics*, **64**, 1045-1097. http://dx.doi.org/10.1103/RevModPhys.64.1045

[15]   Troullier, N. and Martins, J.L. (1991) Efficient Pseudopotentials for Plane-Wave Calculations. *Physical Review B*, **43**, 1993-2006. http://dx.doi.org/10.1103/PhysRevB.43.1993

[16]   Ueno, M., Onodera, A., Shimomura, O. and Takemura, K. (1992) X-Ray Observation of the Structural Phase Transition of Aluminum Nitride under High Pressure. *Physical Review B*, **45**, 10123-10126. http://dx.doi.org/10.1103/PhysRevB.45.10123

[17]   Ueno, M., Yoshida, M., Onodera, A., Shimomura, O. and Takemura, K. (1994) Stability of the Wurtzite-Type Structure under High Pressure: GaN and InN. *Physical Review B*, **49**, 14-21. http://dx.doi.org/10.1103/physrevb.49.14

[18]   Manjón, F.J., Errandonea, D., Garro, N., Romero, A.H., Serrano, J. and Kuball, M. (2007) Effect of Pressure on the Raman Scattering of Wurtzite AlN. *Physica Status Solidi* (*b*), **244**, 42-47.

[19]   Wagner, J.M. and Bechstedt, F. (2002) Properties of Strained Wurtzite GaN and AlN: *Ab initio* Studies. *Physical Review B*, **66**, 115202. http://dx.doi.org/10.1103/PhysRevB.66.115202

[20]   Venkataraman, G., Feldkamp, L.A. and Sahni, V.C. (1975) Dynamics of Perfect Crystals. MIT Press, Cambridge, Massachusetts, 202.

[21]   Sanjurjo, J.A., Lopez-Cruz, E., Vogl, P. and Cardona, M. (1983) Dependence on Volume of the Phonon Frequencies and the IR Effective Charges of Several III-V Semiconductors. *Physical Review B*, **28**, 4579-4584. http://dx.doi.org/10.1103/PhysRevB.28.4579

[22]   Xu, Y.N. and Ching, W.Y. (1993) Electronic, Optical, and Structural Properties of Some Wurtzite Crystals. *Physical Review B*, **48**, 4335-4351. http://dx.doi.org/10.1103/PhysRevB.48.4335

[23]   Bungaro, C., Rapcewicz, K. and Bernholc, J. (2000) *Ab initio* Phonon Dispersions of Wurtzite AlN, GaN, and InN. *Physical Review B*, **61**, 6720-6725. http://dx.doi.org/10.1103/PhysRevB.61.6720

[24]   Christensen, N.E. and Gorczyca, I. (1994) Optical and Structural Properties of III-V Nitrides under Pressure. *Physical Review B*, **50**, 4397-4415. http://dx.doi.org/10.1103/PhysRevB.50.4397

[25] Karch, K. and Bechstedt, F. (1997) *Ab initio* Lattice Dynamics of BN and AlN: Covalent versus Ionic Forces. *Physical Review B*, **56**, 7404-7415. http://dx.doi.org/10.1103/PhysRevB.56.7404

[26] Gorczyca, I., Christensen, N.E., Pelzery Blancá, E.L. and Rodriguez, C.O. (1995) Optical Phonon Modes in GaN and AlN. *Physical Review B*, **51**, 11936-11939. http://dx.doi.org/10.1103/PhysRevB.51.11936

12

Structural and Dielectric Properties of Sn Doped Barium Magnesuim Zirconium Titanate Perovskite Ceramics

Sankararao Gattu[1], Kamala Sujani Dasari[2], Venkata Ramesh Kocharlakota[3]*

[1]Department of Physics, MVJ College of Engineering, Near ITPB, Bangalore, India
[2]Department of Physics, A. C. College, Guntur, India
[3]Deptartment of Physics, GITAM Institute of Science, GITAM University, Visakhapatnam, India
Email: *kvramesh11@gmail.com

Abstract

Perovskite type ceramics $(Ba_{0.9}Mg_{0.1})(Sn_xZr_{0.4-x}Ti_{0.6})O_3$ (with x = 0.01, 0.02, 0.03 and 0.04) relaxor composition prepared through solid state reaction route and calcinated at temperature is 1150°C for 5 hrs with intermediate mixing. The room temperature XRD study suggests that all the samples have the single phase cubic symmetry with space group pm 3 m. The pellets were sintered at 1500°C for 4 hrs. Scanning Electron Microscope (SEM) observations revealed enhanced micro structural uniformity and retarded grain growth with decreasing Sn content. The dielectric measurements at constant frequency show that dielectric constant increases with Sn content. Loss factor and dielectric constant decreased with increasing frequency but at very high frequencies it was independent.

Keywords

Sn Doped, Barium Titanate, Dielectric Properties, Perovskite, Lead Free Ceramics

1. Introduction

Barium Titanate (BT) is the most common ferro electric material, which is used to manufacture electronic components such as multilayer capacitors, positive temperature coefficient thermistors, piezo electric transdures, and ferro electric memory, because of its excellent dielectric, piezo electric and ferro electric properties [1] [2].

---

*Corresponding author.

$Pb(Zr,Ti)O_3$, PZT based ceramics has been study more than anyone else ferroelectric because of their excellent dielectric properties [3]. However, the presence of lead in those materials is about 60% in weight [4], reconsidering its use in technical applications, due to its high toxicity of lead for the environment as well as for humans [5]-[9]. The micro structure and dielectric properties of BT can be modified by addition of the dopants such as La^{3+}, Ce^{2+}, Mn^{4+}, Nb^{5+}, Nd^{3+}, Cr^{3+}, Zr^{4+}, Mg^{2+}, Sr^{2+} and Si^{4+} to occupy Ba^{2+} on A sites or Ti^{4+} on B sites to form the solid solution [10]-[26]. It has been reported that [27] with ~15% Zr substitution in $Ba(ZrTi)O_3$ the three transitions (rhombohedra to orthorhombic, orthorhombic to tetragonal and tetragonal to cubic) of BT, come towards the room temperature with enhanced dielectric constant. Further increase in Zr content beyond 15%, a diffuse dielectric anomaly in ceramic has been observed with the decrease in transition temperature [28] and the material showed typical relaxor like behavior in the range 25% - 42% Zr substitution [29]. Unexpectedly the lead free ceramics show the relaxor properties at low temperatures [30]. Several attempts have been made by researchers on these materials to shift the T_c to close to room temperature. It is well known that homovalent and hetrovalent substitution for barium and titanium ions gives rise to various behaviors including the shifting of the transition temperature. A small content of Ba replaced by Mg in BZT solid solution the dielectric peaks has been shifted. But the transition temperature shifted towards lower temperature. The Sn ion is smaller than the Zr ion. If we substitute the Sn in Zr^+ ion site the T_c may be increased to room temperature and Sn^{4+} substituted BZT ceramic exhibits both high piezoelectric properties and good temperature stabilities in common usage temperature range. This inspired to work on effect of Sn on structural and dielectrical properties of $(Ba_{0.9}Mg_{0.1})(Sn_xZr_{0.4-x}Ti_{0.6})O_3$ relaxor composition prepared through solid state reaction route.

2. Experimental

The perovskite samples of pure and Sn doped Barium Magnesium Zirconate Titanate (BMSZT (0.000), (0.010), (0.015), (0.020), (0.025)) were prepared by conventional solid state reaction method. The starting raw materials were $BaCO_3$ (Chen Chems., Chennai), TiO_2 (Loba Chem., Mumbai) and ZrO_2 (Loba Chem., Mumbai), MgO (Chen Chems., Chennai) and SnO_2 (E. Merck India Ltd.). All the powders were having more than 99% purity. The powders were taken in a suitable stachiometry for 20 gm of samples. The powders were thoroughly mixed in an agate mortar in dry and wet mixing with appropriate amount of Acetone for 6 hrs. After proper mixing, mixed powders were calcinated at 1150°C for 5 hrs. and a small amount polyvinyl alcohol was added to the calcinated powder for fabrication of pellets, which was burnt out during high temperature sintering. The circular disc shaped pellets were prepared by applying a uniaxial pressure of 4.5×10^6 N/m^2. The pellets were subsequently sintered at an optimized temperature of 1500°C for 4 hrs. A preliminary study on compound formation and structural parameters was carried out using an X-ray diffraction (XRD) technique with an X-ray powder diffractometer. The XRD pattern of the calcinated powder was recorded at room temperature PANAlytical X'pert pro with CuK_a radiation (1.5405 Å) in a wide range of Bragg's angles 2θ $(15 \leq 2\theta \leq 80)$. Micro structures of sintered pellets were recorded by scanning electron microscope (SEM)(JEOM JSM-6380 LA). The pellets were then electrode with high purity air-drying silver paste and then dried at 500°C for 1 hr. Impedance spectroscopic analysis was done using a Agilent E4980A Precision LCR meter with temperature (150 - 330 K) and frequency (20 Hz - 200 KHz).

3. Results and Discussion

3.1. Structural Analysis

Figure 1 shows the XRD pattern of the pure and Cu doped BMSZT (0.000, 0.010, 0.015, 0.020, 0.025) samples. The XRD analysis provides that the samples are having single perovskite structure. $BaTiO_3$ (BT) has the tetragonal structure at room temperature. The ionic radii of Ba^{2+} and Ti^{4+} are 1.35 Å and 0.605 Å respectively. If we doped BTO with Mg^{2+} and Zr^{4+} whose ionic radii are both 0.72 Å Mg occupies A site and Zr occupies B site of BT.

The pure BMSZT single phased cubic structure when the Mg content is <1.5% at -% [31] and Zr content is < 0.42% at -% [32]. The small amount of Sn has ionic radius 0.69 Å doping to BMSZT. By doping with Sn the diffraction angles are shifted towards the higher angle side indicating the decrease in lattice parameters due to the incorporation of smaller content of Sn in place of Zr.

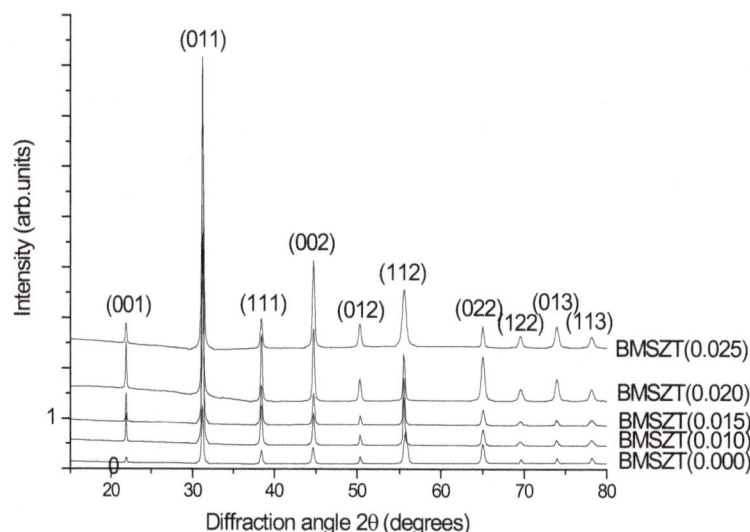

Figure 1. X-ray diffractograms of Sn doped BMSZT (0.000, 0.010, 0.015, 0.020, 0.025) samples.

3.2. Microstructural Analysis

Figure 2 shows The SEM micrographs of pure and Sn dope BMSZT (0.000, 0.010, 0.015, 0.020, 0.025) samples. It is found that the average grain size of samples are ~1.66, ~1.55, ~1.42, ~1.27 and ~1.25 µm as the Sn content decreases from 0% to 0.025%. This decrease is in agreement with our XRD pattern. Moreover the surface observation shows a good density of grains with some porosity.

3.3. Dielectric Properties

3.3.1. Temperature Dependence Dielectric Properties
Figure 3 shows the temperature dependence of the dielectric constant and loss of pure and Sn doped BMSZT samples measured at 1 MHz. The figure shows, the value of dielectric constant increases gradually to a maximum value (ε_m) with increase in temperature up to transition temperature and then decreases indicating a phase transition. It is also found that the Curie temperature Tc of BMSZT samples with Sn dopant of (0.000, 0.010, 0.015, 0.020, 0.025) corresponding to the maximum dielectric constant is 180, 200, 210, 225 and 250 respectively. The results indicates that the curie temperature of BMSZT increased may be due to Zr ions replaced by Sn ions and Sn ionic radius is some small, it can decrease the grain size, again decrease in curie temperature is may be due to occupying the more number of Sn atoms in Zr sites, due to the Sn ions conducts the little current then the dielectric constant may be decreased and curie temperature increased.

According to **Figure 3(a)** the peak value of the dielectric constant of BMSZT samples with the Sn dopant of (0.000, 0.010, 0.015, 0.020, 0.025) is 566, 551, 510, 480 and 462 respectively. The result indicates that the peak value of dielectric constant for undoped sample is the maximum and the peak value decreases with Sn content.

Figure 3(b) shows that the dielectric loss initially increases with temperature reaches maximum. Further increase in temperature loss is decreased but for BMSZT sample of (0.020) it is at lower temperature high value of loss due to the presence of all types of polarisation and may be due to the contribution of finite resistivity of the materials. Further increase in temperature increase in ionic conductivity resulting from the disordering of mobile cations in the oxygen octahedral skeleton [32].

3.3.2. Frequency Dependence Dielectric Properties
As shown in **Figure 4(a)** first, it is found that the dielectric constant of BMSZT samples decreased with frequency. Second, it is also found the dielectric constant of BMSZT (0.015) decreased rapidly at low frequencies. At very high temperatures dielectric constant is very low and it maintains constant value. It may be due to there must be defects with opposite charges (dipoles) to preserve charge neutrality. Theses dipoles could be oriented to align the direction of the applied electric field. When the frequency increases, the dipoles do not catch up with

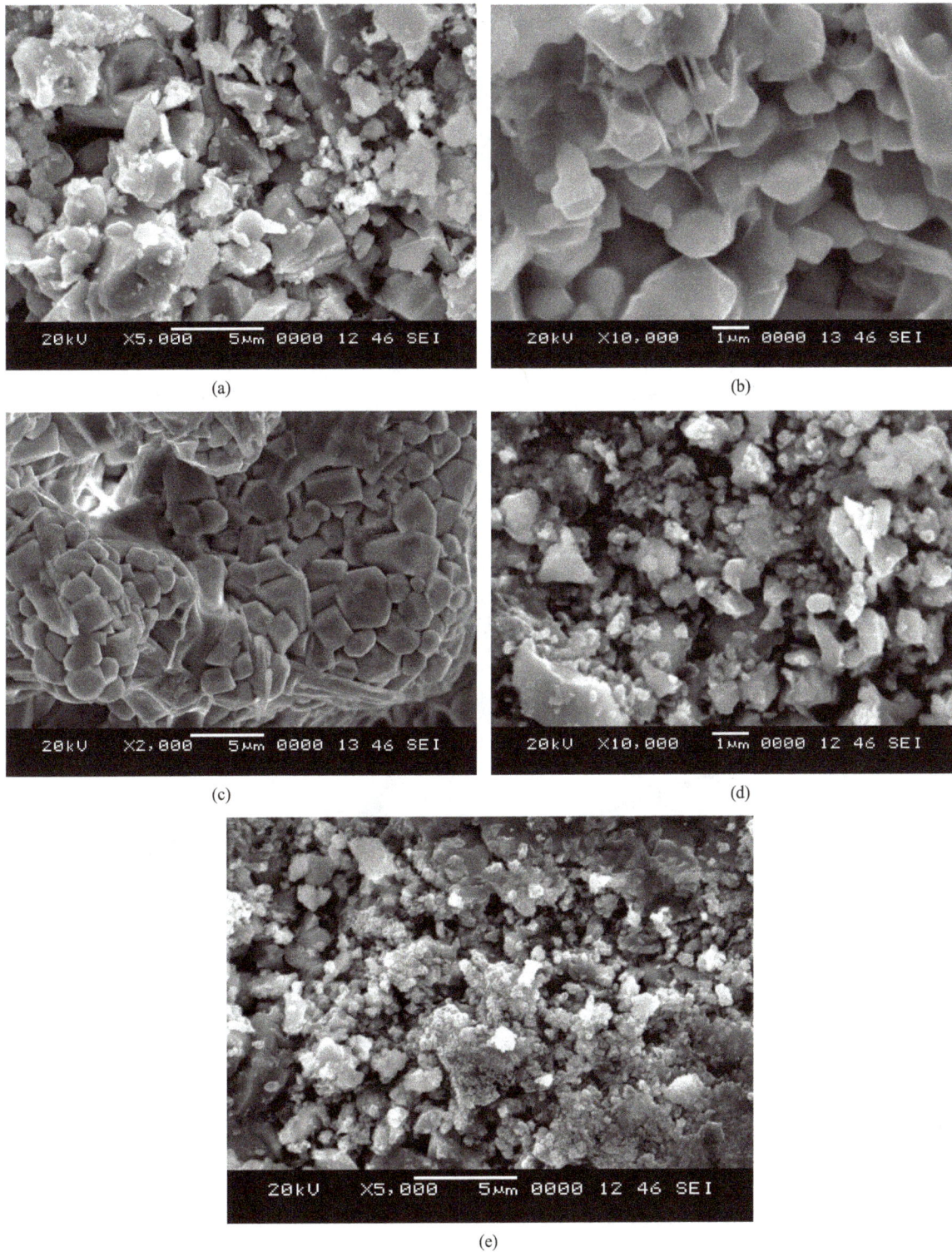

(a)

(b)

(c)

(d)

(e)

Figure 2. SEM micrograph of pure and Sn dope BMSZT ceramics. (a) 0.000; (b) 0.010; (c) 0.01; (d) 0.020; (e) 0.025.

the change of the electric field to complete polarisation so that the dielectric constant decreases.

In the **Figure 4(b)** the dielectric losses were a combined result of electrical conduction and orientational polarisation of the matter. The energy losses, which occur in dielectrics due to dc conductivity and dipole relaxation. The loss factor of a dielectric material is a useful indicator of the energy loss as heat.

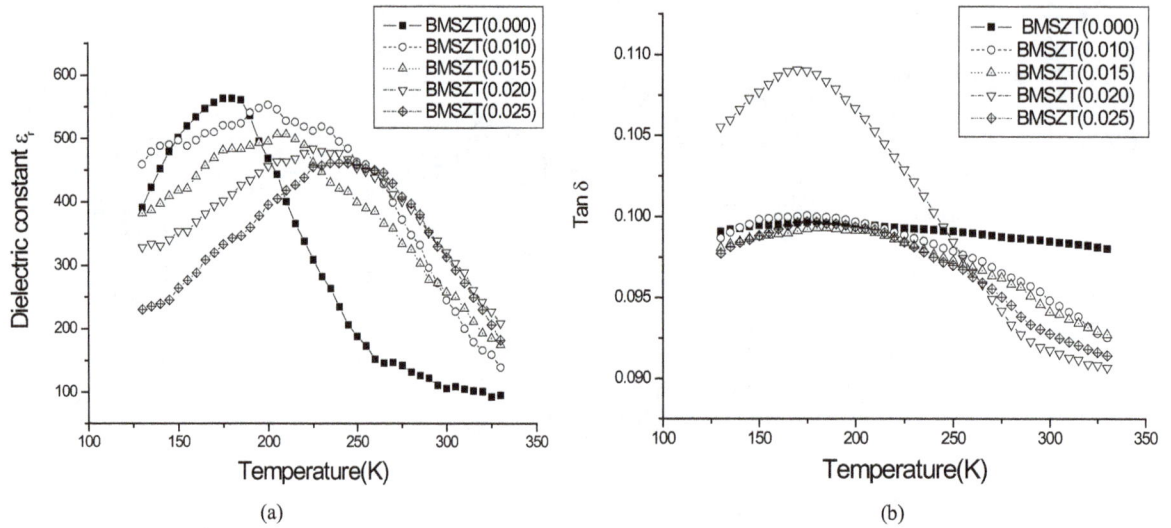

(a)   (b)

Figure 3. Temperature dependence of (a) Dielectric constant (b) Dielectric loss of pure and Sn dope BMSZT (0.000, 0.010, 0.015, 0.020, 0.025) samples.

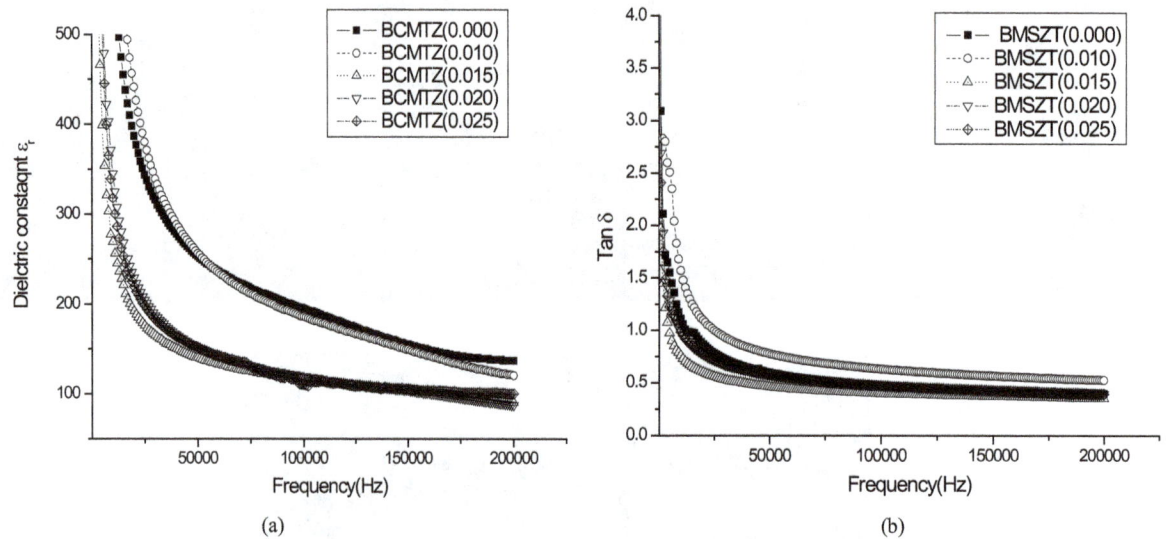

(a)   (b)

Figure 4. Frequency dependence of (a) Dielectric constant (b) Dielectric loss of Sn doped BMSZT (0.000, 0.010, 0.015, 0.020, 0.025) samples.

4. Conclusion

Perovskite types $(Ba_{0.9}Mg_{0.1})(Sn_xZr_{0.4-x}Ti_{0.6})O_3$ (with x = 0.000, 0.010, 0.015, 0.020, 0.025) ceramics have prepared through solid state reaction route. The room temperature XRD study suggests that the compositions have single phase cubic symmetry with space group pm-3m. The dielectric study reveals that the material undergoes a diffuse type ferroelectric phase transition. The transition temperature increased with Sn content and the dielectric constant decreased with Sn content.

References

[1]   Mitic, V.V., Nikolic, Z.S., Pavlovic, V.B., Paunovic, V., Miljkovic, M., Jordovic, B. and Zivkovic, L. (2010) Influence of Rareearth Dopants on Barium Titanate Ceramics Microstructure and Corresponding Electrical Properties. *Journal of the American Ceramic Society*, **93**, 132-137. http://dx.doi.org/10.1111/j.1551-2916.2009.03309.x

[2]   Jung, W.S., Kim, J.H., Kim, H.T. and Yoon, D.H. (2010) Effect of Temperature Schedule on the Particle Size of Barium Titanate during Solid-State Reaction. *Materials Letters*, **64**, 170-172.

http://dx.doi.org/10.1016/j.matlet.2009.10.035

[3]   Takenaka, T. and Nagata, H. (2005) Current Status and Prospects of Lead-Free Piezoelectric Ceramics. *Journal of the European Ceramic Society*, **25**, 2693-2700. http://dx.doi.org/10.1016/j.jeurceramsoc.2005.03.125

[4]   Saito, Y., Takao, H., Tani, T., *et al.* (2004) Lead-Free Piezoceramics. Lead-Free Piezoceramics. *Nature*, **432**, 84-87. http://dx.doi.org/10.1038/nature03028

[5]   Dixit, A., Majumder, S.B., Katiyar, R.S. and Bhalla, A.S. (2003) Relaxor Behavior in Sol-Gel-Derived $BaZr_{(0.40)}Ti_{(0.60)}O_3$ Thin Films. *Applied Physics Letters*, **82**, 2679-2681. http://dx.doi.org/10.1063/1.1568166

[6]   Dobal, P.S., Katiyar, R.S. and Raman, J. (2002) Ferroelectric Perovskites and Bi-Layered Compounds Using Micro-Raman Spectroscopy. *Spectroscopy*, **33**, 405. http://dx.doi.org/10.1002/jrs.876

[7]   Dixit, A., Majumder, S.B., Savvinov, A., Katiyar, R.S., Guo, R. and Bhalla, A.S. (2002) Investigations on the Sol-Gel-Derived Barium Zirconium Titanate Thin Films. *Materials Letters*, **56**, 933-940. http://dx.doi.org/10.1016/S0167-577X(02)00640-7

[8]   Paik, D.S., Park, S.E., Wada, S., Liu, S.F. and Shrout, T.R. (1999) E-Field Induced Phase Transition in <001>-Oriented Rhombohedral $0.92Pb(Zn_{1/3}Nb_{2/3})O_3$-$0.08PbTiO_3$ Crystals. *Journal of Applied Physics*, **85**, 1080. http://dx.doi.org/10.1063/1.369252

[9]   Yu, Z., Ang, C., Guo, R. and Bhalla, A.S. (2002) Dielectric Properties and High Tunability of $Ba(Ti_{0.7}Zr_{0.3})O_3$ Ceramics under dc Electric Field. *Applied Physics Letters*, **81**, 1285. http://dx.doi.org/10.1063/1.1498496

[10]   Parkash, O., Kumar, D., Dwivedi, R.K., Srivastava, K.K., Singh, P. and Singh, S. (2007) Effect of Simultaneous Substitution of La and Mn on Dielectric Behavior of Barium Titanate Ceramic. *Journal of Materials Science*, **42**, 5490-5496. http://dx.doi.org/10.1007/s10853-006-0985-8

[11]   Langhammer, H.T., Müller, T., Böttcher, R. and Abicht, H.P. (2008) Structural and Optical Properties of Chromium-Doped Hexagonal Barium Titanate Ceramics. *Journal of Physics*: *Condensed Matter*, **20**, Article ID: 085206. http://dx.doi.org/10.1088/0953-8984/20/8/085206

[12]   Lu, D.Y., Toda, M. and Sugano, M. (2006) High-Permittivity Double Rare Earth-Doped Barium Titanate Ceramics with Diffuse Phase Transition. *Journal of the American Ceramic Society*, **89**, 3112-3123. http://dx.doi.org/10.1111/j.1551-2916.2006.00893.x

[13]   Chen, Z.W. and Chu, J.Q. (2008) Piezoelectric and Dielectric Properties of $Bi_{0.5}(Na_{0.84}K_{0.16})_{0.5}TiO_3$-$Ba(Zr_{0.04}Ti_{0.96})O_3$ Lead Free Piezoelectric Ceramics. *Advances in Applied Ceramics*, **107**, 222-226. http://dx.doi.org/10.1179/174367608X263403

[14]   Fu, C.L., Cai, W., Chen, H.W., Feng, S.C., Pan, F.S. and Yang, C.R. (2008) Voltage Tunable $Ba_{0.6}Sr_{0.4}TiO_3$ Thin Films and Coplanar Phase Shifters. *Thin Solid Films*, **516**, 5258-5261. http://dx.doi.org/10.1016/j.tsf.2007.07.059

[15]   Cai, W., Fu, C.L., Gao, J.C. and Chen, H.Q. (2009) Effects of Grain Size on Domain Structure and Ferroelectric Properties of Barium Zirconate Titanate Ceramics. *Journal of Alloys and Compounds*, **480**, 870-873. http://dx.doi.org/10.1016/j.jallcom.2009.02.049

[16]   Du, F.T., Yu, P.F., Cui, B., Cheng, H.O. and Chang, Z.G. (2009) Preparation and Characterization of Monodisperse Ag Nanoparticles Doped Barium Titanate Ceramics. *Journal of Alloys and Compounds*, **478**, 620-623. http://dx.doi.org/10.1016/j.jallcom.2008.11.099

[17]   Yuan, Y., Zhang, S.R., Zhou, X.H. and Tang, B. (2009) Effects of Nb_2O_5 Doping on the Microstructure and the Dielectric Temperature Characteristics of Barium Titanate Ceramics. *Journal of Materials Science*, **44**, 3751-3757. http://dx.doi.org/10.1007/s10853-009-3502-z

[18]   Xiao, S.X. and Yan, X.P. (2009) Preparation and Characterization of Si-Doped Barium Titanate Nanopowders and Ceramics. *Microelectronic Engineering*, **86**, 387-391. http://dx.doi.org/10.1016/j.mee.2008.11.042

[19]   Rath, M.K., Pradhan, G.K., Pandey, B., Verma, H.C., Roul, B.K. and Anand, S. (2008) Synthesis, Characterization and Dielectric Properties of Europium-Doped Barium Titanate Nanopowders. *Materials Letters*, **62**, 2136-2139. http://dx.doi.org/10.1016/j.matlet.2007.11.033

[20]   Gulwade, D. and Gopalan, P. (2008) Diffuse Phase Transition in La and Ga Doped Barium Titanate. *Solid State Communications*, **146**, 340-344. http://dx.doi.org/10.1016/j.ssc.2008.02.018

[21]   Unruan, M., Sareein, T., Tangsritrakul, J., Prasetpalichatr, S., Ngamjarurojana, A., Anata, S. and Yimnirun, R. (2008) Changes in Dielectric and Ferroelectric Properties of Fe^{3+}/Nb^{5+} Hybrid-Doped Barium Titanate Ceramics under Compressive Stress. *Journal of Applied Physics*, **104**, Article ID: 124102. http://dx.doi.org/10.1063/1.3042228

[22]   Yaseen, H., Baltianski, S. and Tsur, Y. (2006) Effect of Incorporating Method of Niobium on the Properties of Doped Barium Titanate Ceramics. *Journal of the American Ceramic Society*, **89**, 1584-1589. http://dx.doi.org/10.1111/j.1551-2916.2006.00966.x

[23]   Cha, S.H. and Han, Y.H. (2006) Effects of Mn Doping on Dielectric Properties of Mg-Doped $BaTiO_3$. *Journal of Ap-*

plied Physics, **100**, Article ID: 104102. http://dx.doi.org/10.1063/1.2386924

[24] Shen, Z.J., Chen, W.P., Qi, J.Q., Wang, Y., Chan, H.L.W., Chen, Y. and Jiang, X.P. (2009) Dielectric Properties of Barium Titanate Ceramics Modified by SiO_2 and by BaO-SiO_2. *Physica B: Condensed Matter*, **404**, 2374-2376. http://dx.doi.org/10.1016/j.physb.2009.04.039

[25] Kirianov, A., Hagiwara, T., Kishi, H. and Ohsato, H. (2002) Effect of Ho/Mg Ratio on Formation of Core-Shell Structure in $BaTiO_3$ and on Dielectric Properties of $BaTiO_3$ Ceramics. *Japanese Journal of Applied Physics*, **41**, 6934-6937. http://dx.doi.org/10.1143/JJAP.41.6934

[26] Wang, S., Zhang, S.R., Zhou, X.H., Li, B. and Chen, Z. (2005) Effect of Sintering Atmospheres on the Microstructure and Dielectric Properties of Yb/Mg Co-Doped $BaTiO_3$ Ceramics. *Materials Letters*, **59**, 2457-2460. http://dx.doi.org/10.1016/j.matlet.2005.03.016

[27] Henning, D., Schnell, A. and Simon, G. (1982) Diffuse Ferroelectric Phase Transitions in $Ba(Ti_{1-y}Zr_y)O_3$ Ceramics. *Journal of the American Ceramic Society*, **65**, 539-544. http://dx.doi.org/10.1111/j.1151-2916.1982.tb10778.x

[28] Yu, Z., Guo, R. and Bhalla, A.S. (2000) Dielectric Behavior of $Ba(Ti_{1-x}Zr_x)O_3$ Single Crystals. *Journal of Applied Physics*, **88**, 410. http://dx.doi.org/10.1063/1.373674

[29] Yu, Z., Guo, R. and Bhalla, A.S. (2002) Dielectric Properties and High Tunability of $Ba(Ti_{0.7}Zr_{0.3})O_3$ Ceramics under DC Electric Field. *Applied Physics Letters*, **81**, 1285. http://dx.doi.org/10.1063/1.1498496

[30] Dixit, A., Majumder, S.B., Katiyar, R.S. and Bhalla, A.S. (2003) Relaxor Behavior in Sol-Gel-Derived $BaZr_{(0.40)}Ti_{(0.60)}O_3$ Thin Films. *Applied Physics Letters*, **82**, 2679. http://dx.doi.org/10.1063/1.1568166

[31] Cai, W., Fu, C.L., Gao, J.C. and Zhao, C.X. (2011) Dielectric Properties and Microstructure of Mg Doped Barium Titanate Ceramics. *Advances in Applied Ceramics*, **110**, 181-185. http://dx.doi.org/10.1179/1743676110Y.0000000019

[32] Ravez, J. and Simon, A. (1997) Temperature and Frequency Dielectric Response of Ferroelectric Ceramics with Composition $Ba(Ti_{1-x}Zr_x)O_3$. *European Journal of Solid State and Inorganic Chemistry*, **34**, 1199.

13

Structure and Dielectric Relaxation Behaviour of [Pb$_{0.94}$Sr$_{0.06}$][(Mn$_{1/3}$Sb$_{2/3}$)$_{0.05}$(Zr$_{0.49}$Ti$_{0.51}$)$_{0.95}$]O$_3$ Ceramics

Kumar Brajesh[1]*, Kiran Kumari[2]

[1]Materials Engineering, Indian Institute of Science, Bangalore, India
[2]P G Department of Physics, R N College Hajipur (Vaishali), Bihar, India
Email: *kmrbrjsh9@gmail.com

Abstract

The field dependences of the dielectric response and conductivity are measured in a frequency range from 100 Hz to 1 MHz and in a temperature range from 300 K to about 775 K. The dielectric measurements (real and imaginary parts) of this composition with temperature (300 K - 775 K) at different frequencies (100 Hz - 1 MHz) unambiguously point towards relaxor behaviour of the material. The real part of the dielectric constant is found to decrease with increasing frequency at different temperatures while the position of dielectric loss peak shifts to higher frequencies with increasing temperature indicating a strong dispersion beyond the transition temperature, a feature known for relaxational systems such as dipole glasses. The frequency dependence of the loss peak obeys an Arrhenius law with activation energy of 0.15 eV. The distribution of relaxation times is confirmed by Cole-Cole plots as well as the scaling behavior of the imaginary part of the electric modulus. The frequency-dependent electrical data are also analyzed in the framework of the conductivity and modulus formalisms. Both these formalisms yield qualitative similarities in the relaxation times. The Rietveld analysis conforms that the materials exhibits tetragonal structure. The SEM photographs of the sintered specimens present the homogenous structures and well-grown grains with a sharp grain boundary.

Keywords

Perovskite Oxide, Dielectric Relaxation, Rietveld Analysis

---

*Corresponding author.

1. Introduction

The group of materials with ABO_3 type perovskite structure is also very important due to their attractive electrical and magnetic properties for technological applications and richness of physical and chemical aspects they possess. The perovskite lead zirconate titanate Pb $(Zr_xTi_{1-x})O_3$ abbreviated as PZT is known to have excellent piezoelectric properties and used as transformers, ultrasonic motors and electromechanical transducers; it is desirable to combine high mechanical quality factor (Q_m) with large coupling factor (k_p) and low dielectric loss as well as high mechanical strength [1]-[3]. Relaxors are mainly lead-based perovskite solid solutions and exhibit a stronger piezoelectric effect, high permittivity over a broad temperature range, unique dielectric response with strong frequency dispersion, and anomalous phonon dispersion relation [4]-[6]. Relaxors are used in transducers and capacitors due to the excellent dielectric and piezoelectric properties. As like, prototypical $(1-x)Pb(Mg_{1/3},Nb_{2/3})O_3$ relaxors form the core of the state-of-the-art ultrasound medical imaging probes. However, in SONAR transducers, conventional crystalline PZT is widely used despite the fact that relaxors show about four to ten times larger the d_{33} coefficients. This is due to loss of piezoelectric performance below unsuitably low Curie temperature and low coercive fields in the relaxors compared with PZT. The diffusive and dispersive characteristics of the inverse dielectric response are signatures of relaxor behavior. The diffuse phase transitions that take place gradually are called relaxed phase transitions, giving rise to the term "relaxor". Currently, experimental and theoretical investigations of relaxors are ongoing due to the technological applications and scientific interest of the modified PZT materials [7]-[9].

These compositional modifications may be incorporated either by chemical substitution at A-sites or B-sites of the perovskite structure or by using off-valent element as an additive. In the miniaturization era of technological advancement, there is ever a pressing need for light, efficient, reliable, and long lasting devices for power supply. This requires new generation of electric components like transformer, capacitor, transistor etc. The most commonly used tiny piezoelectric transformers are based on PT and PZT compositions. The hard and soft PZTs have their own advantages. The former has a low dielectric loss factor and a high mechanical quality factor (Q_m) while the latter has a high piezoelectric constant and a high coupling coefficient. Usually hard piezoelectric ceramics are found to be useful for transformer applications, because of their high mechanical quality factor (Q_m) [10]-[13]. Modified PZT ceramics have found applications in high power and transmitting components which demand high mechanical, dielectric and piezoelectric properties. In order to obtain proper ceramics which combine the advantages of both hard and soft PZTs, different modifications have been investigated. Mn^{2+} like Fe^{3+} is generally known as hard additive to generate O^{2-} vacancies. Sr^{2+} cation replaces Pb^{2+} on the A-site. In the same way, Mn^{2+} cation replaces Zr^{4+} or Ti^{4+} on the B-site. O-site vacancies lead to contraction of the grain body. In this paper, we have therefore concentrated on a hard piezoelectric ceramic composition $[Pb_{0.94}Sr_{0.06}][(Mn_{1/3}Sb_{2/3})_{0.05}(Zr_{0.49}Ti_{0.51})_{0.95}]O_3$ as a case study to examine the phenomenon and mechanisms of dielectric relaxation. It is also known that the piezoelectric properties of Sr^{2+} substituted PZT are most pronounced for $x = 0.06$. However, no attempt has so for been made to prepare 6% Sr^{2+} substituted PZT ceramics with lead manganese antimonite as an additional hardener dopant. The present work aims to fill this gap. The distribution of relaxation times [13]-[15] will be confirmed by Cole-Cole plots and scaling behavior of imaginary part of the electric modulus. The frequency-dependent electrical data will also be analyzed in the framework of the conductivity and modulus formalisms.

2. Experimental

Stoichiometric amounts of the $PbCO_3$ and $SrCO_3$ powders were taken for the preparation of $[Pb_{0.94}Sr_{0.06}]CO_3$. Saturated solution of ammonium carbonate was added in the solution of $PbCO_3$ and $SrCO_3$ in dilute nitric acid to obtain the precipitate of $[Pb_{0.94}Sr_{0.06}]CO_3$. The precipitate was washed with distilled water until ammonia was removed and then dried in an oven. Now stoichiometric amounts of Sb_2O_5 (99% purity) and $MnCO_3$ (99.9% purity) were mixed with the help of mortar and pestle for 6 hours to obtain the intimate mixture. In order to obtain the preparation of the final composition $[Pb_{0.94}Sr_{0.06}][(Mn_{1/3}Sb_{2/3})_{0.05}(Zr_{0.49}Ti_{0.51})_{0.95}]O_3$, the stoichiometric amounts of $[Pb_{0.94}Sr_{0.06}]CO_3$, and manganese antimonite ($MnSb_2O_6$), ZrO_2 (99% purity) and TiO_2 (99.5% purity) were mixed with the help of mortar and pestle for 6 hours to get an intimate homogeneous mixture. This mixture was dried in air and then calcined at 800°C for 6 hours. Heat treatments involving calcinations and sintering (1170°C) were carried out with the help of a high temperature Globar furnace. Sintering was performed in PbO atmosphere to avoid lead loss. In the present work, the pellets were sintered in PbO environment created by lead zirconate as a spacer powder in a covered alumina crucible using MgO powder as a sealing agent. This arrangement reduces the lead oxide losses at higher sintering temperature effectively. The duration of calcinations (800°C) and sintering (1170°C) each was kept for 6 hours. The powder compaction was done using a cylindrical die of 8 mm diameter and a hydraulic press. Before cold compaction of calcined powder, a few drops of 2% PVA aqueous solution were added to serve as a binder. For the electrical characterizations, the sintered ceramic pellets were electroded using the silver paste. After applying the paste the pellets were then dried at about 150°C in an oven. The silver paste coated pellets were fired at 500°C for five minutes. The room temperature dielectric measurements of these electroded pellets were done using Hioki LCR meter.

3. Rietveld Refinement Details

Rietveld refinement was carried out using the XRD data with the help of the DBWS-9411 program [9]. The background was fitted with 6-Cofficients polynomials function, while the peak shapes were described by pseudo-Voigt profiles. In all the refinements, scale factor, lattice parameters, positional coordinates (x, y, z) and thermal parameters were varied. Occupancy parameters of all the ions were kept fixed during refinement. No correlation between the positional and thermal parameters was observed during refinement and as such it was possible to refine all the parameters together.

4. Results and Discussion

Figure 1 shows the XRD patterns of the pure perovskite phase of $[Pb_{0.94}Sr_{0.06}][(Mn_{1/3}Sb_{2/3})_{0.05}(Zr_{0.49}Ti_{0.51})_{0.95}]O_3$ ceramics. All the reflection peaks of the X-ray profile are indexed. As shown in the inset of **Figure 2**, the magnified Bragg profiles of the $\{200\}_c$ and $\{111\}_c$ pseudocubic reflections. From virtual inspection of the shape of Bragg's profile we noted that the $\{200\}_c$ is split into two and $\{111\}_c$ is singlet. This is compatible with a tetragonal (P4mm) structure. The Pb^{2+}/Sr^{2+} ions occupy 1 (a) sites at (0, 0, z), $Mn^{2+}/Sb^{5+}/Zr^{4+}/Ti^{4+}$ and O_I^{2-} occupy 1 (b) sites at (1/2, 1/2, z), and O_{II}^{2-} occupy 2(c) sites at (1/2, 0, z). For the refinement, the initial values of the lattice parameters were obtained from our XRD data by least squares method, whereas the values of the structural parameters were taken from Noheda et al. [16] [17]. In this structure, Pb^{2+}/Sr^{2+} coordinates were fixed at (0, 0, 0) in our refinement. **Figure 2** also depicts the observed, calculated and difference profiles for the refined structure. The fit is quite good. The refined structural parameters and the positional coordinates of this composition are given in **Table 1**. SEM studies (**Figure 3**) carried out on the sintered specimens reveal that the average grain size is 7.5 μm with homogenous structures and well-grown grains having a sharp grain boundary. The angular frequency $\omega(= 2\pi\upsilon)$ dependent plots of the real (ε') and imaginary (ε'') parts of complex dielectric permittivity (ε^*) of the given composition at several temperatures from room temperature (303 K) to 775 K are shown in **Figure 4**. A relaxation is observed in the entire temperature range as a gradual decrease in $\varepsilon'(\omega)$ and as a broad peak in $\varepsilon''(\omega)$. Relaxation phenomena in dielectric materials are associated with a frequency-dependent orientational polarization. At low frequency, the permanent dipoles align themselves along the field and contribute fully to the total polarization of the dielectric. At higher frequency, the variation in the field is too rapid for

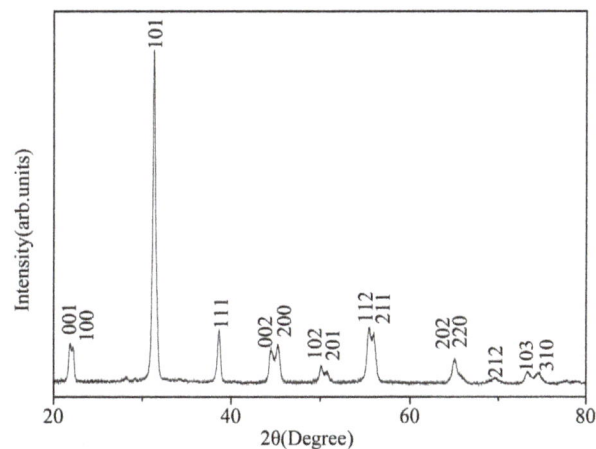

Figure 1. XRD pattern of powder at room temperature.

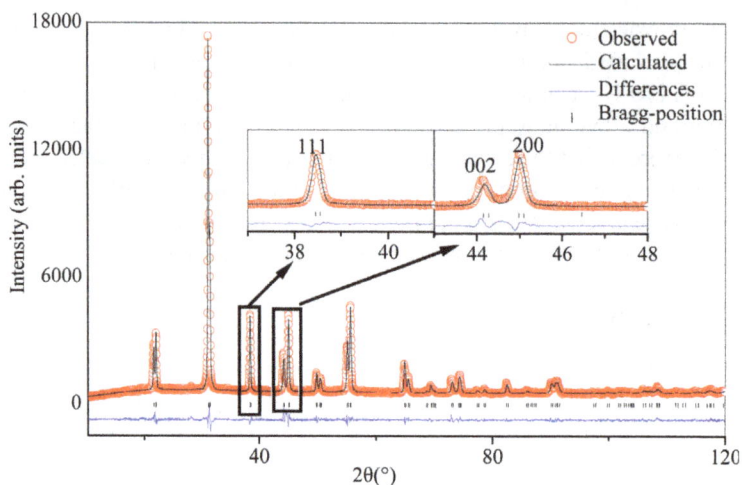

Figure 2. Rietveld fit of the X-ray powder diffraction of the given composition. The dot represent the observed data, the continuous line is the fitted pattern, the vertical bars show the Bragg peak position. The arrows highlight the pseudocubic {111}$_{pc}$ and {200}$_{pc}$ Bragg profiles.

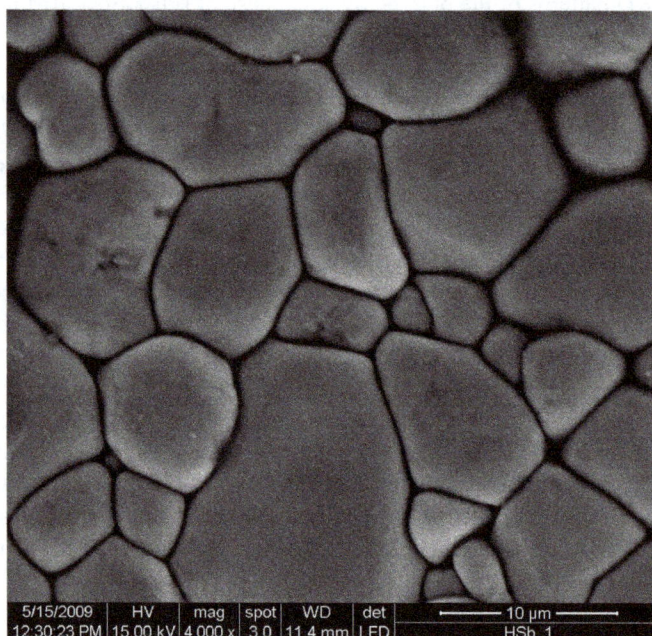

Figure 3. Scanning electron micrographs of sintered composition.

Table 1. Refined structural parameters of $[Pb_{0.94}Sr_{0.06}][(Mn_{1/3}Sb_{2/3})_{0.05}(Zr_{0.49}Ti_{0.51})_{0.95}]O_3$ using tetragonal (space group; P4mm) model.

Ions	Positional coordinates			Thermal parameters U (Å^2)
	x	y	z	
Pb^{2+}/Sr^{2+}	0.00	0.00	0.00	$U_{11} = U_{22} = 0.0374\ (5), U_{33} = 0.008\ (5)$
$Mn^{2+}/Sb^{5+}/Zr^{4+}/Ti^{4+}$	0.50	0.50	0.545 (5)	$U_{11} = U_{22} = 0.0014\ (5), U_{33} = 0.004\ (5)$
O_I^{2-}	0.00	0.50	0.592 (5)	$U_{iso} = 0.093\ (2)$
O_{II}^{2-}	0.50	0.50	0.071 (5)	$U_{iso} = 0.0065\ (1)$

a = b = 4.027 (2) Å, c = 4.098 (2) Å, R_P = 16.5, R_{wp} = 13.1, R_{exp} = 9.59, R_f = 2.57, R_B = 3.13, Vol = 66.475 (5) and χ^2 = 1.86.

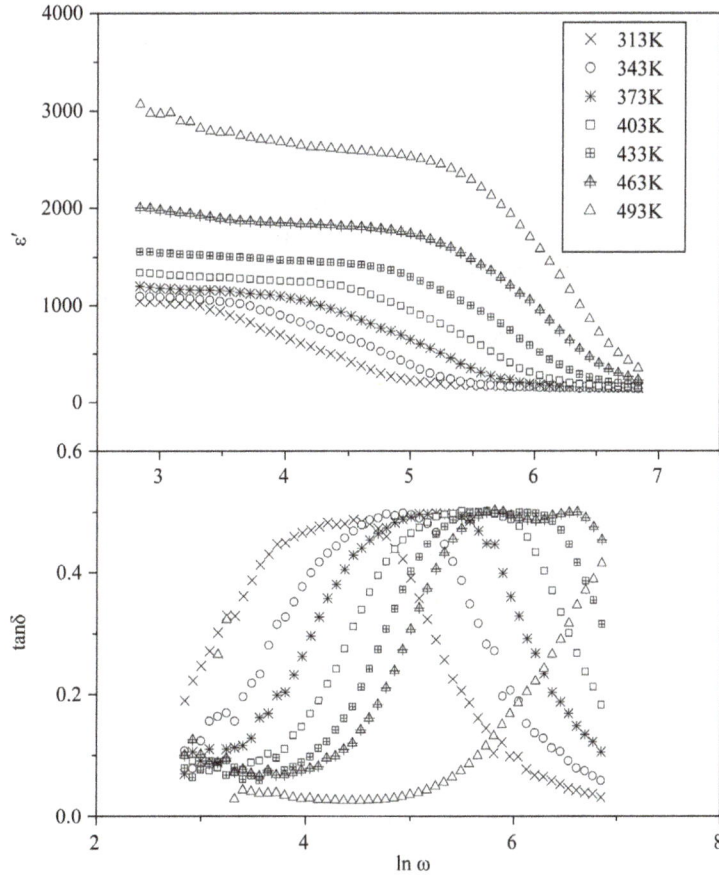

Figure 4. Frequency dependence of the (a) ε' and (b) tanδ of the material at various temperatures.

the dipoles to align themselves, so their contribution to the polarization and hence to the dielectric permittivity can become negligible. Therefore the dielectric permittivity $\varepsilon'(\omega)$ decreases with increasing frequency. It is evident from **Figure 3(b)** that the peak position of ε'' (centered at the dispersion region of ε') shifts to higher frequency with increasing temperature and that a strong dispersion of ε'' exists in present material. The increase in the peak value of ε'' with the increase in temperature indicates an increase in charge carriers in given material by thermal activation (showing the semiconducting behavior of the sample). It is evident that the width of the loss peaks in **Figure 3(b)** cannot be accounted for in terms of a monodispersive relaxation process but points towards the possibility of a distribution of relaxation times. If $g(\tau, T)$ is the temperature dependent distribution function for relaxation time, the complex dielectric constant can be expressed as [18]-[21]

$$\varepsilon^* - \varepsilon_\infty = \varepsilon(0,T)\int_0^\infty g(\tau,T)\frac{\mathrm{d}(\ln\tau)}{1-i\omega\tau}, \tag{1}$$

giving

$$\varepsilon'' = \varepsilon(0,T)\int_0^\infty g(\tau,T)\frac{\mathrm{d}(\omega\tau)}{1+\omega^2\tau^2}, \tag{2}$$

where $\varepsilon(0,T)$ is the low-frequency dielectric constant. Thus, the spectrum of dielectric loss gives direct information about $g(\tau,T)$. One of the most convenient ways of checking the polydispersive nature of dielectric relaxation is through complex Argand plane plots between ε'' and ε', usually called Cole-Cole plots [22] [23]. For a pure monodispersive Debye process, one expects semicircular plots with a centre located on the ε' axis, whereas, for polydispersive relaxation, these Argand plane plots are close to circular arcs with end points on the axis of real and a centre below this axis. The complex dielectric constant in such situations is known to be described by the empirical relation

$$\varepsilon^* = \varepsilon' - i\varepsilon'' = \varepsilon_\infty + \frac{\left(\varepsilon_s - \varepsilon_\infty\right)}{1 + \left(i\omega\tau\right)^{1-\alpha}} \tag{3}$$

where ε_s and ε_∞ are the low- and high-frequency values of ε', respectively, and α is a measure of the distribution of relaxation times. The parameter α can be determined from the location of the centre of the Cole-Cole circles of which only an arc lies above the ε'-axis. **Figure 4(a)** and **Figure 4(b)** depict two such representative plots for T = 313 K and 423 K. It is evident from these plots that the relaxation process differs from the monodispersive Debye process (for which $\alpha = 0$). The parameter α, as determined from the angle subtended by the radius of the circle with the ε' axis passing through the origin of the ε'' axis, shows a very small increase in the interval (0.56, 0.66) with the decrease of temperature from 423 K to 313 K, implying a slight increase in the distribution of the relaxation time with decreasing temperature.

The Cole-Cole plots confirm the polydispersive nature of dielectric relaxation of the titled material. However the small variation in α with decreasing temperature is not convincing enough, keeping in mind the uncertainties in fitting the circle, which was done through a visual fit to the observed data points. **Figure 5** and **Figure 6** are shown the variation of dielectric constants (real and imaginary parts) with frequencies (100 Hz - 1 MHz) at different temperatures (303 K - 583 K). The real part of the dielectric constant is found to decrease with increasing frequency at different temperatures while the position of dielectric loss peak shifts to higher frequencies with increasing temperature indicating a strong dispersion beyond the transition temperature, a feature known for relaxational systems such as dipole glasses. The magnitude of dielectric constant decreases with increasing frequency which is a typical characteristic of ferroelectric material. It is evident from **Figure 5** that the position of the loss peak shifts to higher frequencies with increasing temperature and that a strong dispersion of $\left(\varepsilon''\right)$ exists similar to what is known for relaxational systems such as dipole glasses [24]. At temperature far above T_m in two frequencies (100 Hz & 1 KHz) a monotonous increase in the value of ε'' caused by electrical conduction is observed. The frequency dependence of $M'(\omega)$ and $M''(\omega)$ for our material at various temperatures is shown in **Figure 7**. $M'(\omega)$ show a dispersion tending toward M_∞ (the asymptotic value of $M'(\omega)$ at higher frequencies (**Figure 7(a)**), while $M''(\omega)$ exhibits a maximum $\left(M''_{max}\right)$ (**Figure 7(b)**) centered at the dispersion region of $M'(\omega)$. It may be noted from **Figure 7(b)** the frequency region below peak maximum $\left(M''_m\right)$ determines the range in which charge carriers are mobile on long distances. At frequencies above peak maximum, the carriers are confined to potential wells, being mobile on short distances. The value of M''_m increases with increasing temperature though slightly and shifts to the higher frequency side. At any temperature T, the most probable relaxation time corresponding to the peak position in $\tan\delta$ versus $\ln\omega$ and M'' versus $\ln\omega$ is proportional to $\exp\left(-E_a/k_BT\right)$ (Arrhenius law) with activation energies ≈ 0.15 and 0.21 eV, respectively, as shown in **Figure 8**. Such a value of activation energy indicates that the conduction mechanism for present material may be due to the polaron hopping based on the electron carriers. In the hopping process, the electron disorders its surroundings by moving its neighboring atoms from their equilibrium positions, causing structural defects in the B perovskite sites of the system.

We have scaled each M'' by M''_{max} and each frequency by ω_{max}, where ω_m corresponds to the frequency of the peak position of M'' in the M'' versus $\ln\omega$ plots in **Figure 9**. The overlap of the curves at different temperatures

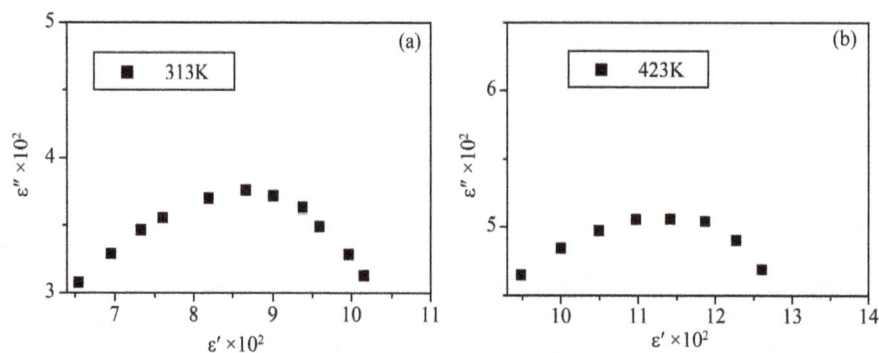

Figure 5. Cole-cole plot at (a) 313 K and (b) 423 K.

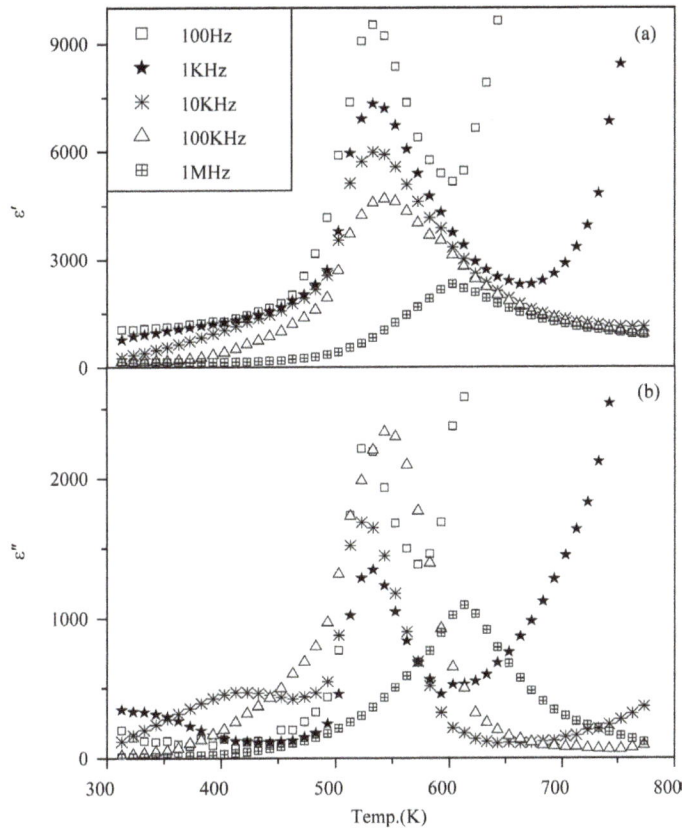

Figure 6. Temperature dependence of the (a) ε' and (b) ε'' of the material at various frequencies.

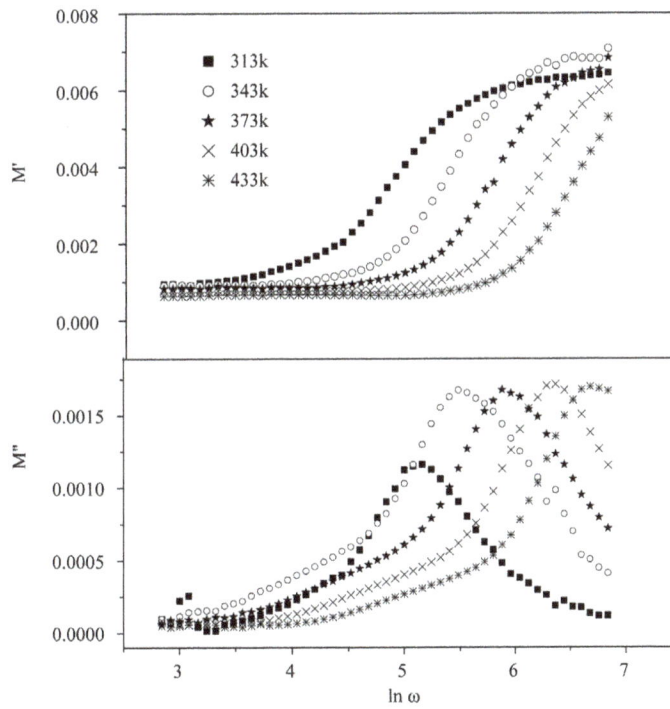

Figure 7. Frequency dependence of the (a) M' and (b) M'' of the material at various temperatures.

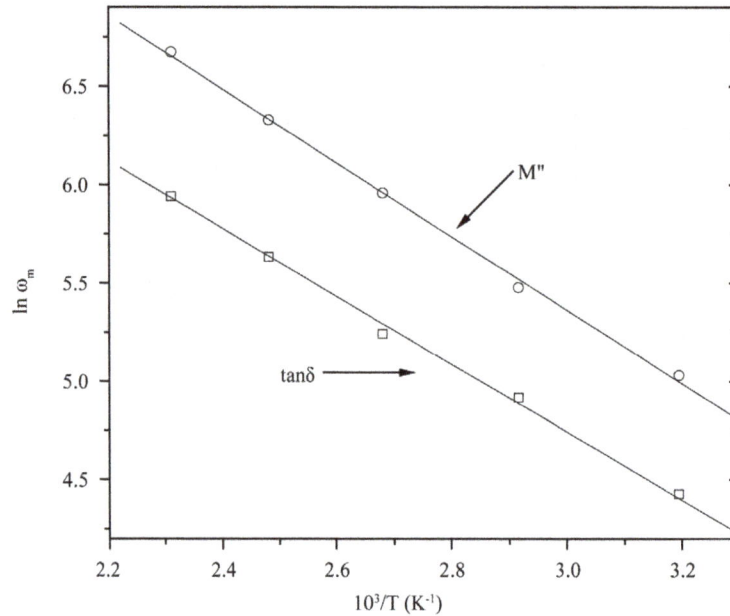

Figure 8. The Arrhenius plot of ω_m corresponding to $\tan\delta$ and corresponding to M'.

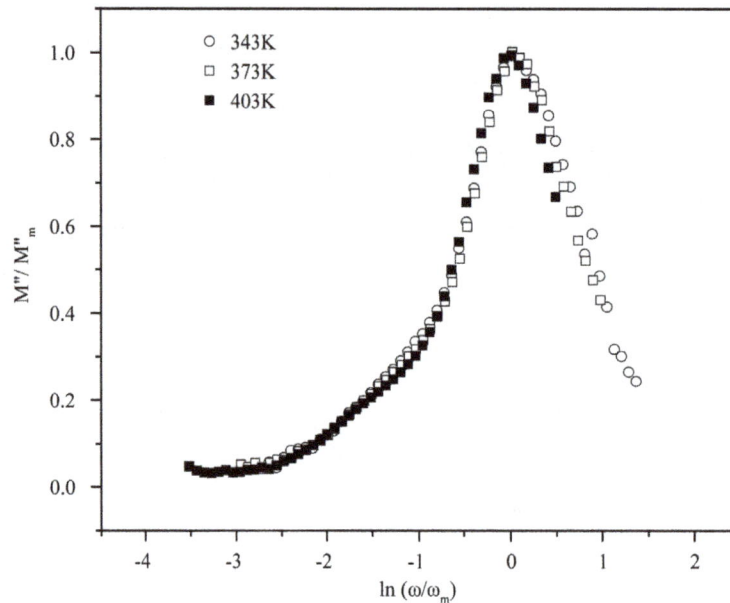

Figure 9. Scalling behavior of M'' at various temperatures for the given composition.

into a single master curve indicates that the relaxation describes the same mechanism at various temperatures. **Figure 10** shows the frequency dependence of the ac conductivity $\sigma(\omega)$ for given composition at different measuring temperaturs. The conductivity shows a dispersion which shifts to higher frequencies with an increase in temperature. In **Figure 11**, the variations of the normalized parameter M''/M''_{max} and $\tan\delta/\tan\delta_{max}$ as a function of logarithmic frequency measured at 403 K for present material are shown. For delocalized or long range conduction, the peak position of two curves should overlap [25] [26]. However, for the present system, the M''/M''_{max} and $\tan\delta/\tan\delta_{max}$ peaks do not overlap, suggesting the components from both long range and localized relaxation. In order to mobilize the localized electron, the aid of lattice oscillation is required. In these circumstances, electrons are considered not to move by themselves but by hopping motion activated by lattice

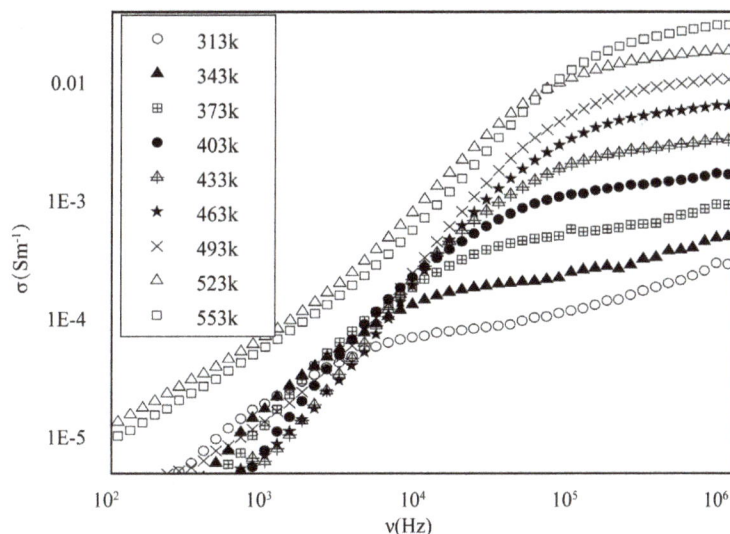

Figure 10. Frequency spectra of the conductivity for the given composition at various temperatures.

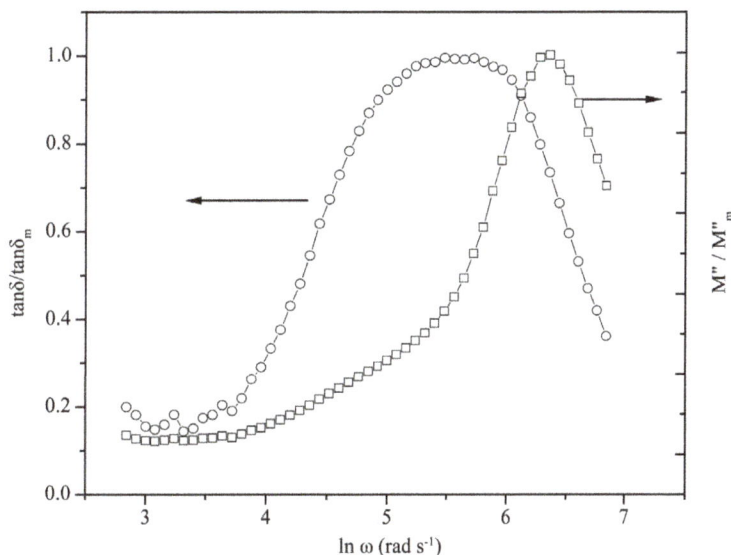

Figure 11. Frequency dependence of normalized peaks, M''/M''_{max} and $\tan\delta/\tan\delta_{max}$ at 403 K.

oscillation, *i.e.*, the conduction mechanism should be considered as phonon-assisted hopping of small polaron between localized states. In addition, the magnitude of the activation energy suggests that the carrier transport is due to the hopping conduction.

5. Conclusion

The frequency dependent dielectric relaxation and conductivity of the $[Pb_{0.94}Sr_{0.06}][(Mn_{1/3}Sb_{2/3})_{0.05}(Zr_{0.49}Ti_{0.51})_{0.95}]$ ceramics synthesized by a solid-state reaction technique are investigated in the temperature range from 313 to 773 K. The SEM of the sample also confirms the formation of single phase of the material. The frequency dependence of the loss peak is found to obey an Arrhenius law with activation energy of 0.15 eV. This value of activation energy suggests that the bulk conduction in present material may be due to polaron hopping based on the electron carriers. The distribution of relaxation times is confirmed by Cole-Cole plots. The frequency-dependent electrical data are also analyzed in the framework of the conductivity and modulus formalisms. These formalisms yield qualitative similarities in the relaxation times. The presence of peak in the temperature dependence of the imaginary part of the dielectric constant $\left[\varepsilon''(T)\right]$ indicates that the hopping of charge carriers is responsible for the dielectric relaxation. The scaling behavior of the imaginary part of the electric modulus $\left(M''\right)$ suggests that the relaxation describes the same mechanism at various temperatures. In conclusion, we have thus here confirmed the polydispersive nature of dielectric relaxation of the present composition through the Cole-Cole plots. We have also used the conductivity as well as the modulus formalism to settle the relaxation mechanisms in the present material.

References

[1] Yoo, J., Lee, Y., Yoon, K., Hwang, S., Suh, S., Kim, J. and Yoo, C. (2001) Microstructural, Electrical Properties and Temperature Stability of Resonant Frequency in $Pb(Ni_{1/2}W_{1/2})O_3$-$Pb(Mn_{1/3}Nb_{2/3})O_3$-$Pb(Zr, Ti)O_3$ Ceramics for High-Power Piezoelectric Transformer. *Japanese Journal of Applied Physics*, **40**, 3256. http://dx.doi.org/10.1143/JJAP.40.3256

[2] Yoo, J., Yoon, K., Lee, Y., Suh, S., Kim, J. and Yoo, C. (2000) Electrical Characteristics of the Contour-Vibration-Mode Piezoelectric Transformer with Ring/Dot Electrode Area Ratio. *Japanese Journal of Applied Physics*, **39**, 2680. http://dx.doi.org/10.1143/JJAP.39.2680

[3] Gao, Y.K., Chen, Y.H., Ryu, J.G., Uchino, K.J. and Viehland, D. (2001) Eu and Yb Substituent Effects on the Properties of $Pb(Zr_{0.52}Ti_{0.48})O_3$-$Pb(Mn_{1/3}Sb_{2/3})O_3$ Ceramics: Development of a New High-Power Piezoelectric with Enhanced Vibrational Velocity. *Japanese Journal of Applied Physics*, **40**, 687. http://dx.doi.org/10.1143/JJAP.40.687

[4] Park, S.-E. and Shrout, T.R. (1997) Ultrahigh Strain and Piezoelectric Behavior in Relaxor Based Ferroelectric Single Crystals. *Journal of Applied Physics*, **82**, 1804. http://dx.doi.org/10.1063/1.365983

[5] Mischenko, A.S., Zhang, Q., Whatmore, R.W., Scott, J.F. and Mathur, N.D. (2006) Giant Electrocaloric Effect in the Thin Film Relaxor Ferroelectric 0.9 $PbMg_{1/3}Nb_{2/3}O_3$-0.1 $PbTiO_3$ near Room Temperature. *Applied Physics Letters*, **89**, Article ID: 242912. http://dx.doi.org/10.1063/1.2405889

[6] Kutnjak, Z., Petzelt, J. and Blinc, R. (2006) The Giant Electromechanical Response in Ferroelectric Relaxors as a Critical Phenomenon. *Nature*, **441**, 956-959. http://dx.doi.org/10.1038/nature04854

[7] Blinc, R., Laguta, V. and Zalar, B. (2003) Field Cooled and Zero Field Cooled Pb_{207} NMR and the Local Structure of Relaxor $PbMg_{1/3}Nb_{2/3}O_3$. *Physical Review Letters*, **91**, Article ID: 247601. http://dx.doi.org/10.1103/PhysRevLett.91.247601

[8] Scott, J.F. (2007) Application of Modern Ferroelectrics. *Science*, **315**, 954-959. http://dx.doi.org/10.1126/Science.1129564

[9] Gehring, P.M., Park, S.-E. and Shirane, G. (2000) Soft Phonon Anomalies in the Relaxor Ferroelectric $Pb(Zn_{1/3}Nb_{2/3})_{0.92}Ti_{0.08}O_3$. *Physical Review Letters*, **84**, 5216. http://dx.doi.org/10.1103/PhysRevLett.84.5216

[10] Jaffe, B., Cook, W.R. and Jaffe, H. (1971) Piezoelectric Ceramics. Academic press, London/New York.

[11] Fu, H. and Cohen, R.E. (2000) Polarization Rotation Mechanism for Ultrahigh Electromechanical Response in Single-Crystal Piezoelectrics. *Nature*, **403**, 281-283. http://dx.doi.org/10.1038/35002022

[12] Lal, R., Krishanan, R. and Ramkrishanan, P. (1988) Transition between Tetragonal and Rhombohedral Phases of PZT Ceramics Prepared from Spray-Dried Powders. *British Ceramic Transactions and Journal*, **87**, 99-102.

[13] Mishra, S.K., Singh, A.P. and Pandey, D. (1997) Effect of Phase Coexistence at Morphotropic Phase Boundary on the Properties of $Pb(Zr_xTi_{1-x})O_3$ Ceramics. *Applied Physics Letters*, **69**, 1707. http://dx.doi.org/10.1063/1.118004

[14] Kutnjak, Z., Petzelt, J. and Blinc, R. (2006) The Giant Electromechanical Response in Ferroelectric Relaxors as a Critical Phenomenon. *Nature*, **441**, 956-959. http://dx.doi.org/10.1038/nature04854

[15] Singh, S., Singh, S.P. and Pandey, D. (2008) A Succession of Relaxor Ferroelectric Transitions in $Ba_{0.55}Sr_{0.45}TiO_3$. *Journal of Applied Physics*, **103**, Article ID: 016107. http://dx.doi.org/10.1063/1.2827506

[16] Noheda, B., Gonzalo, I.A., Guo, R., Park, S.-E., Cross, L.E., Cox, D.E. and Shirane, G. (2000) Tetragonal-to-Monoclinic Phase Transition in a Ferroelectric Perovskite: The Structure of $PbZr_{0.52}Ti_{0.48}O_3$. *Physical Review B*, **61**, 8687. http://dx.doi.org/10.1103/PhysRevB.61.8687

[17] Brajesh, K., Himanshu, A.K., Sharma, H., Kumari, K., Ranjan, R., Bandyopadhayay, S.K. and Sinha, T.P. (2012) Structural, Dielectric Relaxation and Piezoelectric Characterization of Sr^{2+} Substituted Modified PMS-PZT Ceramic. *Physica B: Condensed Matter*, **407**, 635-641. http://dx.doi.org/10.1016/j.physb.2011.11.048

[18] Singh, S.P., Singh, A.K. and Pandey, D. (2007) Dielectric Relaxation and Phase Transitions at Cryogenic Temperatures in $0.65[Pb(Ni_{1/3}Nb_{2/3})O_3]$-$0.35PbTiO_3$ Ceramics. *Physical Review B*, **76**, Article ID: 054102. http://dx.doi.org/10.1103/PhysRevB.76.054102

[19] Courtens, E. (1986) Scaling Dielectric Data on $Rb_{1-x}(NH_4)_xH_2PO_4$ Structural Glasses and Their Deuterated Isomorphs. *Physical Review B*, **33**, 2975-2978. http://dx.doi.org/10.1103/PhysRevB.33.2975

[20] Courtens, E. (1984) Vogel-Fulcher Scaling of the Susceptibility in a Mixed-Crystal Proton. *Physical Review Letters*, **52**, 69-72. http://dx.doi.org/10.1103/PhysRevLett.52.69

[21] Ginzburg, S.L. (1989) Irreversible Phenomena of Spin Glasses. Nauka, Moscow.

[22] Lindgren, L., Svedlindh, P. and Beckman, O.J. (1981) Measurement of Complex Susceptibility on a Metallic Spin Glass with Broad Relaxation Spectrum. *Journal of Magnetism and Magnetic Materials*, **25**, 33-38. http://dx.doi.org/10.1016/0304-8853(81)90144-X

[23] Cole, K.S. and Cole, R.H. (1941) Dispersion and Absorption in Dielectrics I. Alternating Current Characteristics. *The Journal of Chemical Physics*, **9**, 341. http://dx.doi.org/10.1063/1.1750906

[24] Hochli, U.T., Knorr, K. and Loidl, A. (1990) Publication Models and Dates Explained. *Advances in Physics*, **39**, 405-615. http://dx.doi.org/10.1080/00018739000101521

[25] Gerhardt, R. (1994) Impedance and Dielectric Spectroscopy Revisited: Distinguishing Localized Relaxation from Long-Range Conductivity. *Journal of Physics and Chemistry of Solids*, **55**, 1491-1506. http://dx.doi.org/10.1016/0022-3697(94)90575-4

[26] Brajesh, K., Kumar, P., Himanshu, A.K., Ranjan, R., Bandyopadhayay, S.K., Sinha, T.P. and Singh, N.K. (2014) Dielectric Relaxation, Phase Transition and Rietveld Studies of Perovskite $[Pb_{0.94}Sr_{0.06}][(Mn_{1/3}Sb_{2/3})_{0.05}(Zr_{0.52}Ti_{0.48})_{0.95}]O_3$ Ceramics. *Journal of Alloys and Compounds*, **589**, 443-447. http://dx.doi.org/10.1016/j.jallcom.2013.11.170

14

Fabrication, Mechanical and Dielectric Characterization of 3D Orthogonal Woven Basalt Reinforced Thermoplastic Polyimide Composites

Shuna Hou, Jianfei Xie, Ye Kuang, Xianhong Zheng, Lan Yao*, Yiping Qiu

Key Laboratory of Textile Science and Technology, Ministry of Education, College of Textiles, Donghua University, Shanghai, China
Email: *yaolan@dhu.edu.cn

Abstract

The 3D orthogonal woven basalt fiber reinforced polyimide (PI) composites were fabricated and characterized in this study. The PI film was firstly prepared to determine PI processing parameters. Fourier transform infrared (FTIR) analysis showed that 300°C was the suitable imidization temperature. Thermal gravimetric analysis (TGA) and differential scanning calorimetry (DSC) results showed relatively good thermal properties of the PI film. In the fabrication of composites, the multi-step impregnation method was applied. The bending properties of 3 mm-thick composite showed increasing trend in all and the second-time impregnated composite had much higher value than the first-time impregnated composite. Moreover, the bending fracture mode photos showed obvious creases except for the first-time impregnated materials, which agreed well with the bending property values. The dielectric constants for the composites were complex because they had not regular value following the mixing rule of the composites, which was mainly due to the interfacial polarization and other effects in the fabrication processing.

Keywords

3D Orthogonal Woven Structure, Basalt Fiber, Thermoplastic Polyimide, Bending Properties, Dielectric Properties

---

*Corresponding author.

1. Introduction

For recent years, much attention has been given to polyimide (PI) and its composites due to their high mechanical strength, good thermal stability, high stability under vacuum, good anti-radiation, and good solvent resistance. Investigations on the fiber reinforced thermoplastic polyimide composites have been made by many researchers [1]-[7]. In [3], the effect of rare earth (RE) solution surface modification of poly-p-phenylenebenzobisoxazole (PBO) fibers on the tensile property of PBO fiber-reinforced thermoplastic polyimide (PBO/PI) composites have been investigated. Experimental results revealed that RE surface treatment could effectively improve the interfacial adhesion between PBO fibers and PI matrix. In [5], they compared the effects of rare earth solution (RES) treatment and air oxidation surface treatment on the mechanical and tribological properties of carbon fiber-reinforced polyimide composites and found that the RES surface treatment was superior to air oxidation treatment in promoting interfacial adhesion between carbon fiber and PI matrix. Paplham et al. [6] have investigated the effect of crystallization of the thermoplastic polyimide upon the microhardness values of the resin. The results showed that the addition of carbon fibers to the neat resins greatly increased the microhardness and thus the yield stress of the composite.

The emergence of 3D woven composites is aimed to improve the weaknesses of traditional laminated structures, namely delamination. Among the different technologies to produce 3D fiber architecture, 3D orthogonal woven preforms have gained industrial acceptance [8]-[15]. However, little has been reported on the properties of 3D fiber reinforced thermoplastic polyimide composites though the 3D composites have been more and more frequently used because of their damage tolerance and anti-delamination properties.

In this study, 3D basalt fiber reinforced PI composites have been fabricated. The selection of basalt fibers is due to its desirable properties such as high tensile strength, high tensile modulus, and excellent heat resistance. It is produced in a similar way as glass fibers using basalt rock which is an over-ground, effusive volcanic rock with 45% - 52% SiO_2 [16]-[18]. The whole paper was arranged as follows. The thermoplastic PI film was firstly prepared by imidization process of polyamic acid. Then the measurements on the film including surface chemical analysis and thermal property analysis were investigated. Finally, the 3D composite materials were fabricated and the bending properties as well as the dielectric properties were characterized and discussed.

2. Experimental

2.1. Preparation of Polyamic Acid (PAA) and PI Film

The polyimide was obtained through the imidization process of the PAA. The PAA was synthesized by reacting 3,3', 4,4'-benzophenonetetracarboxylic dianhydride (BTDA) with 4,4-oxydianiline (4,4-ODA) .The molar fraction of them was 1.02:1. The N, N-dimethyl formamide (DMF) was used as the solvent of 4,4'-ODA. **Figure 1** shows the synthesis process of PAA and PI. The PAA film was cast onto a clean glass, where thickness was controlled by the diameter of copper wire around glass rod. The PAA film was imidized according to the procedure in a FIR radiation oven as shown in **Figure 2**. The final yellow imidized film was peeled off the glass plate by soaking in hot water and dried in the oven.

2.2. Characterization of the PI Film

Fourier transform infrared (FTIR) spectra obtained on a Nicolet 5700 ConTinu μm Fourier Transform Infrared Microscopy was used to analyze the surface chemical properties of PI films imidized at different temperatures. Thermal gravimetric analysis (TGA) was performed on a Shimadzu DT-40 thermal analysis system at a heating rate of $10°C \cdot min^{-1}$ in a nitrogen atmosphere at a flow rate of $20 \ cm^3 \cdot min^{-1}$. Dfferential scanning calorimetry (DSC) was performed on a Perkin Elmer Pyris 1 DSC Differential Scanning Calorimeter in a nitrogen atmosphere at a flow rate of $50 \ cm^3 \cdot min^{-1}$. The glass transition temperature (T_g) was determined by the inflection point of the heat flow versus temperature curve.

2.3. Composite Manufacturing

In the manufacturing process of the composite, the multi-step impregnation method was applied. In the first-time impregnation, the 3D orthogonal woven basalt fabric was immersed into the PAA, and then was put into the vacuum oven to remove the air bubbles. Padding was also necessary to remove the superfluous liquid on the fabric.

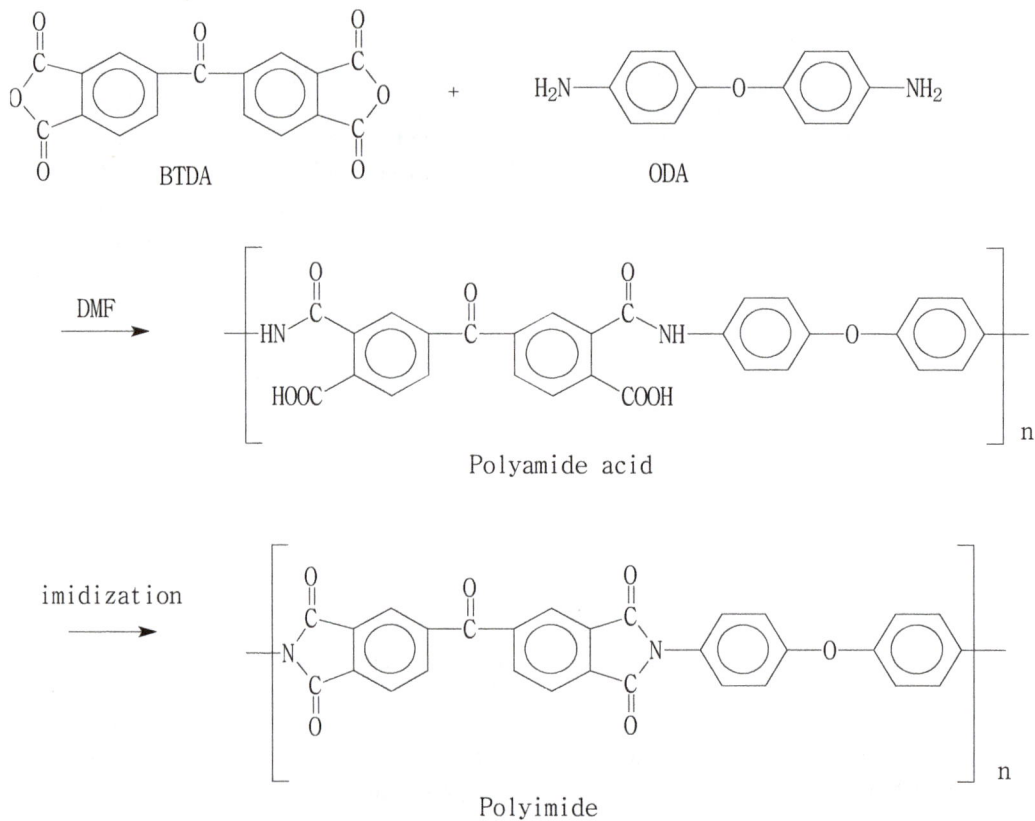

Figure 1. The synthesis process of PAA and PI.

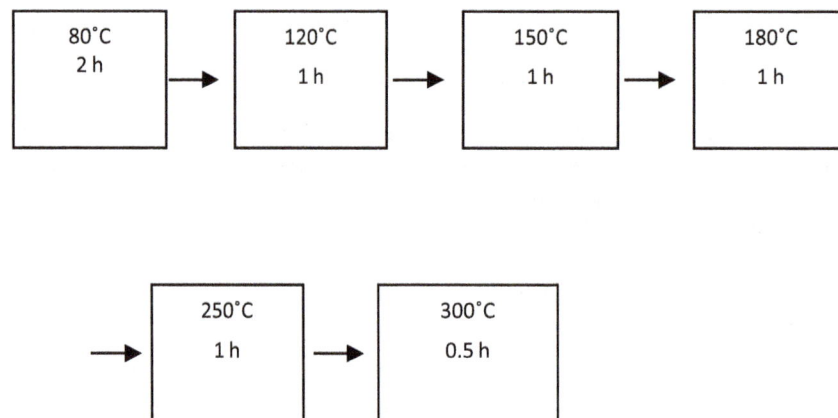

Figure 2. The processing of thermal imidization.

Under the pressure of 0.08 - 0.09 MPa, the 3D fabric was dried and pre-imidized at 80/1 h, 120/1 h, 150/1 h and 180/1 h. In the second-time impregnation process, the first-time prepregs were immersed into the PAA and being treated following the same process as the first-time impregnation. Thus impregnation for five times was performed totally and after each impregnation, the samples were saved and the resin content was calculated. In the very final step, the prepregs were molded into composites using the press vulcanizer at the following sequence shown in **Figure 3**. After cooled to the room temperature, the five compact 3D orthogonal woven basalt /PI composites were obtained. The 3 mm and 5 mm thickness of composite were fabricated finally. For convenience, the composite samples of 3 mm thickness were named as S1, S2, S3, S4 and S5, representing one to five times of impregnation respectively. Using the 3D orthogonal woven structure fabric as the preform, the molding method applied in this study was new and totally different with other PI composites processing methods [2].

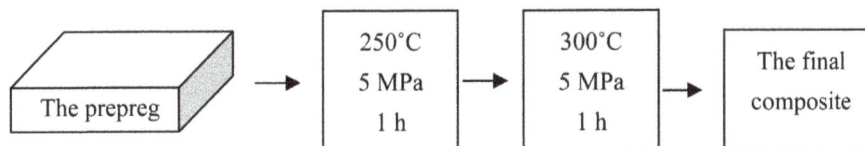

Figure 3. The post-impregnation process for the final composite panels.

2.4. The Photographs

To observe different appearances of the prepregs for different times of impregnation, the microscopic photos were taken using the KH-1000 3D digital video microscopic measurement system.

2.5. Bending Test

The 3-point bending tests were conducted using WDW-20 Computerized Electronic Universal Testing Machine on regular specimens of 250×25 mm, with the span length fixed at 150 mm. The specimens were cut with the length being in the directions of the warp and weft yarns. The specimens were deflected until rupture occurred in the outer fiber. The failure modes for the 3-point bending test of the composites were analyzed using the manual camera.

2.6. Dielectric Property Test

The dielectric property tests were performed on Agilent 4291 B 1.8 GHz Impedance/Material Analyzer. All the specimens were dried before measuring. The dielectric constant and loss tangent of the composites were obtained in the frequency range from 1 MHz to 1000 MHz.

3. Results and Discussion

3.1. The Properties of PI Films

FTIR of the PI films at different imidization temperatures are shown in **Figure 4**. We have calculated the ratio of the absorbance at 1720 or 1380 cm^{-1} to that at 1500 cm^{-1}, that is C=O/C=C or C-N/C=C. The changing trend of the ratios versus the imidization temperature can be used to characterize the imidization process indirectly (shown in **Figure 5**). It can be seen that the imidization degree had the increasing trend with increasing temperature in all. At 300°C, the imidization degree of the PI film reached almost the maximum value. Therefore, 300°C was used determined as the final imidization temperature of PI films.

Figure 6 and **Figure 7** show DSC and TGA curves of the PI film imidized at 300°C. The PI film had T_g of 261.87°C and decomposed temperature of 531.0°C in the nitrogen, indicating the adequate thermal properties of the film.

3.2. The Resin Content of the Composites

Figure 8 shows the resin content versus times of impregnation. The resin content increased as the times were increased for the fabrics of 3 and 5 mm thickness. The 3 mm-thick composite had higher resin content compared with the 5-mm thick composite at each-time impregnation. The lower thickness made the resin easier infuse into the fabric under the same processing parameters. After five-time impregnation, the 3 mm-thick composite had 33.61% resin content which was much higher than that of 5 mm-thick composite (18.30%).

3.3. Bending Test

Table 1 lists the bending properties of the 3 mm-thick composite. In all, the bending strength and modulus increased as the resin content was increased, which was mainly due to the improved face properties of the composite as the PI resin was infused. It is also noticed that composites showed significantly better bending properties after the second impregnation processing was finished indicating the positive effect of multi-step impregnation processing. In addition, the bending properties in the weft direction were higher than those in the warp direction due to the higher volume fraction of the weft yarns.

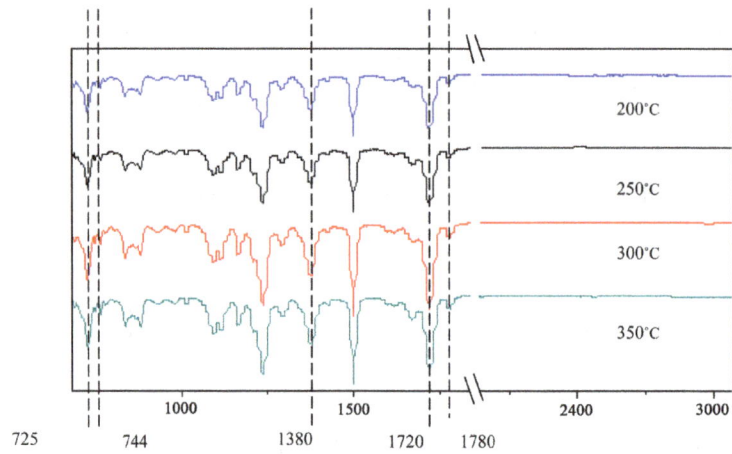

Figure 4. FTIR of PI films at different temperatures.

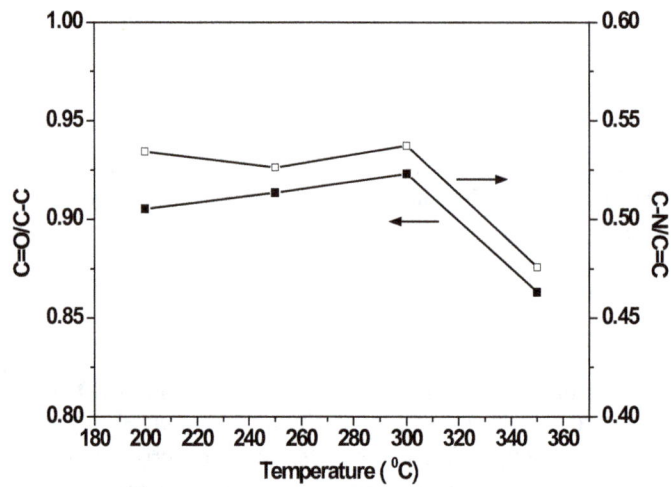

Figure 5. Effect of temperature on imidization degree of PI.

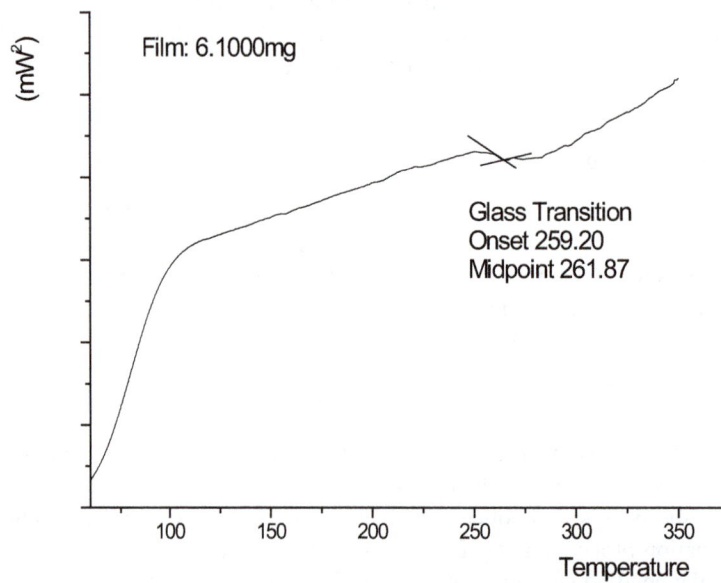

Figure 6. DSC curve of PI film imidized at 300 degree.

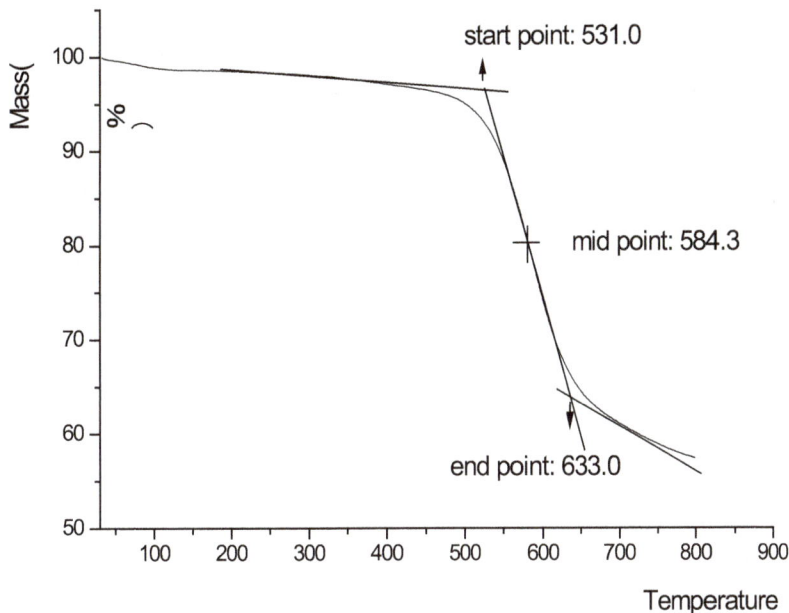

Figure 7. TGA curve of the PI film imidized at 300 degree.

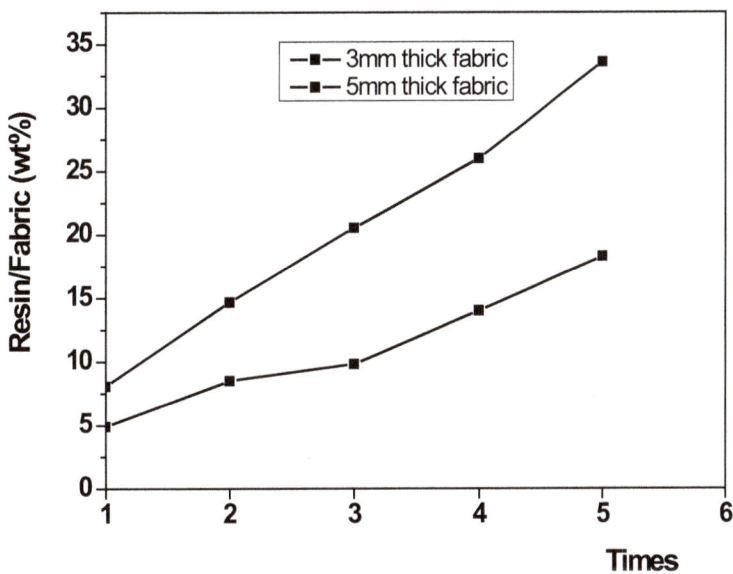

Figure 8. The resin content of the composites.

Table 1. The bending properties of the composite with different times of impregnation.

Samples	Bending strength (MPa)		Bending modulus (GPa)	
	Warp direction	Weft direction	Warp direction	Weft direction
S1	57.20	65.53	1.98	6.72
S2	179.77	161.54	7.96	13.26
S3	168.32	190.02	8.96	14.10
S4	183.76	214.97	8.73	13.83
S5	214.14	236.24	8.16	14.49

Figure 9 shows the bending failure modes of the composite materials. It can be seen that all the materials had creases after the bending damage expect for the one-time impregnated material which had the lowest resin content. These damage modes illustrated the bending properties and agreed with the bending properties in **Table 1**.

3.4. Dielectric Properties

Frequency dependence of dielectric constant and dielectric loss are shown in **Figure 10** and **Figure 11**. For each type of the composites, the dielectric constant remained relatively constant but a little decrease as the frequency was increased indicating dielectric stability under frequencies of the PI composites. In addition, the resin content of the five composites were decreasing in the order S5 > S4 > S3 > S2 > S1, however, the dielectric constant for them are found to be decreasing in the order S3 > S2 > S5 > S4 > S1. This does not agree with the well-known mixing rule for dielectric constant of a multi-component material. Similar results have been reported in our previous study, in which the dielectric constants of the five aramid/glass hybrid composite structures were found not to be in the order of the fiber content [19]. The dielectric losses of the five composites were very close in **Figure 11** and showed constant value as the frequency was increased.

4. Discussions

The photos in **Figure 12** and **Figure 13** show the typical surface and inner images of the 3 mm and 5 mm thick composites. It has been mentioned that the surface resin content increased as the times were increased; however, it was not the case for the inner part of the composites. When the temperature was raised, the solvent evaporating leaded to the viscosity rising with a film forming at the outer surface of the composite. The film hindered the inner solvent evaporating outwards and thus air bubbles were produced. Due to the lower thickness, the 3 mm thick composite had some yellow polyimide resin in the inner part. However, for the 5 mm thick composite, relatively less polyimide could be observed.

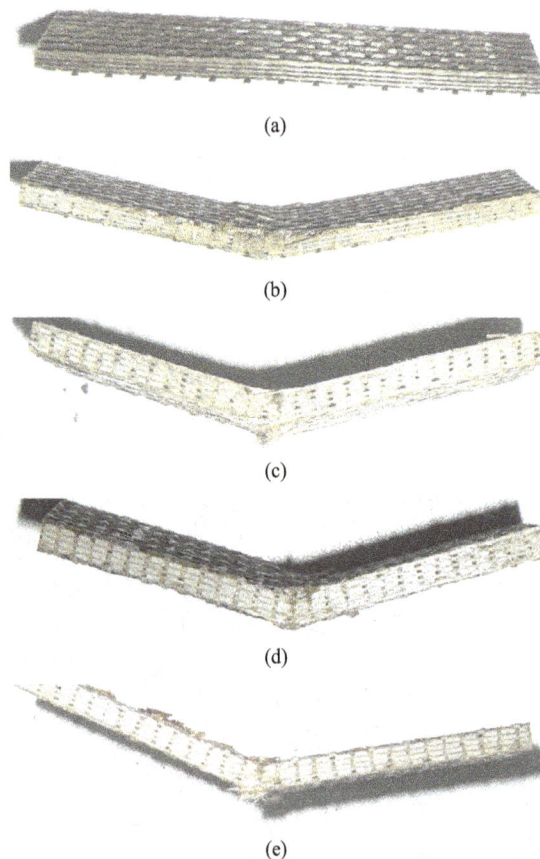

(a)

(b)

(c)

(d)

(e)

Figure 9. The bending failure modes of (a) S1 - (e) S5.

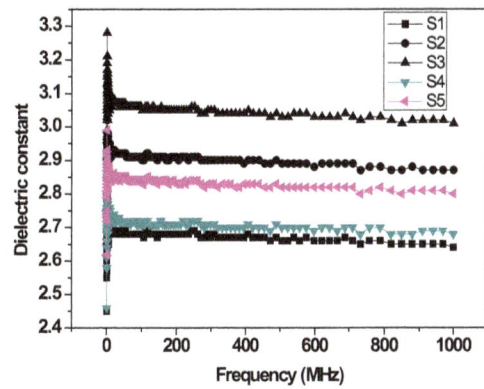

Figure 10. Frequency dependence of dielectric constant of the composites.

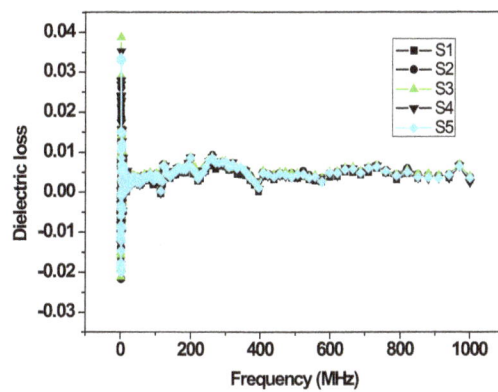

Figure 11. Frequency dependence of dielectric loss of the composites.

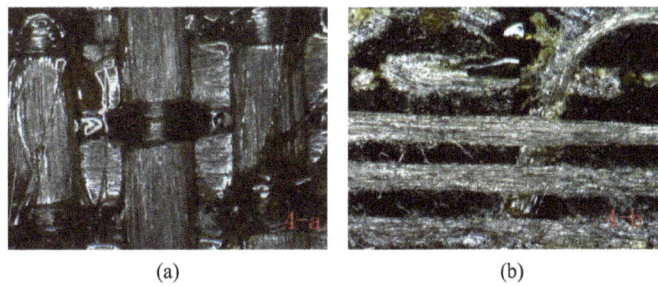

| (a) | (b) |

Figure 12. Typical images for (a) the outer surface and (b) the inner section of four-time impregnated composite with 3 mm thickness.

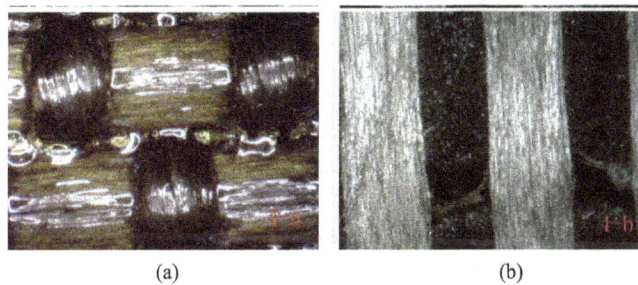

| (a) | (b) |

Figure 13. Typical images for (a) the outer surface and (b) the inner section of four-time impregnated composite with 5 mm thickness.

Dielectric constant is relatively a complex value in the composites, which is related to the material component, volume fraction, manufacturing method and other effects. In this study, the phenomenon of not agreeing with the mixing rule was mainly due to the existence of interface polarization between fiber and resin, which made the final dielectric constant complicated. For further studies, interface polarization of the basalt fiber and PI needs to be investigated to find the influence rule.

5. Conclusion

The PI film was firstly prepared to obtain the imidization temperature for the thermoplastic polyimide. To fabricate the 3D orthogonal woven basalt PI composite, the multi-step impregnation method was applied. After five-time impregnation, the 3 mm and 5 mm thick composites had 33.61% and 18.30% resin content respectively indicating that the lower thickness was easier for the resin to infuse. In the bending test for 3 mm-thick composite, the bending strength and modulus increased as the resin content was increased and the significant increase could be observed after the second impregnation processing. For the dielectric properties, the dielectric constant and loss had relatively stable value under different frequencies; however, the dielectric constants of the five composites did not change with the resin content regularly.

Acknowledgements

This research was funded by the National High Technology Research and Development Program of China (No. 2007AA03Z101), the State Key Program of National Natural Science of China (No. 51035003), the China Natural Science Foundation (Grant No. 50803010), the Shanghai Natural Science Foundation (Grant No. 12ZR1440500 and 14ZR1400100), the Doctoral Scientific Fund Project of the Ministry of Education of China (Grant No. 20120075120016) and the Fundamental Research Funds for the Central Universities.

References

[1]   Sheng, D., Wang, H., Ying, Z., Wang, H.T. and Ying, Z.H. (2013) Application of Thermoplastic Polyimide Composite in Aerospace Field. *Plastics*, **42**, 46-48.

[2]   Xu, H.Y., Yang, H.X., Tao, L.M., Liu, J.G., Fan, L. and Yang, S.Y. (2010) Preparation and Properties of Glass Cloth-Reinforced Meltable Thermoplastic Polyimide Composites for Microelectronic Packaging Substrates. *High Performance Polymers*, **22**, 581-597. http://dx.doi.org/10.1177/0954008309354609

[3]   Yu, L. and Cheng, X. (2013) Tensile Property of Surface-Treated Poly-p-phenylenebenzobisoxazole (PBO) Fiber-Reinforced Thermoplastic Polyimide Composite. *Journal of Thermoplastic Composite Materials*, **26**, 307-321. http://dx.doi.org/10.1177/0892705711423290

[4]   Li, J. and Cheng, X.H. (2007) Effect of Rare Earth Solution on Mechanical and Tribological Properties of Carbon Fiber Reinforced Thermoplastic Polyimide Composite. *Tribology Letters*, **25**, 207-214. http://dx.doi.org/10.1007/s11249-006-9168-7

[5]   Li, J. and Cheng, X.H. (2008) Evaluation of Tribological Performance of Surface-Treated Carbon Fiber-Reinforced Thermoplastic Polyimide Composite. *Journal of Applied Polymer Science*, **107**, 1147-1153. http://dx.doi.org/10.1002/app.26639

[6]   Paplham, W.P., Seferis, J.C., Calleja, F.J.B. and Zachmann, H.G. (1995) Microhardness of Carbon-Fiber-Reinforced Epoxy and Thermoplastic Polyimide Composites. *Polymer Composites*, **16**, 424-428. http://dx.doi.org/10.1002/pc.750160512

[7]   Rodeffer, C.D., Maybach, A.P. and Ogale, A.A. (1996) Influence of Thermal Aging on the Transverse Tensile Creep Response of a Carbon Fiber Thermoplastic Polyimide Composite. *Journal of Advanced Materials*, **27**, 46-51.

[8]   Pankow, M., Quabili, A. and Yen, C.-F. (2014) Hybrid Three-Dimensional (3-D) Woven Thick Composite Architectures in Bending. *Journal of Materials Science*, **66**, 255-260. http://dx.doi.org/10.1007/s11837-013-0825-7

[9]   Sun, B., Zhang, R., Zhang, Q., Gideon, R. and Gu, B. (2013) Drop-Weight Impact Damage of Three-Dimensional Angle-Interlock Woven Composites. *Journal of Composite Materials*, **47**, 2193-2209. http://dx.doi.org/10.1177/0021998312454904

[10]   Udatha, P., Kumar, C.V.S., Nair, N.S. and Naik, N.K. (2012) High Velocity Impact performance of Three-Dimensional Woven Composites. *Journal of Strain Analysis for Engineering Design*, **47**, 419-431. http://dx.doi.org/10.1177/0309324712448578

[11]   Yao, L., Wang, X., Liang, F., Wu, R., Hu, B., Feng, Y., *et al.* (2008) Modeling and Experimental Verification of Di-

electric Constants for Three-Dimensional Woven Composites. *Composites Science and Technology*, **68**, 1794-1799. http://dx.doi.org/10.1016/j.compscitech.2008.01.014

[12] Callus, P.J., Mouritz, A.P., Bannister, M.K. and Leong, K.H. (1999) Tensile Properties and Failure Mechanisms of 3D Woven GRP Composites. *Composites Part A*: *Applied Science and Manufacturing*, **30**, 1277-1287. http://dx.doi.org/10.1016/S1359-835X(99)00033-0

[13] Cox, B.N., Dadkhah, M.S. and Morris, W.L. (1996) On the Tensile Failure of 3D Woven Composites. *Composites Part A*: *Applied Science and Manufacturing*, **27**, 447-458. http://dx.doi.org/10.1016/1359-835X(95)00053-5

[14] Qiu, Y.P., Xu, W., Wang, Y.J., Zikry, M.A. and Mohamed, M.H. (2001) Fabrication and Characterization of Three-Dimensional Cellular-Matrix Composites Reinforced with Woven Carbon Fabric. *Composites Science and Technology*, **61**, 2425-2435. http://dx.doi.org/10.1016/S0266-3538(01)00164-6

[15] Tan, P., Tong, L.Y., Steven, G.P. and Ishikawa, T. (2000) Behavior of 3D Orthogonal Woven CFRP Composites. Part I. Experimental Investigation. *Composites Part A*: *Applied Science and Manufacturing*, **31**, 259-271. http://dx.doi.org/10.1016/S1359-835X(99)00070-6

[16] Czigany, T. (2005) Basalt Fiber Reinforced Hybrid Polymer Composites. *Materials Science Forum*, **473-474**, 59-66. http://dx.doi.org/10.4028/www.scientific.net/MSF.473-474.59

[17] Subagia, I.D.G.A., Kim, Y., Tijing, L.D., Kim, C.S. and Shon, H.K. (2014) Effect of Stacking Sequence on the Flexural Properties of Hybrid Composites Reinforced with Carbon and Basalt Fibers. *Composites Part B*: *Engineering*, **58**, 251-258. http://dx.doi.org/10.1016/j.compositesb.2013.10.027

[18] Wang, J., Chen, B., Liu, N., Han, G. and Yan, F. (2014) Combined Effects of Fiber/Matrix Interface and Water Absorption on the Tribological Behaviors of Water-Lubricated Polytetrafluoroethylene-Based Composites Reinforced with Carbon and Basalt Fibers. *Composites Part A*: *Applied Science and Manufacturing*, **59**, 85-92. http://dx.doi.org/10.1016/j.compositesa.2014.01.004

[19] Yao, L., Li, W.B., Wang, N., Li, W., Guo, X. and Qiu, Y.P. (2007) Tensile, Impact and Dielectric Properties of Three Dimensional Orthogonal Aramid/Glass Fiber Hybrid Composites. *Journal of Materials Science*, **42**, 6494-6500. http://dx.doi.org/10.1007/s10853-007-1534-9

Composite Materials Damage Modeling Based on Dielectric Properties

Rassel Raihan[1*], Fazle Rabbi[2], Vamsee Vadlamudi[2], Kenneth Reifsnider[1]

[1]University of Texas at Arlington, Arlington, USA
[2]University of South Carolina, Columbia, USA
Email: [*]mdrassel.raihan@uta.edu

Abstract

Composite materials, by nature, are universally dielectric. The distribution of the phases, including voids and cracks, has a major influence on the dielectric properties of the composite materials. The dielectric relaxation behavior measured by Broadband Dielectric Spectroscopy (BbDS) is often caused by interfacial polarization, which is known as Maxwell-Wagner-Sillars polarization that develops because of the heterogeneity of the composite materials. A prominent mechanism in the low frequency range is driven by charge accumulation at the interphases between different constituent phases. In our previous work, we observed *in-situ* changes in dielectric behavior during static tensile testing, and also studied the effects of applied mechanical and ambient environments on composite material damage states based on the evaluation of dielectric spectral analysis parameters. In the present work, a two dimensional conformal computational model was developed using a COMSOL™ multi-physics module to interpret the effective dielectric behavior of the resulting composite as a function of applied frequency spectra, especially the effects of volume fraction, the distribution of the defects inside of the material volume, and the influence of the permittivity and Ohmic conductivity of the host materials and defects.

Keywords

Polymer Matrix Composite Materials, Dielectric Properties, Degradation of Composite Materials, Broadband Dielectric Spectroscopy (BbDS)

1. Introduction

The applications of composite materials are now widespread because of their various advantages over conven-

---

[*]Corresponding author.

tional isotropic materials. These heterogeneous material system's properties can be tailored based on the needs of the application and design. Aerospace and automotive industries are using composite materials to reduce weight to increase fuel efficiency, and also for energy storage and structural stability. The automobile Company Volvo has developed structural composite materials which can store and discharge electrical energy while also being used as a car body structure, fabricated from carbon fibers and a polymer resin [1] [2].

It is necessary to understand the material state changes caused by applied mechanical, thermal, and electrical fields to design and synthesize an effective material system. These complex material systems degrade progressively under combined applied field conditions. To evaluate such material state changes there are many computational tools and methods but most of them do not give a direct and quantitative assessment of the damage state. Numerous experimental techniques and methods have also been developed to measure such material state changes but most of them do not give a direct and quantitative assessment of the damage state.

During the service life of composite materials many degradation processes occur and generally this degradation initiates and evolves by microdamage development, especially matrix microcracking and crack growth, delamination, fiber fracture, fiber-matrix debonding, and microbuckling [1]. In our previous work, **Figure 1**, we have shown that the analysis of the dielectric data gives us information about the types of material state changes throughout the mechanical life of composite materials [2] [3].

Broadband Dielectric Spectroscopy (BbDS) measures the interaction of EMF with a material system over a wide range of frequencies which is shown in **Figure 2**. The response to that broad frequency range typically contains information about molecular and collective dipolar fluctuations, and charge transport and polarization effects that occur at inner and outer boundaries as they affect the form of different dielectric properties of the composite material under study.

Various researchers have used finite element methods (FEM) to model the effective dielectric properties of periodic and random composites containing inclusions of various shapes [4] [5]. In this present work, we used

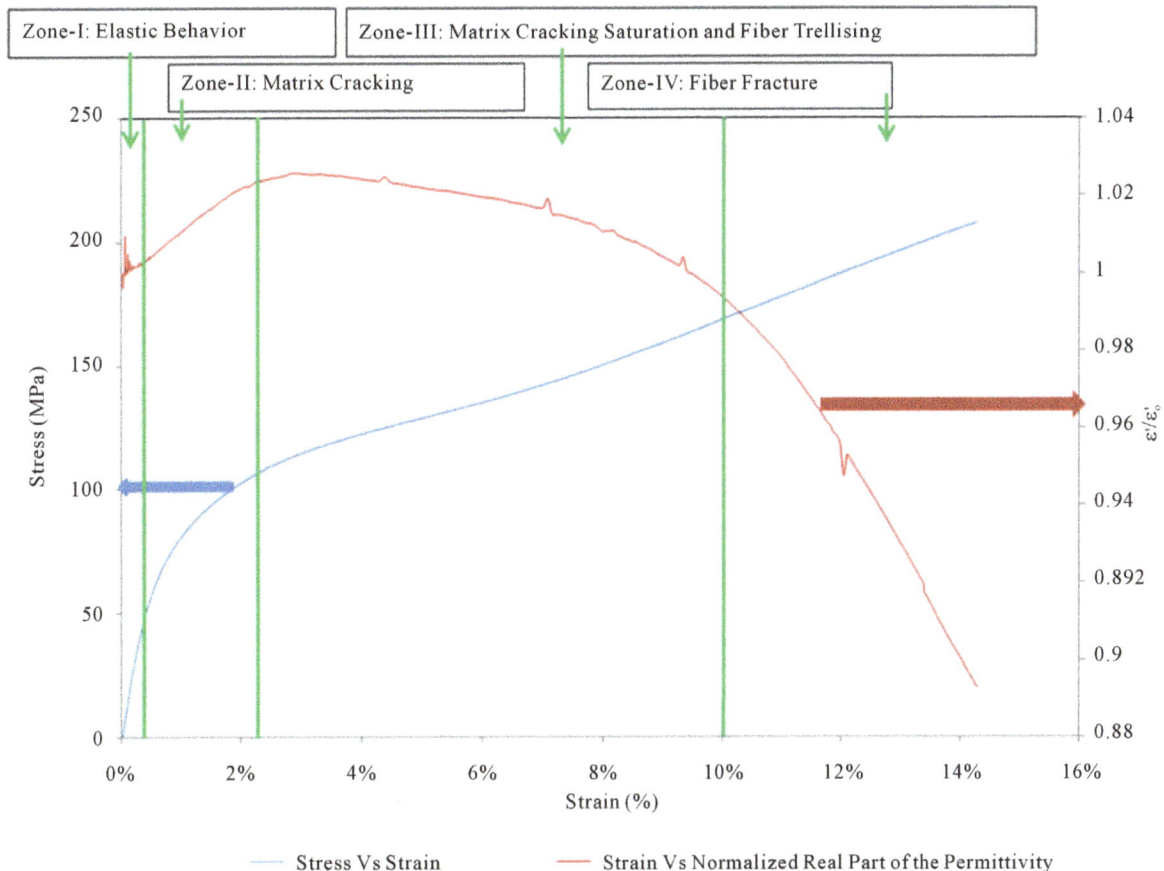

Figure 1. Response of the dielectric property in different zones of damage progression [2].

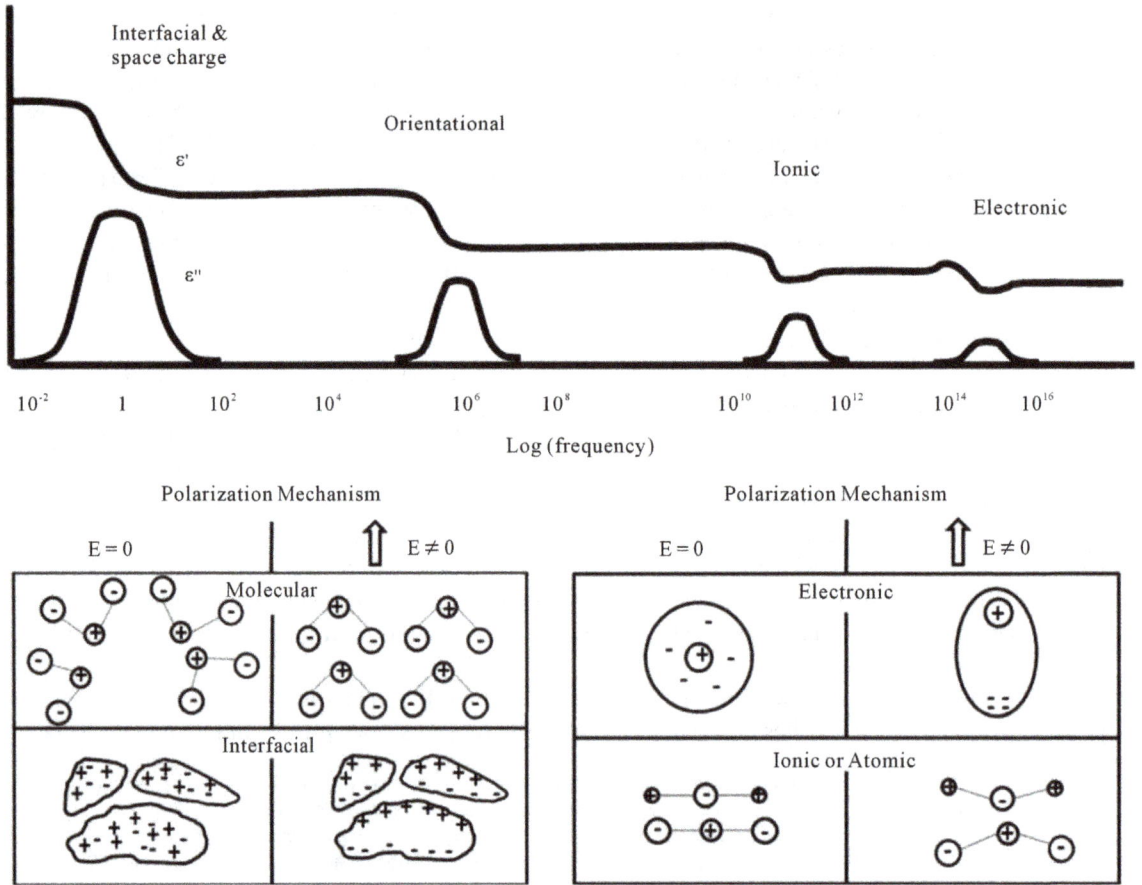

Figure 2. Dielectric responses of material constituents at broad band frequency ranges and different polarization mechanism.

COMSOL Multiphysics™ for conformal modeling, and to reduce the complicacy of the model we only considered the interfacial polarization which is caused by the permittivity and conductivity difference between two constituents. We assumed that the composite materials were homogeneous, and represented defects/cracks as inclusions as shown in **Figure 3**.

2. Basic Equations

In classical dielectrics the relation between the applied electric field E and the dielectric displacement D is linear and can be expressed as [6],

$$D = \varepsilon_o \varepsilon_r E \tag{1}$$

where, ε_o is the permittivity of the free space and ε_r is the relative permittivity of the dielectric material.

If ρ is the charge density, from Maxwell's equations we know that the dielectric displacement follows the following relationship

$$\nabla \cdot D = \rho \tag{2}$$

For current density J we can state the following from the continuity equation

$$\nabla \cdot J = -\frac{d\rho}{dt} \tag{3}$$

From also Ohm's law we know

$$J = \sigma E \tag{4}$$

Here σ is the conductivity of the material.

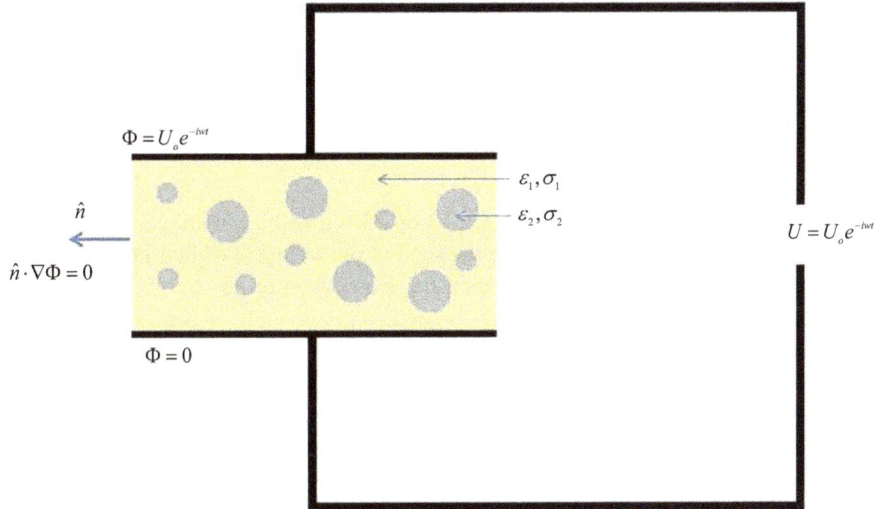

Figure 3. Schematic of the calculation of effective dielectric properties of composites.

So from Equations (2) and (3) we obtain

$$\nabla \cdot \left(\boldsymbol{J} + \frac{d\boldsymbol{D}}{dt} \right) = 0 \tag{5}$$

Now using 1, 4 and 5 we can write the following

$$\nabla \cdot \left(\sigma \boldsymbol{E} + \frac{d(\varepsilon_o \varepsilon_r \boldsymbol{E})}{dt} \right) = 0 \tag{6}$$

In case of a sinusoidal applied electric field $\boldsymbol{E}$ of angular frequency ω

$$\nabla \cdot (\sigma + i\omega\varepsilon_o\varepsilon_r) \boldsymbol{E} = 0 \tag{7}$$

We know

$$\boldsymbol{E} = -\nabla\Phi \tag{8}$$

From Equation (7) and (8) we get

$$\nabla \cdot \left[(\sigma + i\omega\varepsilon_o\varepsilon_r) \nabla\Phi \right] = 0 \tag{9}$$

From Equation (9), we can tell that in a heterogeneous material the product of the physical properties (some form of the conductivity and permittivity) and the slope of the potential must be a constant as we cross material boundaries. For the quasi-static case with harmonic input fields, the gradient of that product vanishes. The interacting field is a result of the charge difference at the interface, and unless the conductivity and permittivity of adjacent material phases are identical, there is a disruption of charge transfer at the material boundary which results in internal polarization.

To solve Equation (9), we set the potential on the top electrode to be,

$$\Phi = U = U_o e^{-i\omega t} \tag{10}$$

And on the bottom electrode,

$$\Phi = 0 \tag{11}$$

Boundary conditions on the interfaces are,

$$\Phi_1 = \Phi_2 \tag{12}$$

$$\varepsilon_1 \hat{\boldsymbol{n}} \cdot \nabla\Phi_1 = \varepsilon_2 \hat{\boldsymbol{n}} \cdot \nabla\Phi_2 \tag{13}$$

where $\hat{\boldsymbol{n}}$ is the unit vector normal to the interface surface. To eliminate fringe effects on the side planes, we set

$$\hat{n} \cdot \nabla \Phi = 0 \qquad\qquad (14)$$

here $\hat{n}$ is the unit vector normal to the side plane.

3. Results and Discussion

3.1. Two Phase Model

For this model, an undamaged composite material is considered to be a homogeneous material and the cracks (here as circular inclusions) are considered to be the second phase inside of that homogeneous material system. Permittivity and ohmic conductivity of the host material were taken to be $\varepsilon_1 = 5$ and $\sigma_1 = 10^{-13}$ S/m and for the inclusion permittivity and ohmic conductivity, $\varepsilon_2 = 2$ and $\sigma_2 = 10^{-15}$ S/m, were chosen which are values close to those of the ambient air permittivity and conductivity [7]. Because of the difference in the permittivities and conductivities of the phases, the accumulation of charge at the interphase boundaries causes an undulation of the space distribution of the potential which is shown in the **Figure 4** and **Figure 5**. **Figure 4** shows the potential distribution around the inclusion and **Figure 5** shows the potential distribution along the horizontal center line. It can be seen that around the boundary of the phases there is a potential nonlinearity (in the figure the nonlinearity is shown inside two ellipses) that is caused by the charge accumulation at the interface between the host material and inclusion. **Figure 6** shows that the space charge accumulation is higher, which is caused by the dissimilarity of the material properties around the inclusion boundary in the presence of the applied electric field.

Computer simulations were performed for different volume fractions of the inclusions. **Figure 7** shows that the space charge density increases with an increase of the inclusion volume fraction. In the frequency range above 1 Hz the space charge density is constant but below 1 Hz a nonlinear increase is observed in the space charge density around the inclusion interface as shown in **Figure 8**.

Figure 9 shows the change of the real and imaginary parts of the global permittivities with the increase of volume fraction of the inclusion as a function of frequency. At a high frequency the period of potential oscillations is not sufficient for charge accumulation but at low frequency the charge has enough time to accumulate around the interface which leads to interfacial polarization (Maxwell-Wagner-Sillar polarization); that is why there is an increase in the real part of the permittivity (shown in **Figure 10**) and dielectric loss at the lower frequencies.

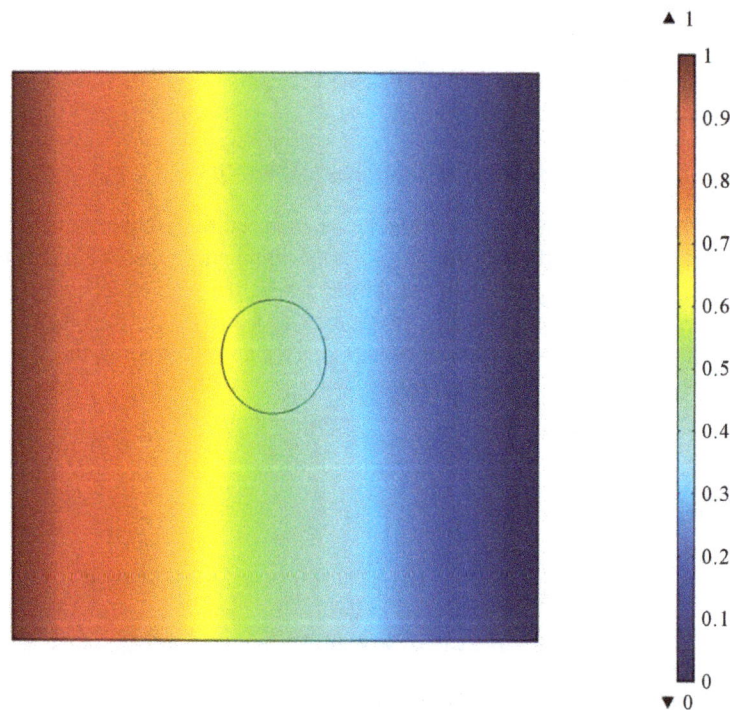

Figure 4. Potential distributions around the inclusion.

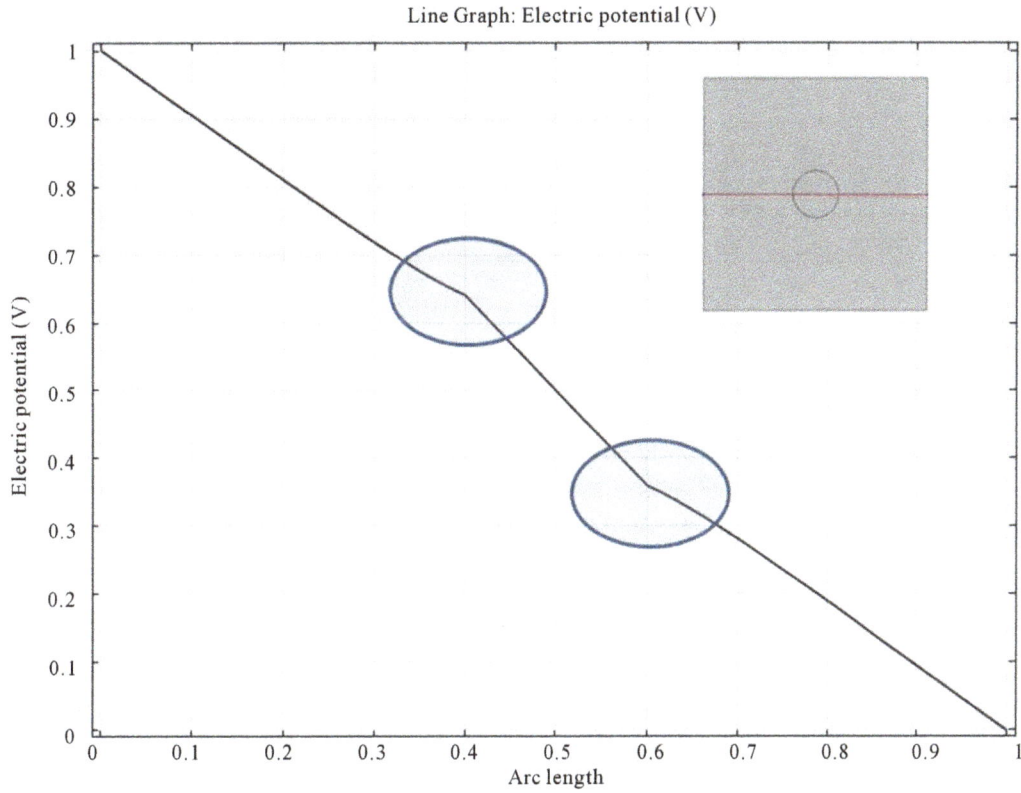

Figure 5. Potential distributions along the line.

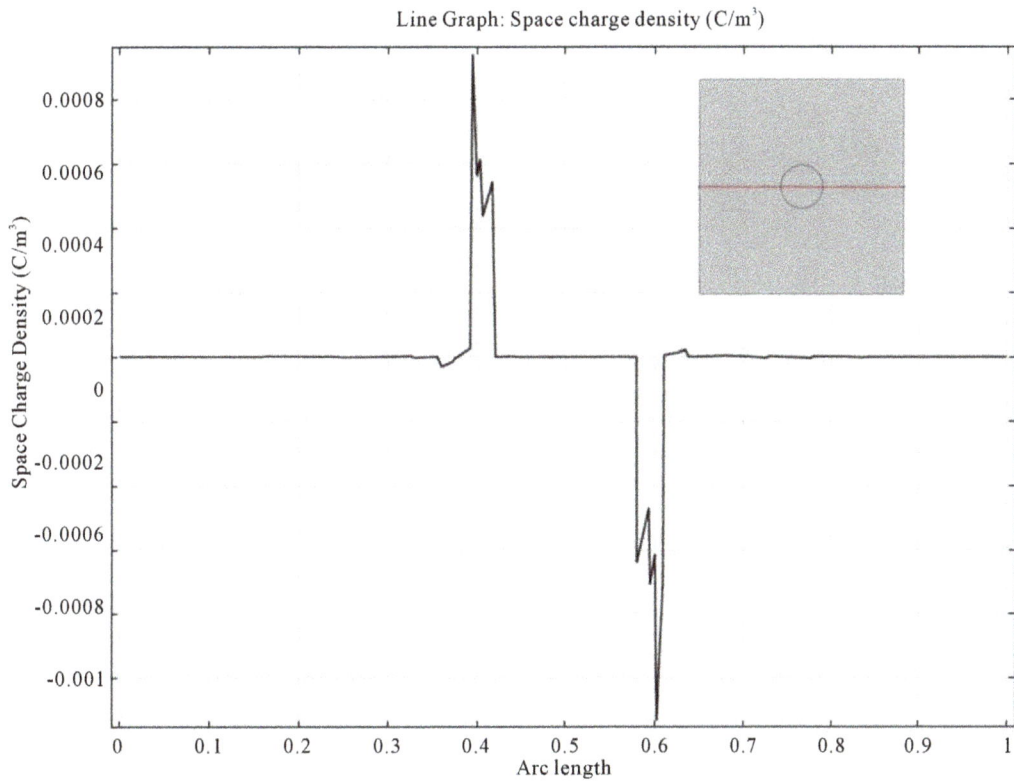

Figure 6. Space charge densities along the line.

Space Charge Density of different volume fraction vs Frequency

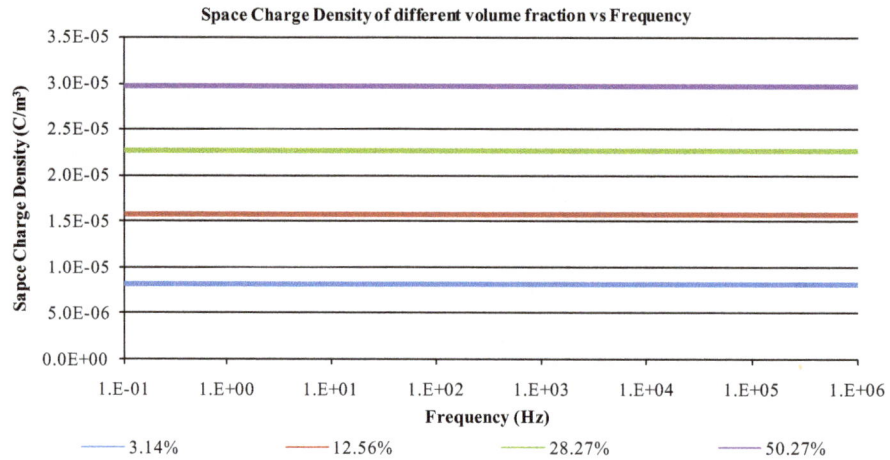

Figure 7. Space charge densities around the inclusion interface of different volume fraction in different frequency.

Space Charge Density of 3.14% volume fraction vs Frequency

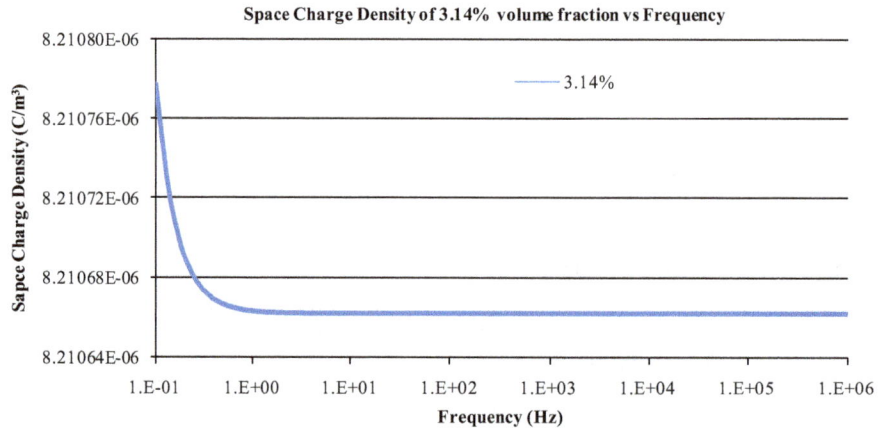

Figure 8. Space charge density change of 3.14% volume fraction of inclusion with frequency.

Real and Imaginary part of Permittivity of different volume fraction vs Frequency

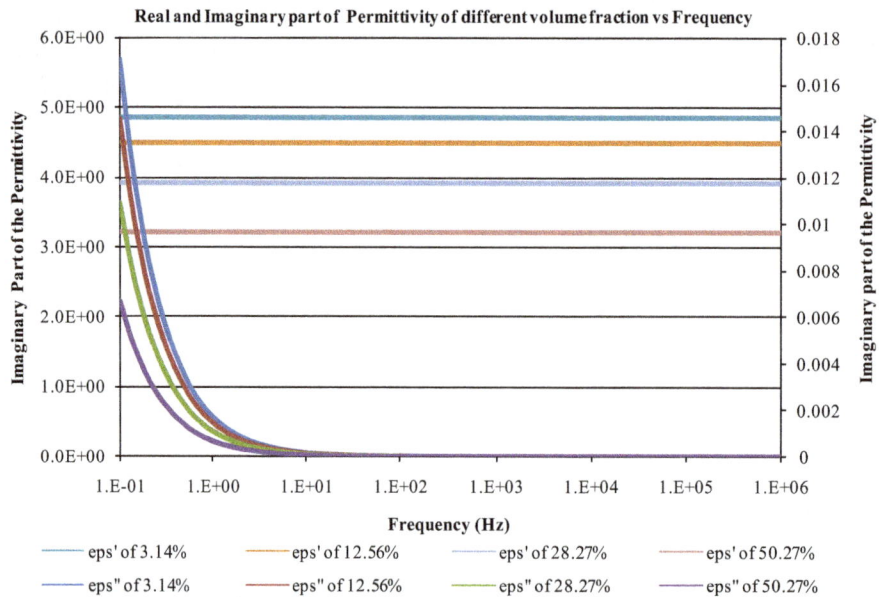

Figure 9. Real and imaginary part of the permittivity with different volume fraction of the inclusion.

For different volume fractions of inclusion, the real part of the permittivity was calculated from the computer simulation for a frequency of 10 Hz. **Figure 11** is the comparison of real part of the permittivity change with increasing volume fraction of the inclusion phase. The relation between the real part of the permittivity and volume fraction is almost linear, but when the real part of the permittivity is plotted with surface area fraction (surface area fraction is the ratio of inclusion surface to the material surface) of the inclusion there is clearly a nonlinear relationship predicted. For low volume fractions the effect of surface area fraction of the inclusion is more dominant than the volume fraction, but for the higher volume fractions it is opposite.

Figure 12 illustrates the comparison between computer simulation results with increasing inclusion volume/surface-area fraction and experimental results. **Figure 12(a)** and **Figure 12(b)** show slight increases in the real part of the impedance for low volume/surface-area fractions. **Figure 12(c)** shows the experimental results for the real part of the impedance change with the strain. The real part of the impedance increases below the low strain (below 5%) for the off axis sample where matrix microcracking is dominant and distributed throughout the material system.

3.2. Three Phase Model

Composite materials are filled with various additive materials to achieve the desired mechanical, thermal and electrical properties. Typical filler materials used for the present modeling are carbon or glass fibers. The use of these fibers as filler materials introduces a water sensitive component into the polymer composites. Glass fibers are well known for their water affinity on their surfaces. Currently, epoxies are widely used matrix materials in

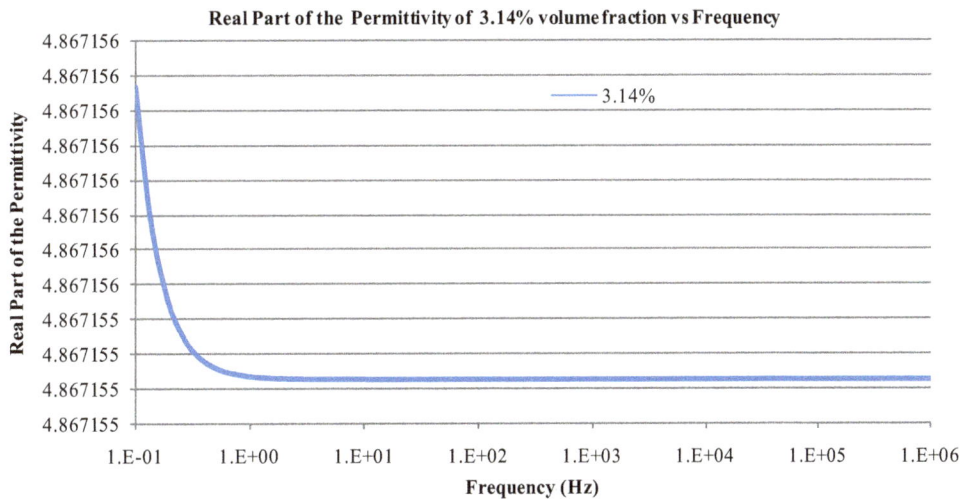

Figure 10. Real part of the permittivity of the material with 3.14% volume fraction of inclusion.

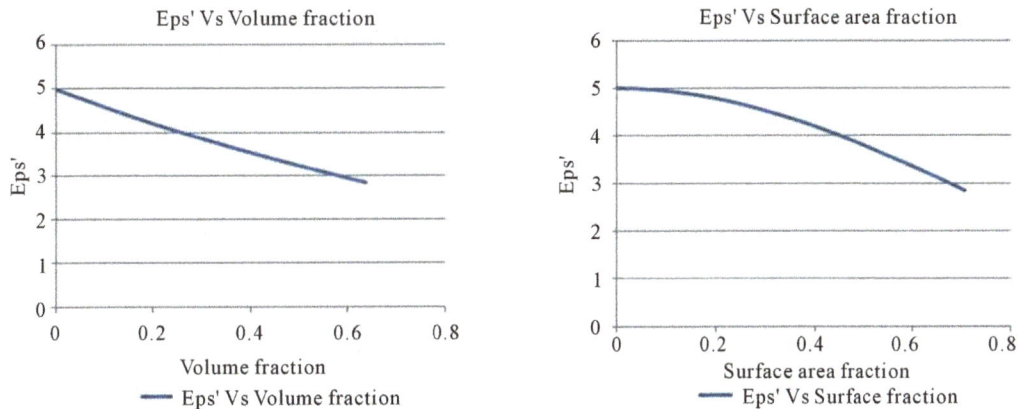

Figure 11. Eps′ comparison with volume fraction and surfacearea fraction of the inclusion.

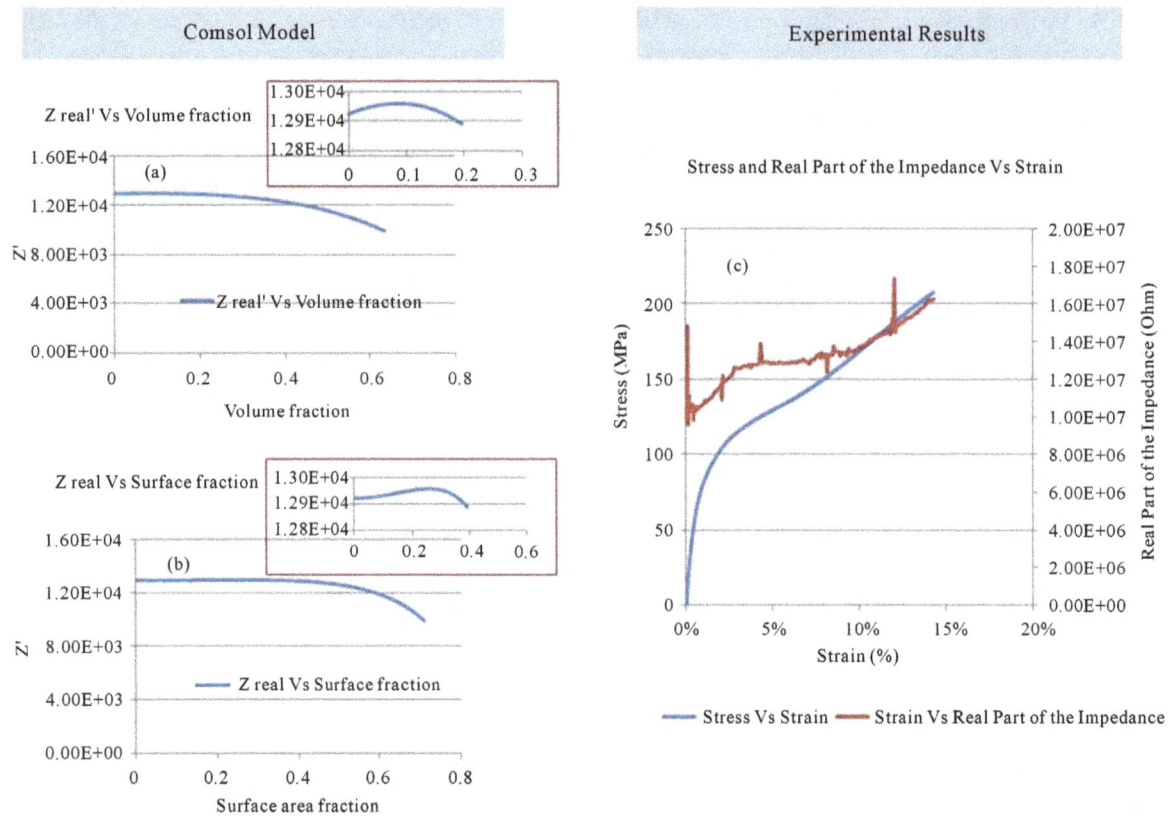

Figure 12. Comparison of Comsol^{TM} simulation result and experimental results.

composite industries, which also have the potential of being sensitive to moist conditions or humid environments. Soles and Yee [8] found that a network of nanopores that is inherent in the epoxy structure helps free water to traverse the epoxy; they found the average size of nanopore diameters to vary from 5 to 6.1 Å and account for 3% - 7% of the total volume of the epoxy material. The approximate diameter of a kinetic water molecule is just 3.0 Å, so via the nanopores network the moisture can easily traverse into the epoxy. They also found that the volume fraction of nanopores does not affect the diffusion coefficient of water and argued that polar groups coincident with the nanopores are the rate-limiting factor in the diffusion process, which could explain why the diffusion coefficient is essentially independent of the nanopore content. In their **Figure 13** they explain how the water transport happens in epoxy networks.

There are many theories about the state of water molecules in polymers. Adamson [9] suggested that moisture can transfer in epoxy resins in the form of either liquid or vapor. It is proposed by Tencer [10] that it is also possible that vapor water molecules undergo a phase transformation and condense to the liquid phase. This condensed moisture was stated to be either in the form of discrete droplets on the surface or in the form of a uniform monolayer [11].

Water has a higher dielectric permittivity and conductivity than the glass fiber and matrix, so it has strong effects on the dielectric properties, *i.e.* relative permittivity and dielectric loss, of the material system. In the literature it is well established that water absorption increases the dielectric constant of the dielectric material [12]-[16]. This dielectric loss is observed in the low frequency range. Water diffuses through the interface and also weakens the interfacial strength of filler and matrix.

When composite materials go through degradation processes, microcracks typically form and these microcracks can also be filled with moist air, and condensed or adsorbed water layers can form on the surface of those defects. In our two phase model we saw an interfacial polarization (Maxwell-Wagner-Sillars polarization) that is present in the low frequency region of the frequency spectra. If a water layer is present on the surface of the defect it will become electrically conductive. Since the host material and defect have low electrical conductivity and permittivity is not significantly high, this will give rise to interfacial polarization.

Figure 14 illustrates the tri-layer computational model used for the study where yellow, gray and blue parts represent respectively the host material, defect, and a conductive layer. The permittivity and ohmic conductivity of the host material were taken to be $\varepsilon_1 = 5$ and $\sigma_1 = 10^{-13}$ S/m and for the inclusion the permittivity and ohmic conductivity had values of $\varepsilon_2 = 2$ and $\sigma_2 = 10^{-15}$ S/m. For the conductive layer a different permittivity ε_3 and conductivity σ_3 (higher than host and defect properties) was used to see the effect on the effective dielectric properties of the material system.

Figure 15 shows the potential distributions around the inclusion of the tri-layer model which is different than the potential difference shown in **Figure 4** for the two phase inclusion model. Because of the conductive layer around the inclusion there is a large undulation of the space distribution of the electric potential.

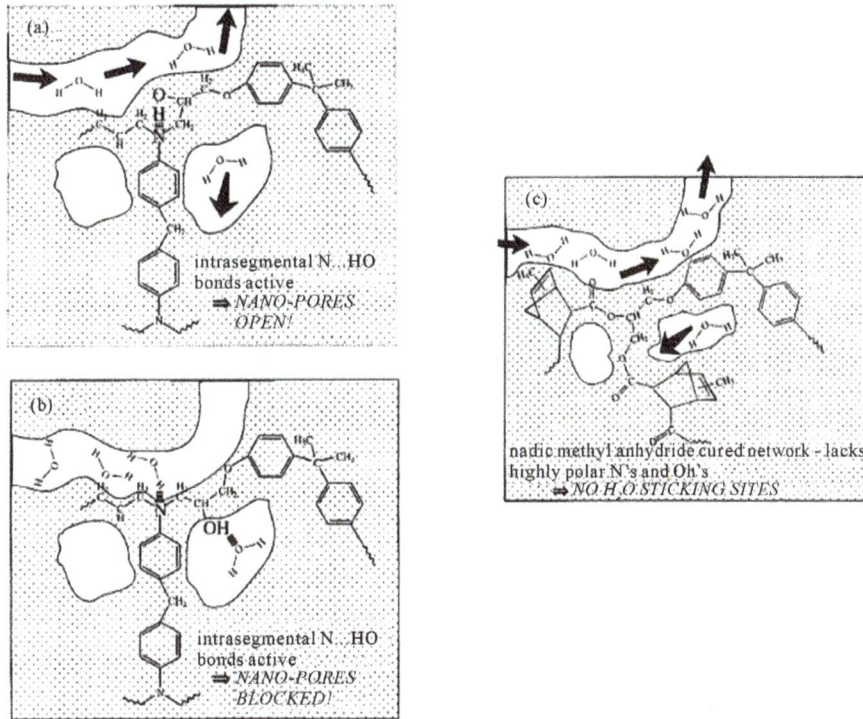

Figure 13. A plausible picture of moisture diffusion through the nanopores of an amine-containing epoxy (Figure from reference [7]).

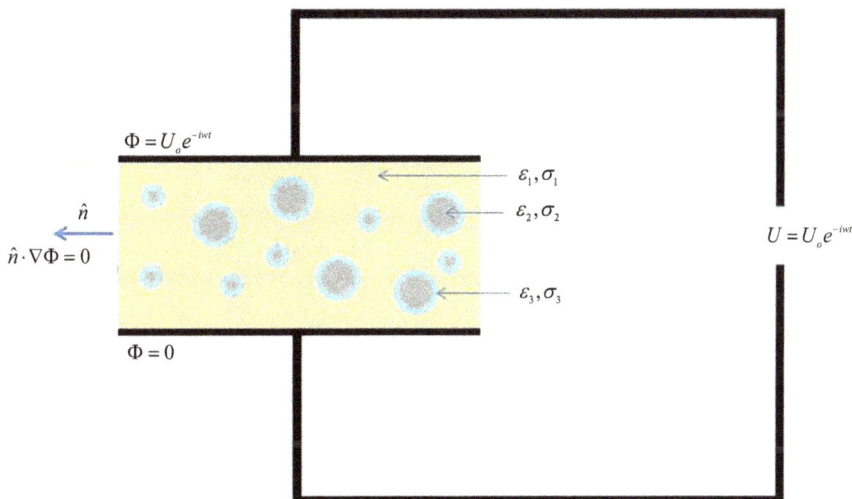

Figure 14. Tri-layer model.

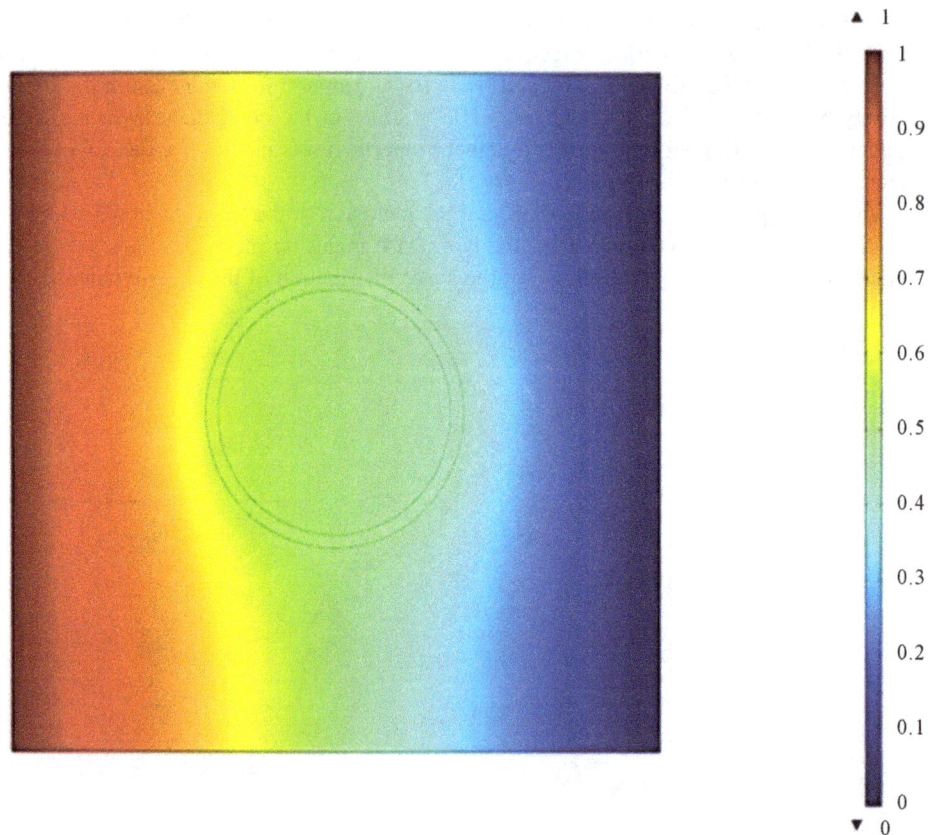

Figure 15. Potential distributions around the inclusion with a conductive layer.

For the tri-layer model, the total volume fraction is the sum of the volume fraction of the defect and the volume fraction of the conductive layer. For all of the cases of tri-layer modelling, the conductive layer thickness was specified as 0.5 micro meter. We observed that for two phase models, the real part of the permittivity was almost linear but in **Figure 16** we can see that for the three phase case there is an increase in real part of the permittivity for lower volume fractions and then a decreasing trend. As there is a conductive layer in between the defect and the host matrix, the interfacial polarization plays a vital role for this type of behavior. The subsequent decrease of the real part of the permittivity for higher volume fractions is caused by the dominance of the volume of the defects which is higher than the interfacial polarization contributed by the conductive layer.

It is also clear from **Figure 17** that for the same conductivity and permittivity values of the conductive layer, for low frequency the real part of the permittivity of the material system is higher than the value for higher frequency.

Figure 18 shows the dependence of dielectric constant on the conductivity of the conductive layer. The real part of the permittivity at different volume fractions for the same permittivity and frequency behave differently for different surface layer conductivities.

As shown in **Figure 19**, for the same volume fraction of the inclusion but variable frequency of the input field, there is a step-like increase of the real part of the permittivity over a narrow frequency range, and the dielectric loss also has the peak in that region, where the Maxell-Wagner-Sillars polarization dominates.

Figure 20 shows the relation of the real part of the permittivity with the frequency for all volume fractions of the inclusion. That simulation was also done for just the matrix material (this is the host material, as we considered it homogeneous). Since there were no other phases presents for that case, there was no chance of charge accumulation and there is no predicted change of the dielectric constant for the matrix-only case. We can see for higher volume fractions the dielectric relaxation strength (difference between the real parts of the permittivity at low frequency and high frequency) also increases. At higher frequency the real part of the permittivity drops as a function of volume fraction because charge accumulation does not occur at the interface at those frequencies.

Real Part of the Permittivity vs Total Volume Fraction of the Inclusion

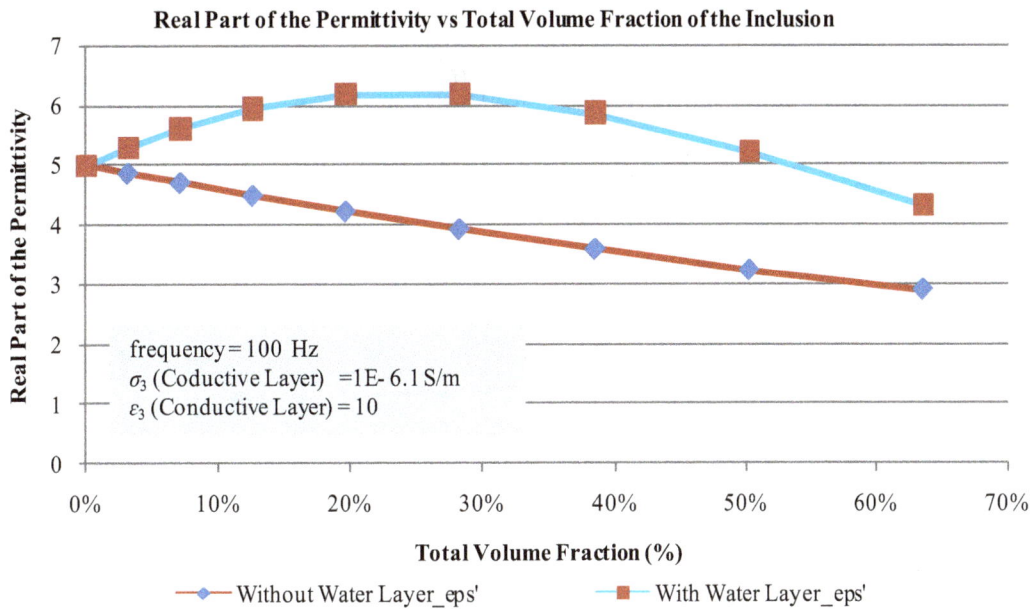

Figure 16. Variation of real part of the permittivity with increasing volume fraction. For the tri-layer model, total volumefraction is the sum of the volume fraction of defect.

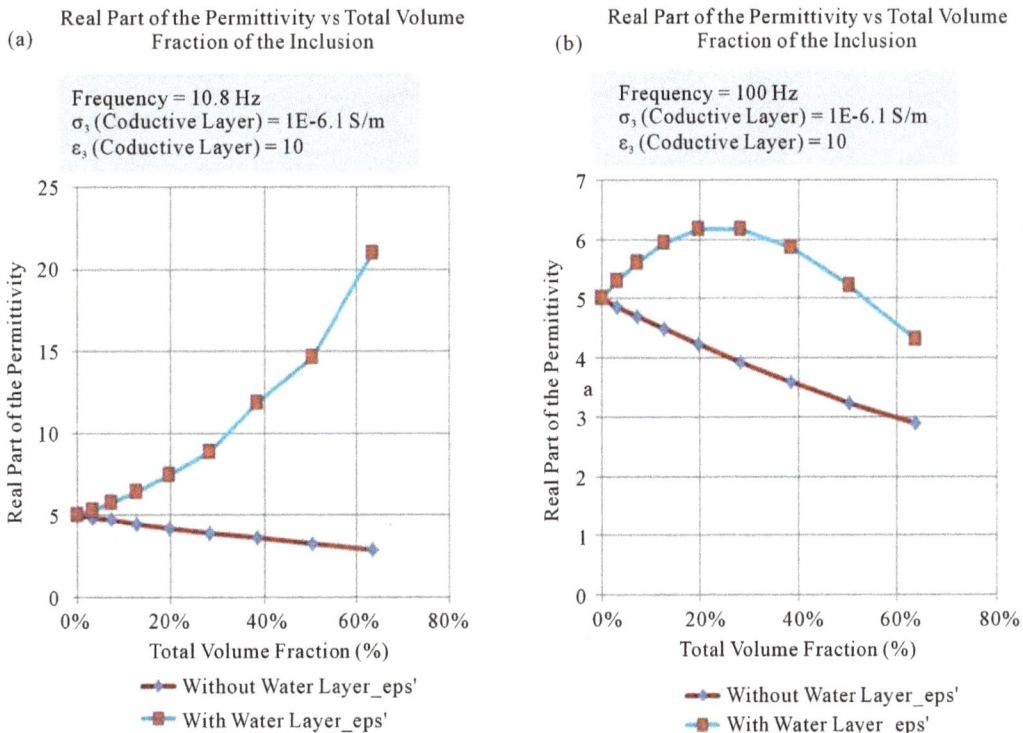

Figure 17. Frequency dependency of the real part of permittivity for different volume fraction.

Dielectric loss (the imaginary part of the permittivity) also varies with volume fraction and it is illustrated in **Figure 21**. For high volume fraction dielectrics, the loss changes somewhat and the peak of the loss also increases.

The corresponding Cole-Cole plot, **Figure 22**, also shows the shift in relaxation for different volume fractions.

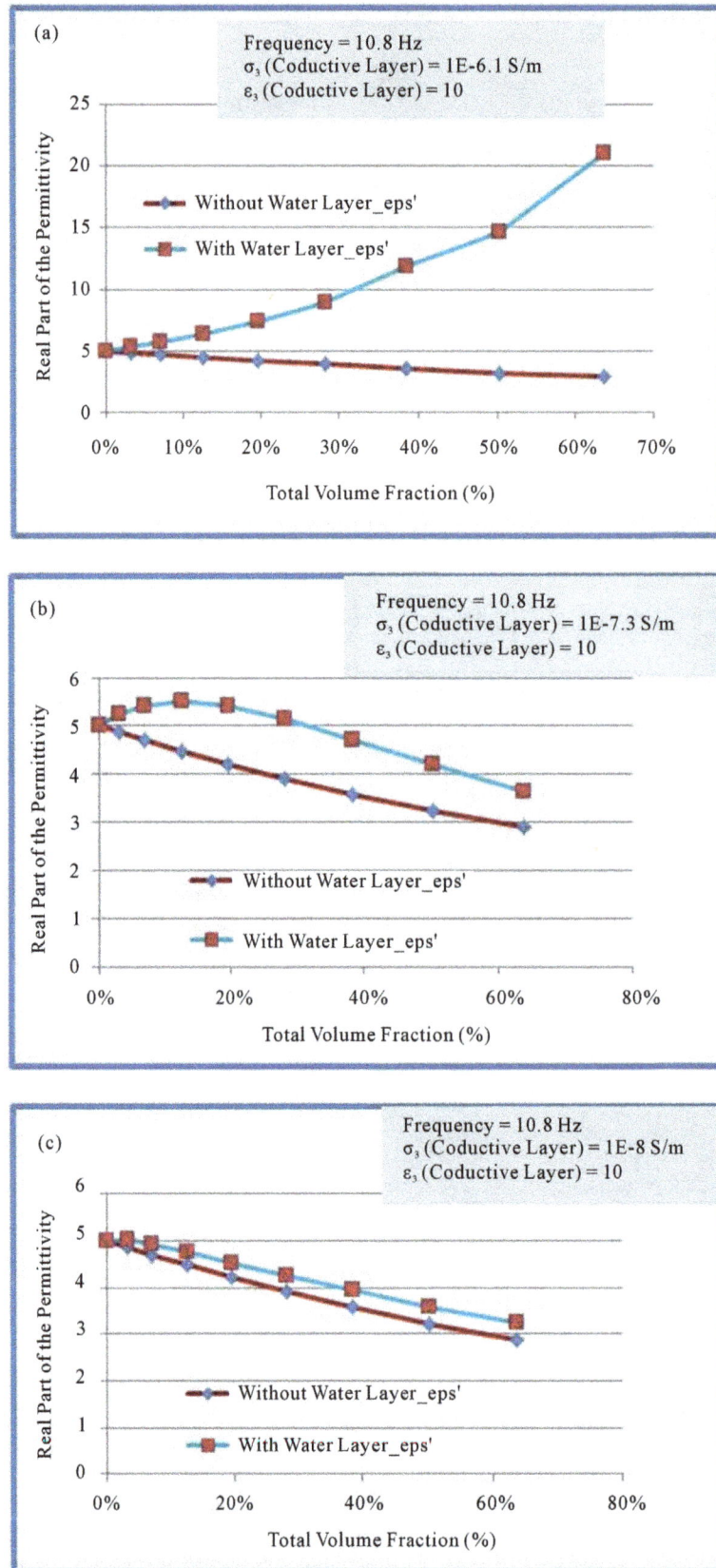

Figure 18. Dielectric properties dependence on the conductivity.

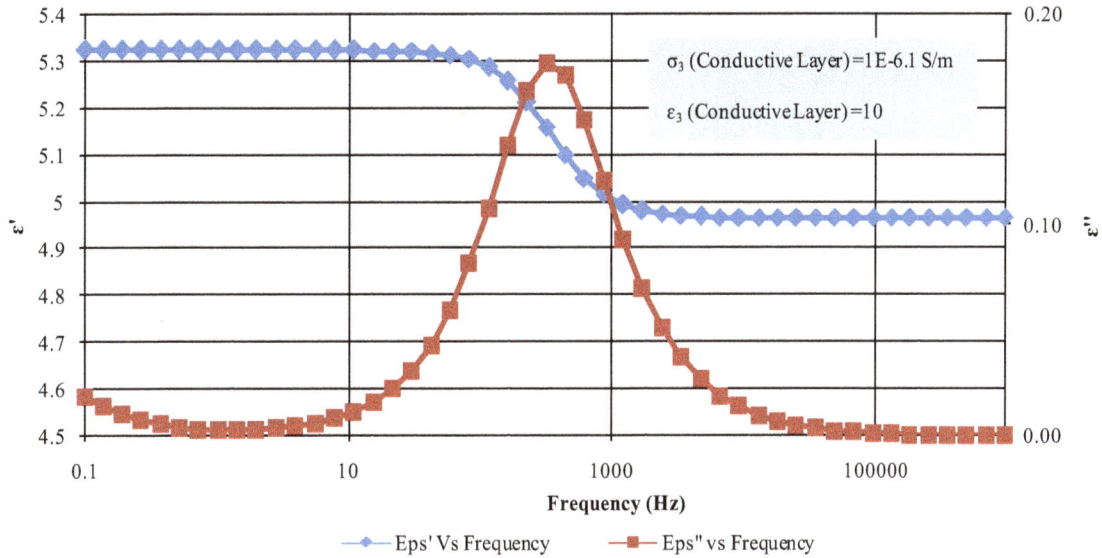

Figure 19. Real and imaginary part of the permittivity Vs frequency for 3.14% total volume fraction of the inclusion.

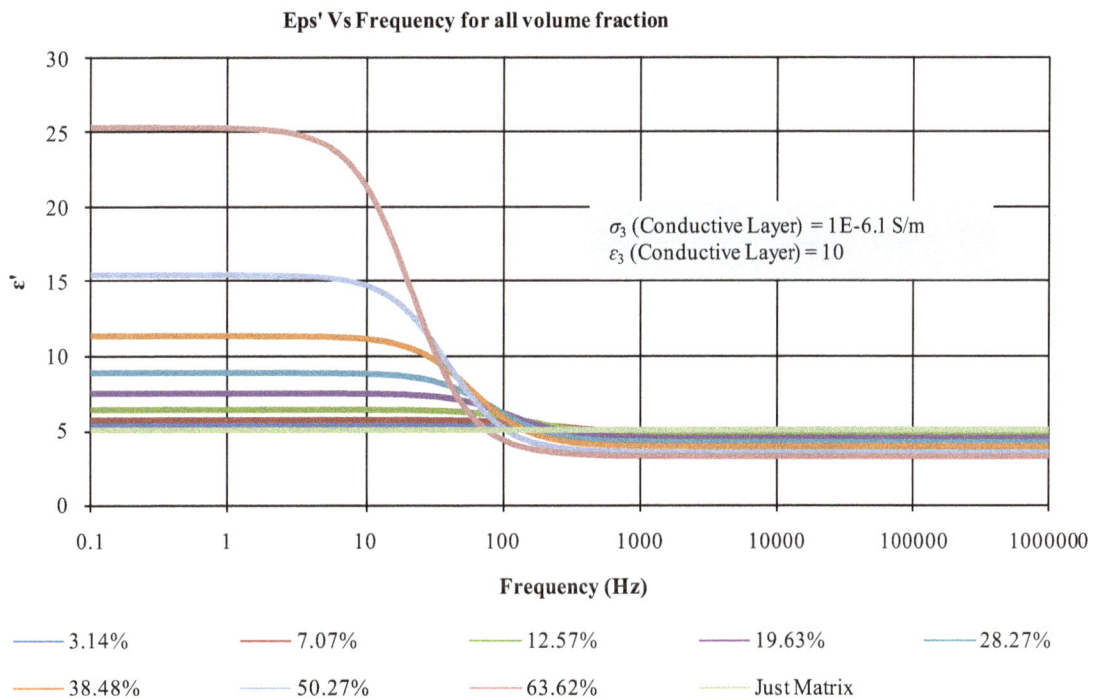

Figure 20. Real part of the permittivity in frequency spectra of all volume fractions.

3.3. Distributed Damage Model

A distributed damage model was created to see the effect of the distribution of the damage. A dielectric study was performed for a certain volume fraction of inclusion, and then that inclusion was divided into 10 inclusions while keeping the total volume fraction the same.

Figures 23-26 show the change of dielectric properties of a single damage volume and distributed damage volumes with the same amount of volume fraction without any conductive layer around the defects. The dielectric loss increased for the distributed damage because of the presence of more interfacial polarization.

Eps" Vs Frequency for all volume fraction

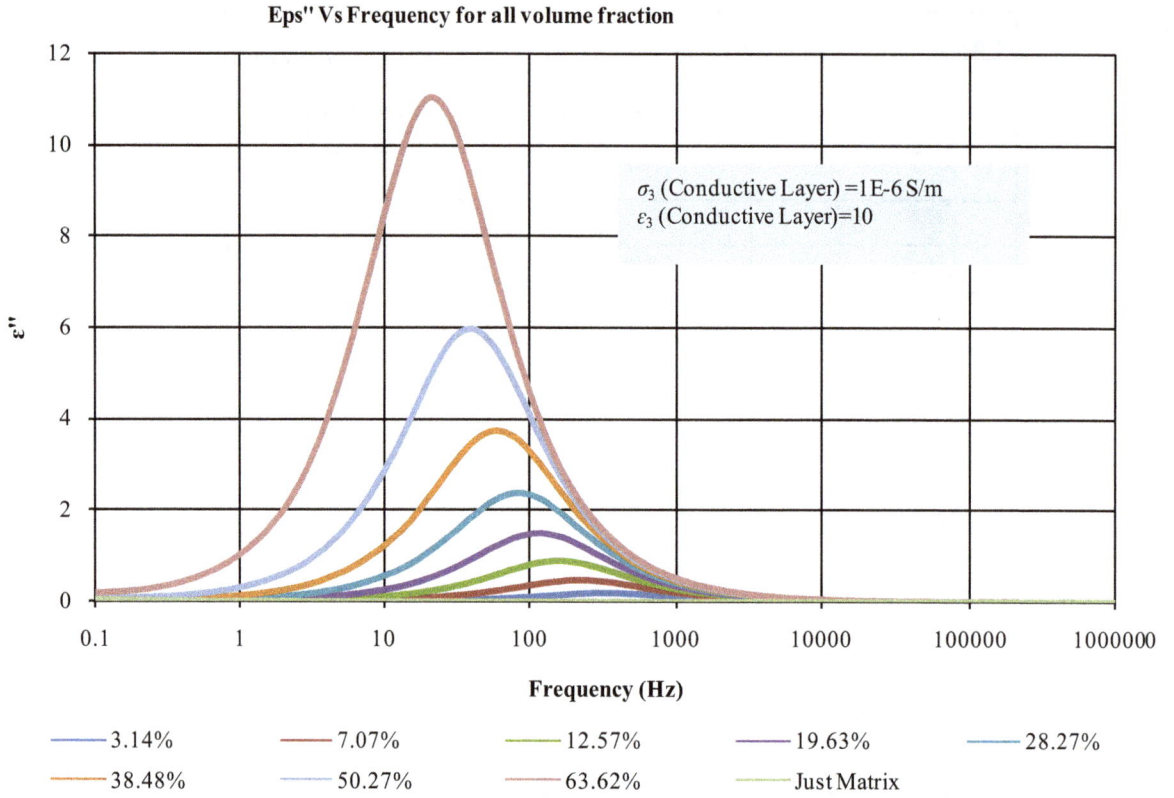

Figure 21. Imaginary part of the permittivity in frequency spectra of all volume fractions.

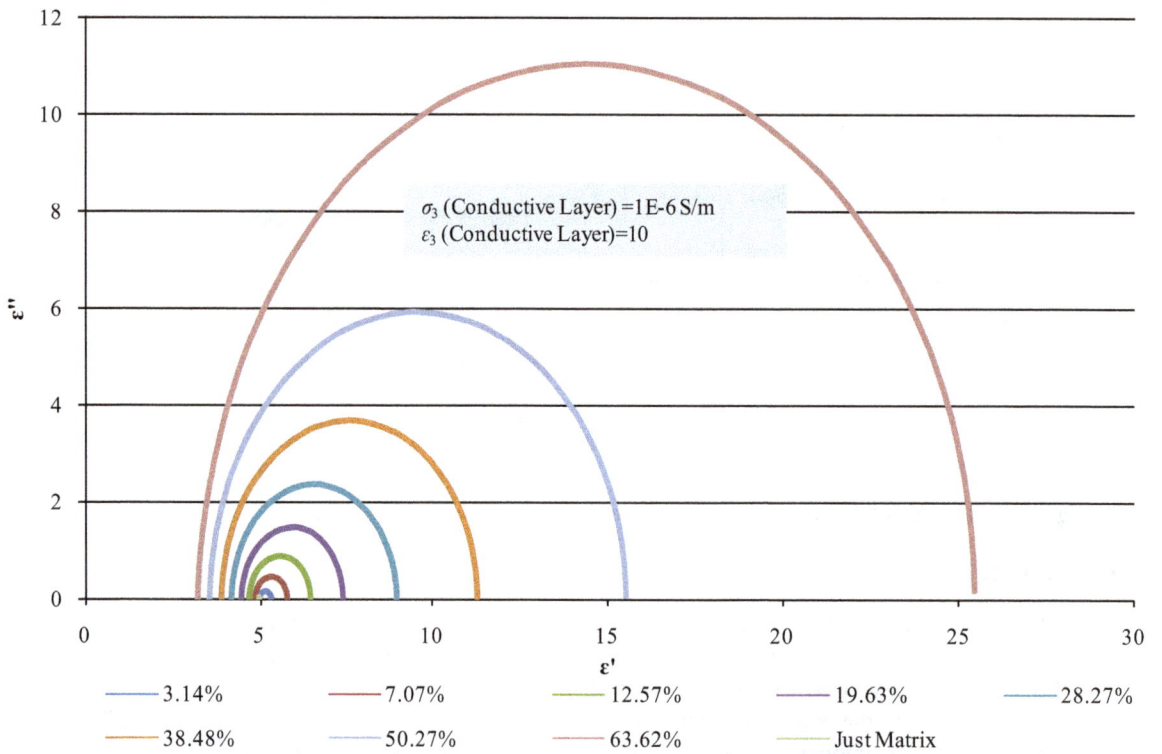

Figure 22. Cole-Cole plot of different volume fraction.

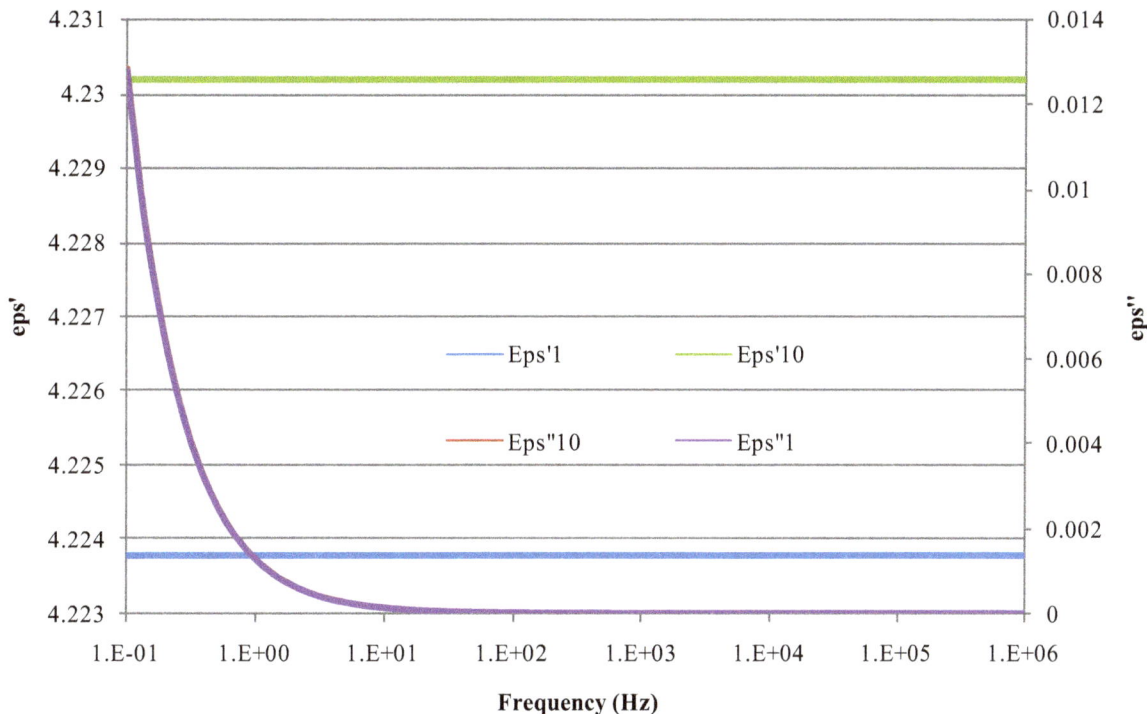

Figure 23. Dielectric Properties without conductive layer for same volume fraction but different number of inclusion.

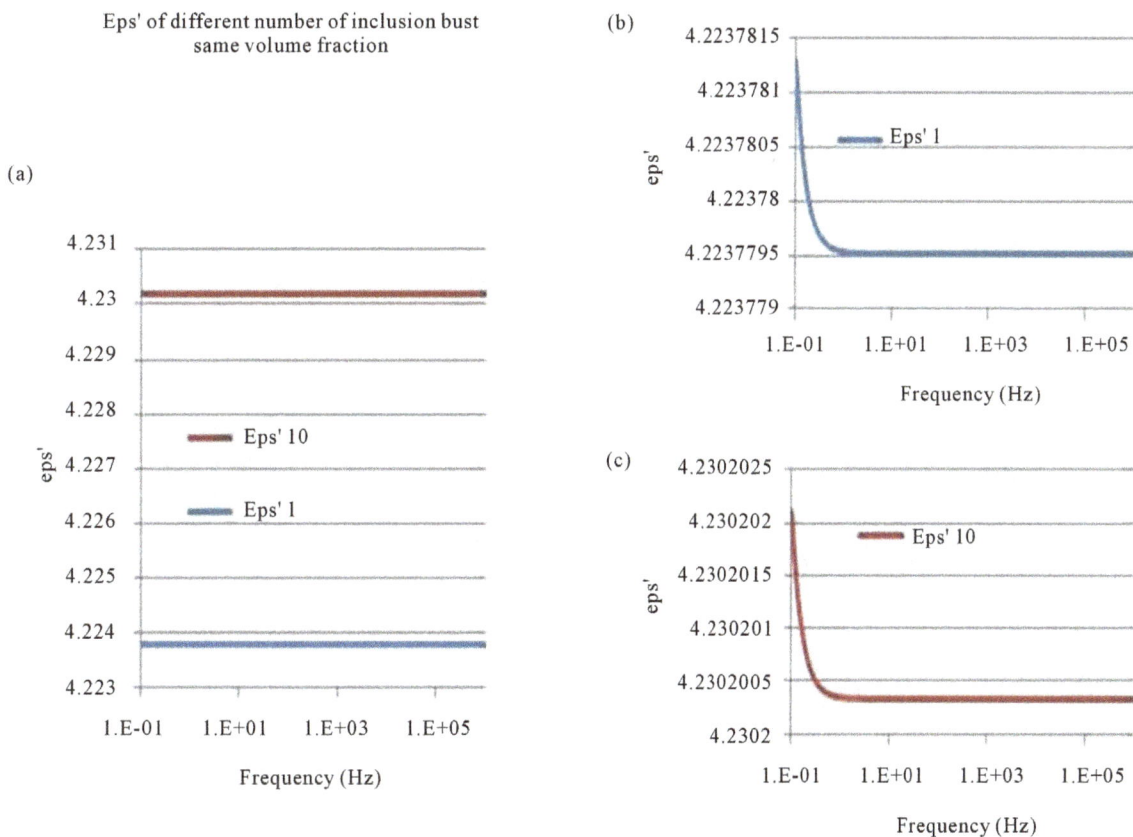

Figure 24. Real Part of the permittivity for different number of inclusion but same volume fraction.

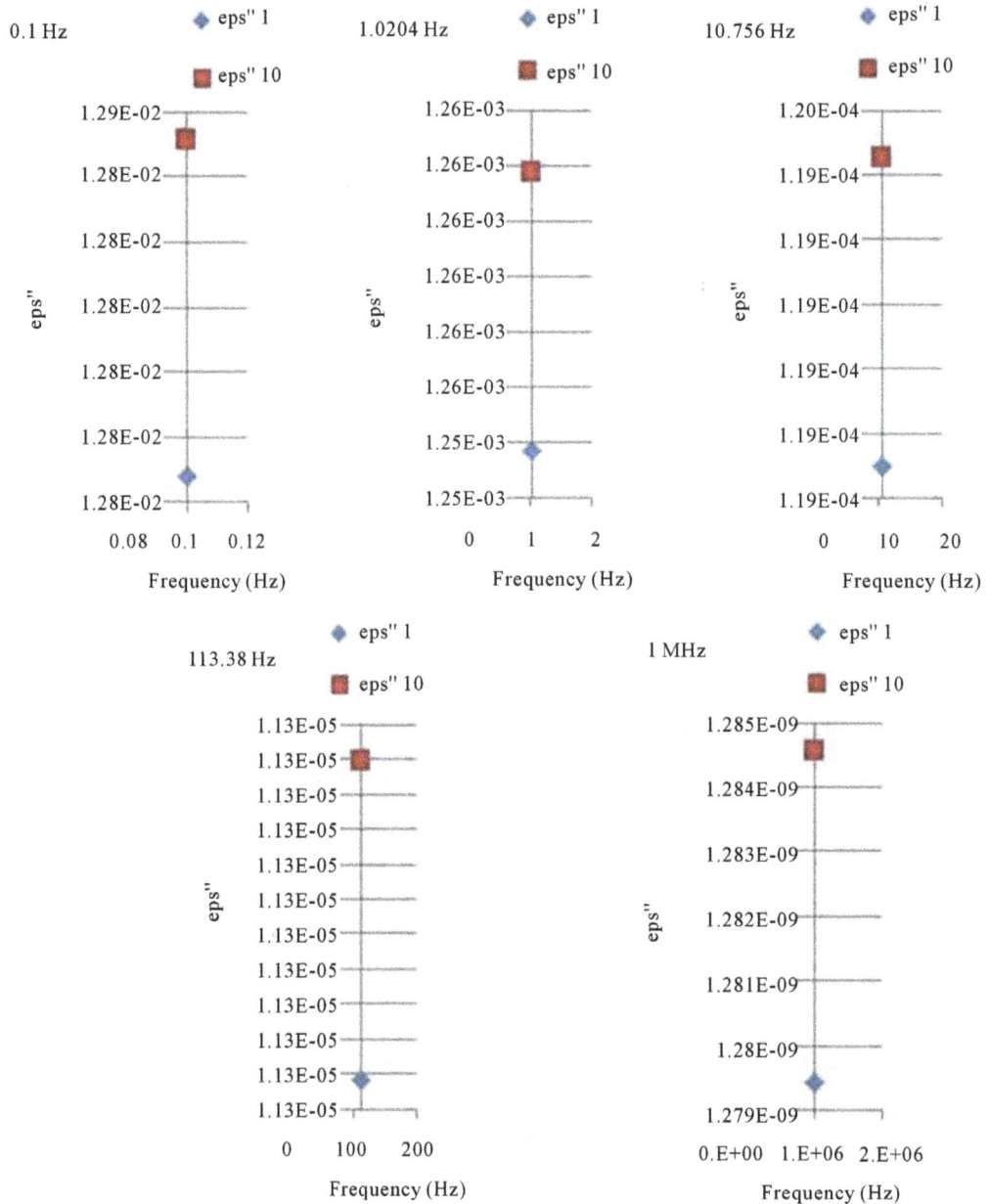

Figure 25. Dielectric losses at different frequency for different number of inclusion but same volume fraction without any conductive layer.

Figures 27-29 show the change of dielectric properties of a single damage phase and distributed damage with the same amount of volume fraction with a conductive layer around the defect. The dielectric loss increased for the distributed damage because of more interfacial polarization and it is more evident than the prior case because the conductive layer around the defect leads to increased interfacial polarization.

The difference between the static permittivity and the limiting high frequency dielectric permittivity is called the Dielectric relaxation strength (DRS), $\Delta\varepsilon$, as given in Equation (15).

$$\Delta\varepsilon = \varepsilon_s - \varepsilon_\infty \qquad (15)$$

where, ε_s is the static permittivity and ε_∞ is the limiting high frequency dielectric constant. To calculate DRS from the experimental data we subtract the value of real part of the permittivity at 1 MHz from the value of real part of the permittivity at 0.1 Hz.

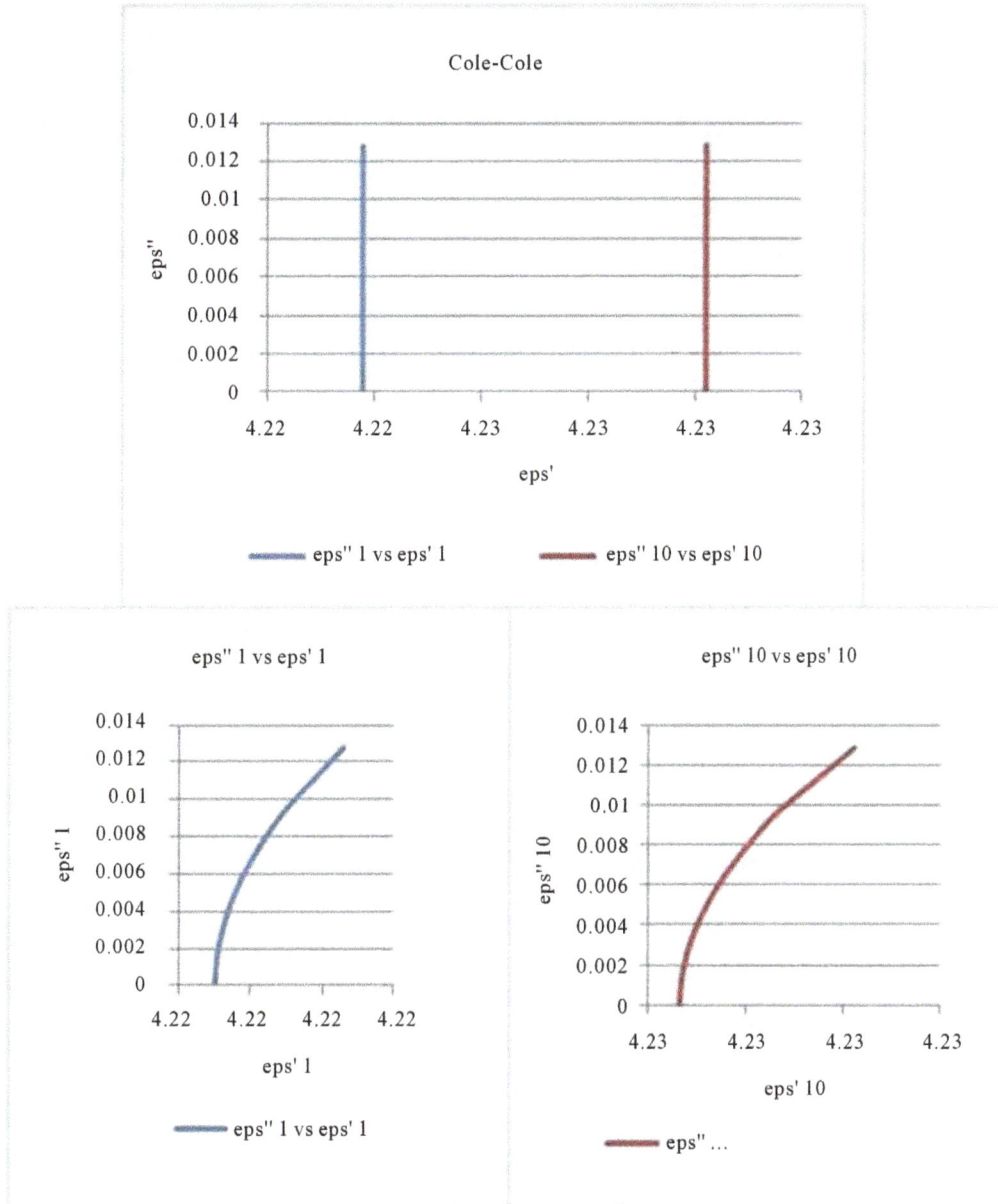

Figure 26. Cole-Cole plot of different number of inclusion without conductive layer.

Figure 30(a) shows the increase of DRS with the increase of damage state defects which is in agreement with what we saw in the computational data (**Figure 20** and **Figure 27**). **Figure 30(b)** shows the increase in dielectric loss with the increase of damage, and this loss increased more in the lower frequency region when the defects had conductive solution layers on their surface, which in also in agreement with the computational model (**Figure 21** and **Figure 28**).

4. Conclusions

In this paper, we have demonstrated a computational model to predict global dielectric property changes caused by increasing defects inside of a materials system. We show that the dielectric character of the defects, their volume, and the morphology of the defect surfaces play an important role in the overall dielectric properties of the materials system during degradation. The data presented here also demonstrate the possibility of using dielectric properties to model and interpret the progressive damage of heterogonous materials systems.

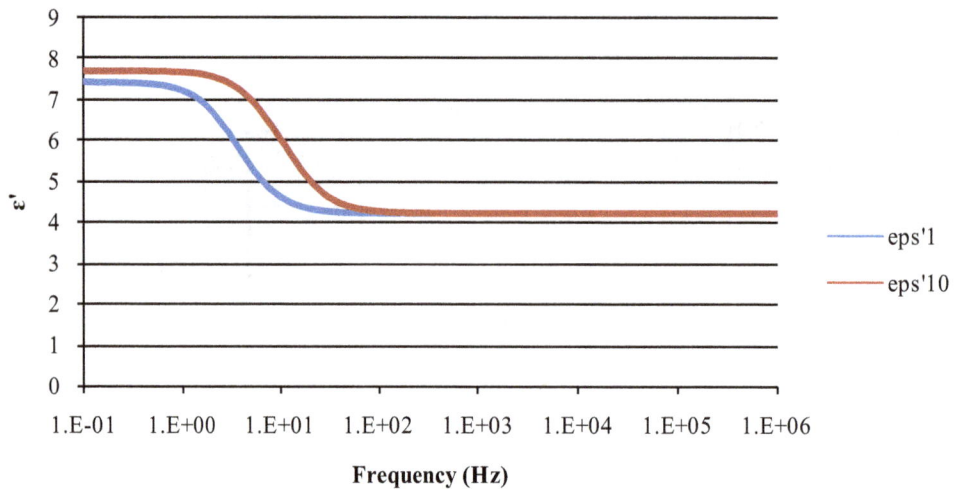

Figure 27. Real part of the Permittivity of different number inclusion but same volume fraction with conductive layer.

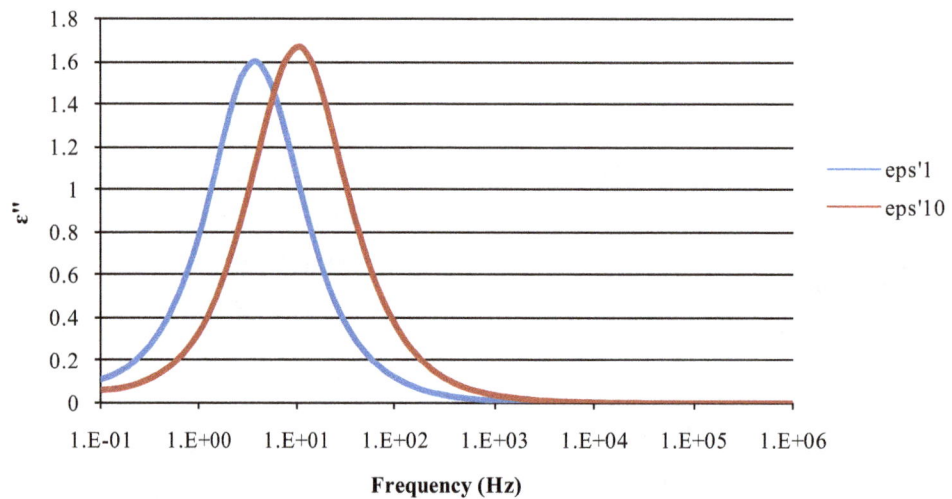

Figure 28. Imaginary part of the Permittivity of different number inclusion but same volume fraction with conductive layer.

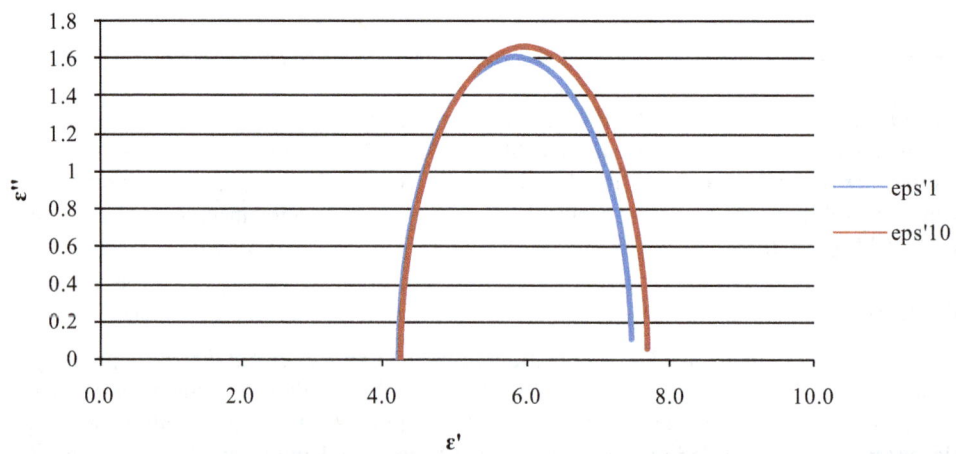

Figure 29. Cole-Cole plot of different number of inclusion with conductive layer.

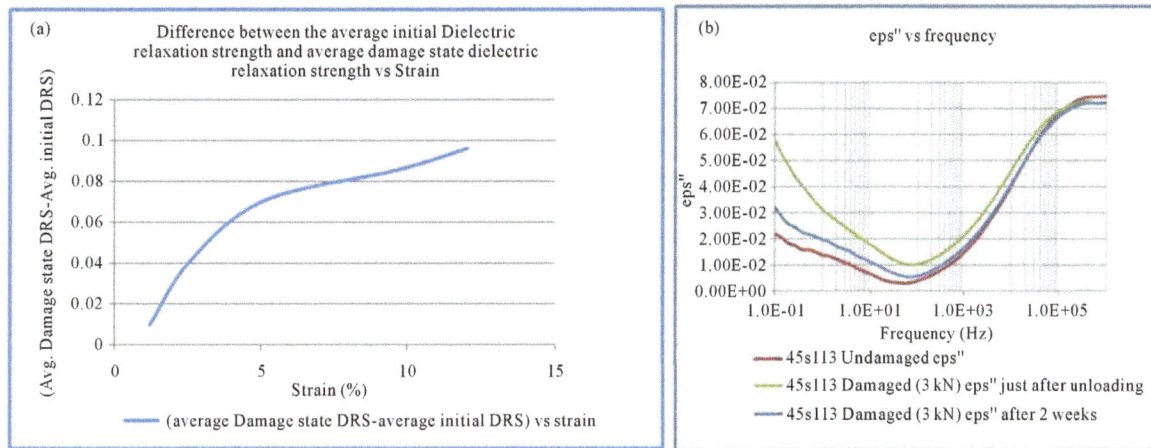

Figure 30. Experimental results of dielectric properties change of damaged composite materials.

In general, we have shown that the dielectric properties of heterogeneous systems are influenced by various physical factors: electrical and structural interactions between particles, heterogeneity of morphological and electrical properties of the constituent phases, frequency dependence of electrical phase parameters, intra-particle structure, particle shape, size, orientation and, volume and surface fraction of the constituent phases. This dependence complicates the determination of the electrical parameters of heterogeneous materials from the observed global dielectric relaxation spectra, but also presents us with an opportunity to recover important information not only about the electrical and structural properties of constituents but also about the interactions between constituents, including the parent materials and damage phases. Further theoretical and experimental investigation is required to fully understand the changes in dielectric spectra associated with many of the specific damage accumulation events and local details in heterogeneous material systems.

From the results presented in this paper, it can be concluded that analysis of the dielectric data gives us information about the type of material state changes throughout the mechanical life of a composite material. It should be emphasized that these changes in the dielectric properties are distinct and measurable changes in material state, and that they are caused by a non-conservative, non-equilibrium material response to the applied fields. Opportunities for further understanding include the identification of the material and physical limitations of this method of characterization, e.g., specimen size, material property ranges, and specimen shapes that are most and least suited to the approach. A robust study of the interpretation of dielectric data associated with specific damage modes and details is also needed.

References

[1] Reifsnider, K.L. and Case, S.W. (2002) Damage Tolerance and Durability of Material Systems. John Wiley and Sons, New York.

[2] Raihan, R., Adkins, J.M., Baker, J., Rabbi, F. and Reifsnider, K. (2014) Relationship of Dielectric Property Change to Composite Material State Degradation. *Composites Science and Technology*, **105**, 160-165. http://dx.doi.org/10.1016/j.compscitech.2014.09.017

[3] Raihan, R., Reifsnider, K., Cacuci, D. and Liu, Q. (2015) Dielectric Signatures and Interpretive Analysis for Changes of State in Composite Materials. *ZAMM-Journal of Applied Mathematics and Mechanics/Zeitschrift für Angewandte Mathematik und Mechanik.* http://dx.doi.org/10.1002/zamm.201400226

[4] Tuncer, E., Serdyuk, Y.V. and Gubanski, S.M. (2001) Dielectric Mixtures—Electrical Properties and Modeling.

[5] Brosseau, C., Beroual, A. and Boudida, A. (2000) How Do Shape Anisotropy and Spatial Orientation of the Constituents Affect the Permittivity of Dielectrich Eterostructures? *Journal of Applied Physics*, **88**, 7278-7288. http://dx.doi.org/10.1063/1.1321779

[6] Baker, J., Adkins, J.M., Rabbi, F., Liu, Q., Reifsnider, K. and Raihan, R. (2014) Meso-Design of Heterogeneous Dielectric Material Systems: Structure Property Relationships. *Journal of Advanced Dielectrics*, **4**, 1450008. http://dx.doi.org/10.1142/S2010135X14500088

[7] Pawar, S.D., Murugavel, P. and Lal, D.M. (2009) Effect of Relative Humidity and Sea Level Pressure on Electrical Conductivity of Air over Indian Ocean. *Journal of Geophysical Research*: *Atmospheres*, 114(D2).

[8]   Soles, C. and Yee, A. (2000) A Discussion of the Molecular Mechanisms of Moisture Transport in Epoxy Resins. *Journal of Polymer Science, Part B: Polymer Physics*, **38**, 792-802. http://dx.doi.org/10.1002/(SICI)1099-0488(20000301)38:5<792::AID-POLB16>3.0.CO;2-H

[9]   Adamson, M.J. (1980) Thermal Expansion and Swelling of Cured Epoxy Resin Used in Graphite/Epoxy Composite Materials. *Journal of Material Science*, **15**, 1736-1745. http://dx.doi.org/10.1007/bf00550593

[10]   Tencer, M. (1994) Moisture Ingress into Nonhermetic Enclosures and Packages—A Quasisteady State Model for Diffusion and Attenuation of Ambient Humidity Variations. *IEEE 44th Electronic Components Technology Conference*, Washington DC.

[11]   Shirangi, M.H. and Michel, B. (2010) Mechanism of Moisture Diffusion, Hygroscopic Swelling, and Adhesion Degradation in Epoxy Molding Compounds. *Moisture Sensitivity of Plastic Packages of IC Devices*, Springer, 29-69. http://dx.doi.org/10.1007/978-1-4419-5719-1_2

[12]   Banhegyi, G. and Karasz, F.E. (1986) The Effect of Adsorbed Water on the Dielectric Properties of $CaCO_3$ Filled Polyethylene Composites. *Journal of Polymer Science Part B: Polymer Physics*, **24**, 209-228. http://dx.doi.org/10.1002/polb.1986.090240201

[13]   Banhegyi, G., Hedvig, P. and Karasz, F.E. (1988) DC Dielectric Study of Polyethylene/$CaCO_3$ Composites. *Colloid and Polymer Science*, **266**, 701-715. http://dx.doi.org/10.1007/BF01410279

[14]   Cotinaud, M., Bonniau, P. and Bunsell, A.R. (1982) The Effect of Water Absorption on the Electrical Properties of Glass-Fibre Reinforced Epoxy Composites. *Journal of Materials Science*, **17**, 867-877. http://dx.doi.org/10.1007/bf00540386

[15]   Reid, J.D., Lawrence, W.H. and Buck, R.P. (1986) Dielectric Properties of an Epoxy Resin and Its Composite I. Moisture Effects on Dipole Relaxation. *Journal of Applied Polymer Science*, **31**, 1771-1784. http://dx.doi.org/10.1002/app.1986.070310622

[16]   Paquin, L., St-Onge, H. and Wertheimer, M.R. (1982) The Complex Permittivity of Polyethylene/Mica Composites. *IEEE Transactions on Electrical Insulation*, **5**, 399-404. http://dx.doi.org/10.1109/TEI.1982.298482

16

Effect of Aluminum Nano-Particles on Microrelief and Dielectric Properties of PE+TlInSe$_2$ Composite Materials

E. M. Gojayev*, Kh. R. Ahmadova, S. I. Safarova, G. S. Djafarova, Sh. M. Mextiyeva

Azerbaijan Technical University, Baku, Azerbaijan
Email: *geldar-04@mail.ru

Abstract

The paper presents the results of studies surface microrelief, frequency-temperature characteristics of the imaginary part of the dielectric permittivity and dielectric loss of PE+TlInSe$_2$ composite materials in 25°C - 150°C temperature and 25 Hz - 1 MHz frequency range before and after application of the aluminum nano-particles with a size of 50 nm. The change in the amount of semiconductor filler TlInSe2 and aluminum nano-particles changes the state of the surface and the frequency-temperature characteristics of composite materials PE+xvol.%TlInSe$_2$<Al>, which allows to obtain composites with the desired dielectric permittivity and dielectric loss.

Keywords

Surface Microrelief, Aluminum Nano-Particles, Dielectric Permeability, Tangent of the Angle Dielectric Loss

1. Introduction

Study of the dielectric properties of polymers over a wide temperature and frequency ranges is one of the most effective ways to establish the characteristics of their structure. However, the "response" of the polymer system to the action of the electric field of a certain frequency is not equivalent to the "mechanical response". Therefore, the method of dielectric loss can be used to identify softening of polymers. The maximum of dielectric loss may differ quite significantly from the temperature of the structural vitrification, as well as the frequency (at a given temperature corresponding to the maximum) may be different from the frequency of the mechanical vitrification. That mismatch of relaxation transitions corresponding to electrical or mechanical effects on the temperature or

---

*Corresponding author.

frequency scale provides additional information about the levels of structural organization of the polymers.

Taking the above mentioned into consideration, this article studied the surface microrelief using atomic force microscope and the temperature and frequency dependences of PE+TlInSe$_2$ and PE+TlInSe$_2$ compositions with aluminum nano-particles.

2. Techniques of Experiment

In scanning, the probe microscope study of surface microrelief and its local properties is carried out with the help of specially prepared probes in the form of needles. The dimension of the working part of such probes is about 10 nm. The typical distance between the probe and the sample surface in probe microscopes is equal 0.1 - 10 nm. Operation of the probe microscopes based on different types of interaction of the probe with the surface [1]. The process of scanning of the surface in scanning probe microscope is similar to the motion of the electron beam across the screen in the TV cathode-ray tube. Probe moves along the line first in the forward direction and then in the opposite one, then moves to the next line. The movement of the probe is carried out using a scanner in small steps under the effect of saw-tooth voltage, generated by digital to analog converters. Registration of the information about surface is carried out, as a rule, on direct pass under two conditions: in the process of scanning probe must touch points on the surface and in each case only one of its points. And if the scanning probe cannot reach some areas of the surface (e.g., when the samples have the highest portions of the relief), the relief only partially is restored. And the greater the number of points touching the probe, the more reliable is possible to reconstruct the surface.

The test samples were prepared as follows: a polymer powder is mixed with a powder of a semiconductor material TlInSe2 and aluminum nano-particles. After that, the mixture placed between aluminum foil sheets is compressed into 100-μm-thick films at the melting temperature of the polymer matrix and a press sure of 15 MPa. The prepared samples with the foil are quenched in water, and the foil is removed. The obtained samples are useful for studying the properties of electrets. Research conducted at the facility described in [2]. Composites with additives investigated x = 0, 1, 3, 5, 7, 10 wt% and y = 3; 5; 7; 10 wt%.

To calculate the dielectric permittivity of the sample thickness and diameter of the upper electrode (for determining an area) is measured automatically using calipers. For this purpose, the sample is placed between two flat electrodes of circular shape. Capacitance and dielectric loss tangent are measured at the same time [3].

The dielectric permittivity of the compositions is calculated from the measured capacitance values, the thickness of the sample and the area of the electrodes. The dielectric permittivity ε is calculated according to the formula

$$\varepsilon = \frac{Cd}{\varepsilon_0 S},$$

where C—measured electrical capacitance of the sample, F; $\varepsilon_0 = 8.85 \times 10^{-12}$ F/m; d—diameter of the sample, m; S—area of the sample, m^2.

Dielectric loss tangent tgδ is measured directly. Thus, for each of the selected dielectrics must be measured capacitance and dielectric loss tangent corresponding to 1 Hz frequency.

The sample is mounted between two electrodes in the measuring cell. Then it is heated in a cell (the heater is mounted in the cover) with a constant rate of 2 K/min. The sample temperature is recorded using a thermocouple and the dielectric losses—with the help of the measuring bridge LCK E7-8 heating at a constant rate achieved by three-LAT-system.

3. Results and Discussion

As a result, we have obtained planar surface image for PE+xvol.%TlInSe$_2$ composites of size $16 \times 10^3 \times 16 \times 10^3$ nm (**Figure 1(a)** and **Figure 1(b)**), respectively. From **Figure 1(a)**, **Figure 1(b)**, which shows 3D (three-dimensional) images of the same sections it is seen that the surface relief (topography) of the compositions varies with the volume amount of the filling nano-particles [4]-[6].

Analysis of AFM images histograms (**Figure 2(a)** and **Figure 2(b)**) shows that the uniformity of the surface varies up to 25 nm. It is seen that in the boundary layer of the compositions yet is observed some rough edges. Rather, they are connected with the fact that the destruction of the binding forces leaves on the surface not individual atoms but their groups-clusters. This is also evidenced from Fourier spectrum obtained by AFM method

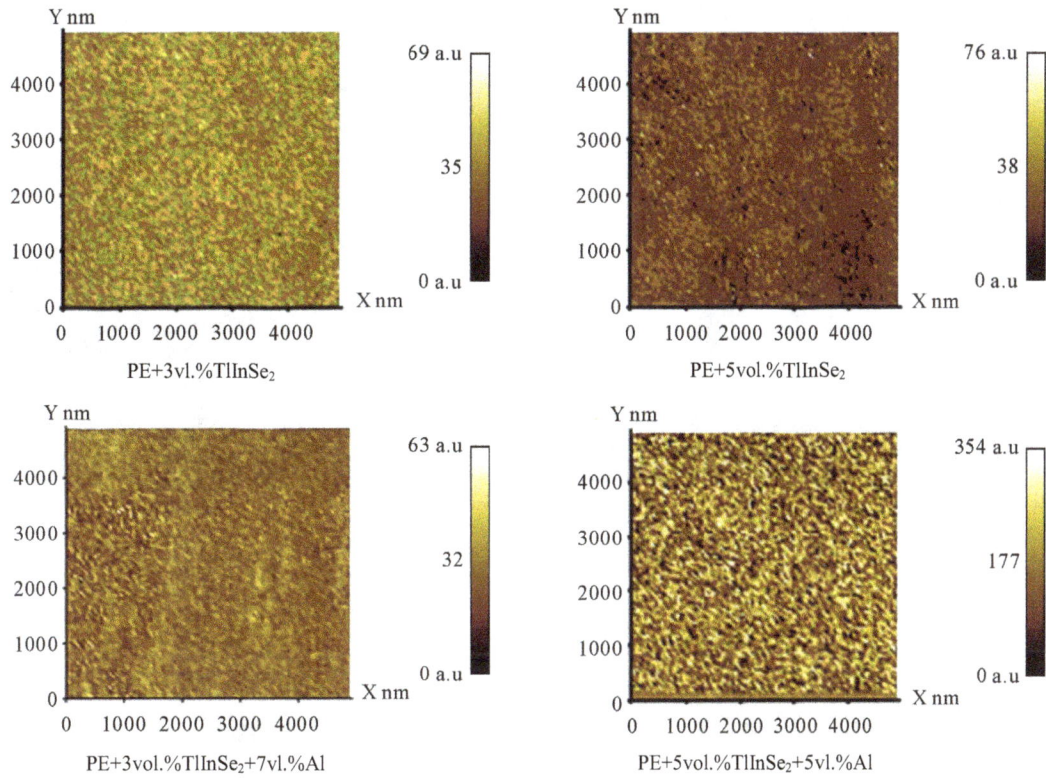

(a)

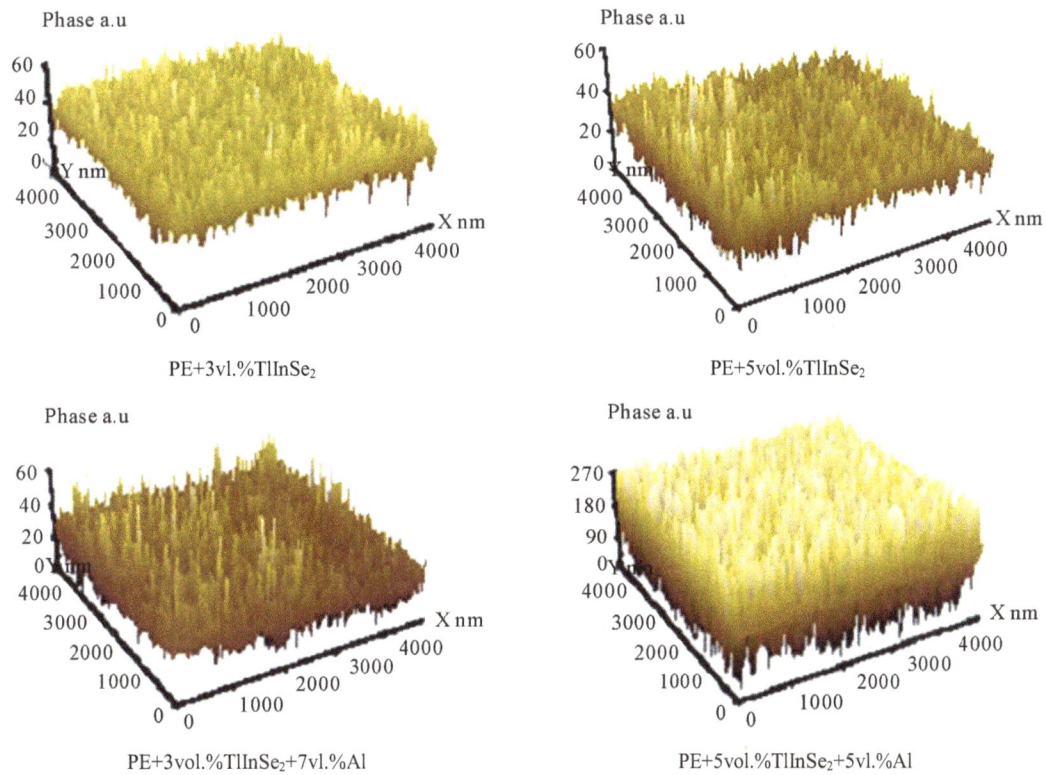

(b)

Figure 1. Planar (a) and 3D (b) images of the surface of composites.

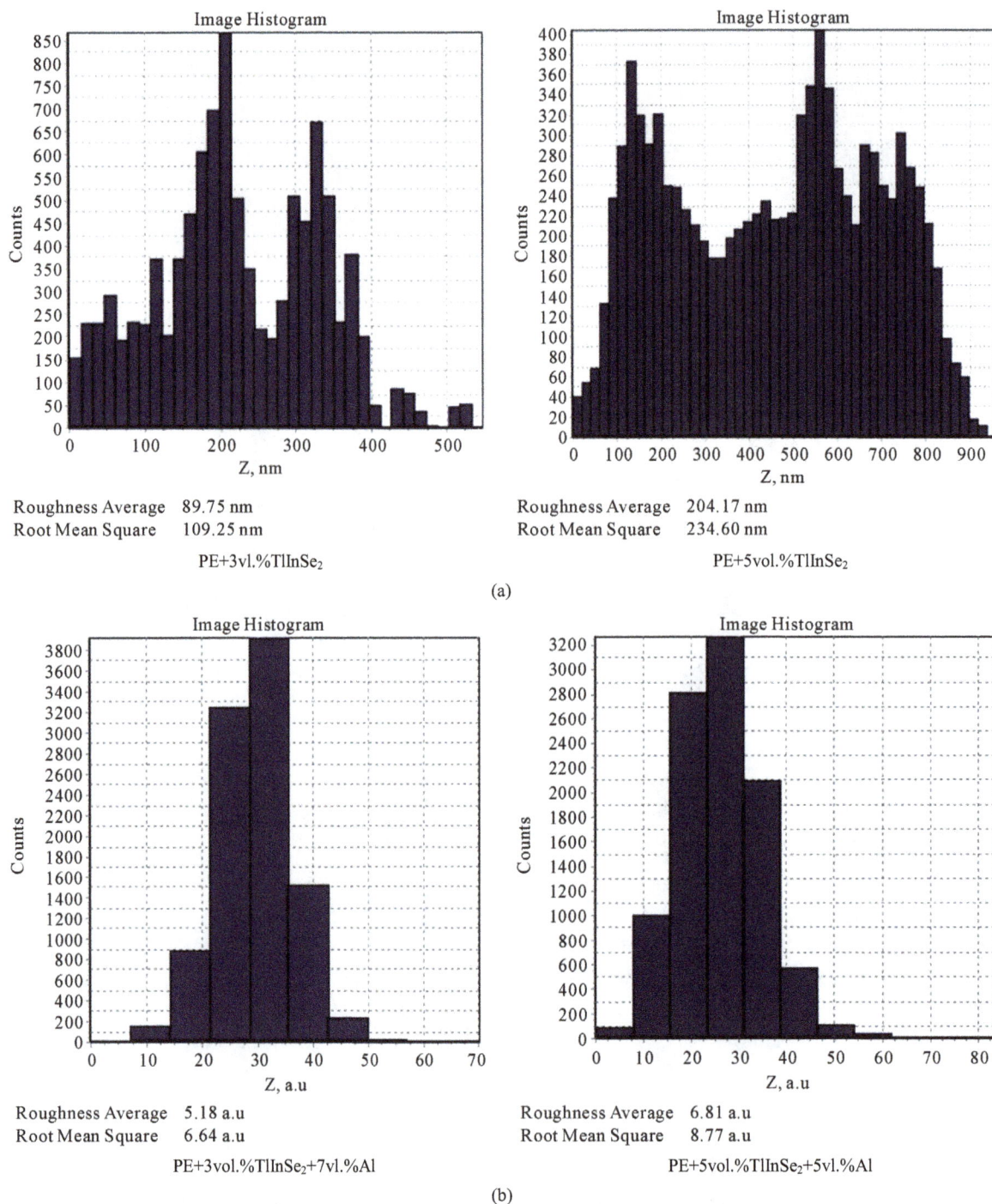

Roughness Average   89.75 nm
Root Mean Square   109.25 nm

PE+3vl.%TlInSe$_2$

Roughness Average   204.17 nm
Root Mean Square   234.60 nm

PE+5vol.%TlInSe$_2$

(a)

Roughness Average   5.18 a.u
Root Mean Square   6.64 a.u

PE+3vol.%TlInSe$_2$+7vl.%Al

Roughness Average   6.81 a.u
Root Mean Square   8.77 a.u

PE+5vol.%TlInSe$_2$+5vl.%Al

(b)

Figure 2. Surface histogram composites.

(**Figure 3(a)** and **Figure 3(b)**). The concentration of the spectrum in the center of the image shows that the surface particles have approximately the same dimensions, *i.e.* are commensurable.

Results of the study of frequency-temperature dependences of the imaginary part of the dielectric permittivity and dielectric loss tangent of the compositions PE+xvol.%TlInSe$_2$ ($1 \leq x \leq 10$) are shown in **Figures 4-7**.

Figure 4 shows the dependence of the imaginary part of the dielectric permittivity with temperature in the range of 20˚C - 150˚C for composites PE+xvol.%TlInSe$_2$ ($1 \leq x \leq 10$).

As follows from **Figure 4**, the imaginary part of the dielectric permittivity increases with temperature in

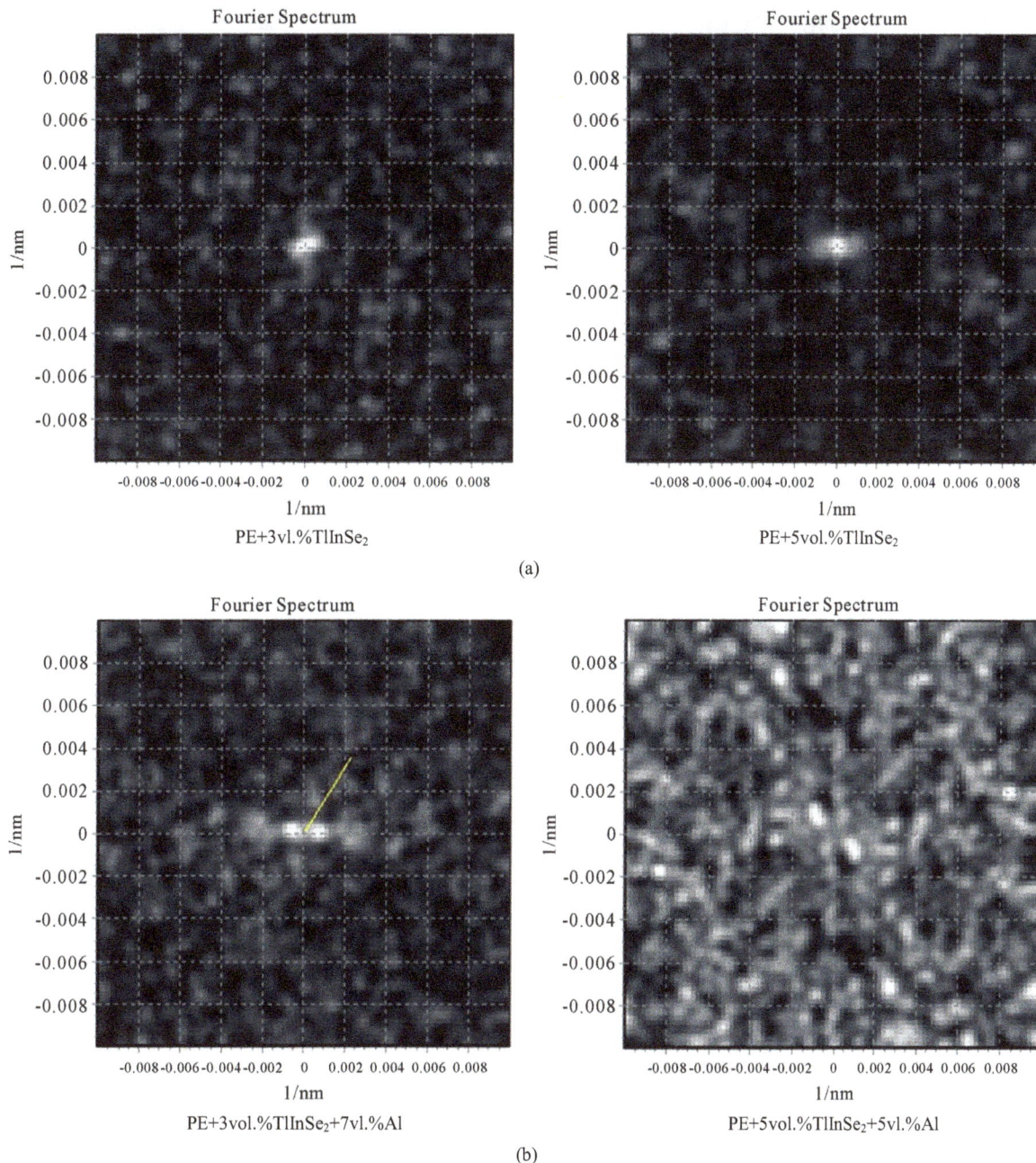

Figure 3. Fourier spectrum of the surface of the composites.

PE+xvol.%TlInSe$_2$ composite materials with an increase in the additive from 1 to 5 vol.%. In general, for investigated composites the variation $\varepsilon''(T)$ is not significantly different. However, for the composite PE+5vol.% TlInSe$_2$ at a temperature 600°C a mild maximum is revealed. Dependence $\varepsilon''(T)$ shows that at low additive content imaginary part of the dielectric permittivity remains substantially constant. And when the amount of additives in TlInSe$_2$ increases up to 5 vol.%, increase in $\varepsilon''(T)$ and the further reduction with increasing of the temperature occurs. Changes in the imaginary part of the dielectric permittivity with temperature for this composite are substantially different. Firstly, at 25°C, imaginary part of the dielectric permittivity increases by three times. İn 60°C - 100°C temperature range, relatively moderate reduction occurs. Further, from 100°C to 125°C sharp decrease in the imaginary part of the dielectric constant takes place. In this temperature range, the imaginary part of the dielectric permittivity reduces by 3.5 times. Subsequently up to 150°C moderate decrease of ε'' occurs.

Results of the study of the temperature dependence of the dielectric loss tangent of composite materials PE+xvol.%TlInSe$_2$ (1≤ x ≤ 10) are shown in **Figure 4(b)**. As can be seen from **Figure 4(b)**, a change of tgδ with temperature for composites with x = 1; 3 in the temperature range investigated does not practically occur. For PE+5vol.%TlInSe$_2$ composite tgδ at low temperatures is relatively high, but with increasing temperature decreases to a stable value (0.0076). For the composite with x = 10 in 250˚C - 100˚C temperature range a slight increase in tgδ, in the range of 100˚C - 125˚C sharp decrease from 0.0373 to 0.0096, and further a slight decrease occurs.

We investigated the frequency and temperature dependence of composite materials PE+xvol.%TlInSe$_2$<Al> with aluminum nano-particles. The results are shown in **Figure 5(a)** and **Figure 5(b)**.

As can be seen from **Figure 5(a)**, in the investigated temperature range of 250˚C - 150˚C the imaginary part of the dielectric permittivity increases with increasing content of additives TlInSe$_2$. But a change in the nature of overall identical for the three composites, where (x = 3, 5, 7), respectively with the addition of the aluminum nano-particles in an amount of y = 7; 5; 3.

Similarly occur changes of ε''(T) for the composite PE+10vol.%TlInSe$_2$+10vl.%Al. However, note that in the entire temperature range ε''(T) in magnitude greater than in other composites. In the temperature range of 850˚C - 100˚C is observed a significant increase, and further moderate decrease in ε''(T).

We investigated the frequency dependence of the imaginary part of the dielectric permittivity and dielectric loss tangent of composites in the frequency range 25 Hz - 1 MHz. As can be seen from **Figure 6(a)**, change of the imaginary part of the dielectric permittivity against the frequency for composites with x = 3; 5; 10 are of the same nature. Is typical of all three composites in the range of 100 kHz - 1 MHz relatively high decrease, slow decrease up to 1 kHz, and then an inversion of the imaginary parts of the permittivity occurs. At 200 Hz a deep minimum is observed and further an increase of the imaginary part of the permittivity occurs. Nature of the change of the imaginary part of the dielectric permittivity of the composite with the addition of x = 1 vol.% before detecting an inversion has identical character. However, at frequencies of 500 Hz, 100 Hz deep minima, and between them at 200 Hz pronounced peak is observed.

The dielectric loss tangent for the above mentioned composites with the addition of TlInSe$_2$ have been investigated.

As can be seen from **Figure 6(b)**, a plot tgδ(v) is identical with the character of ε''(v). For composites with x = 3; 5; 10 is revealed a single deep minimum at a frequency of 200 Hz and a sign inversion of tgδ at the frequency of 1 kHz; for the composite with x = 1 also a sign inversion for tgδ at the same frequency, deep minima at frequencies of 100 and 500 Hz and clear maximum at a frequency of 200 Hz is detected.

Results of the study of the imaginary part of the dielectric permittivity and dielectric loss of composites PE+xvol.%TlInSe$_2$<Al> with aluminum nano-particles are shown in **Figure 7(a)** and **Figure 7(b)**.

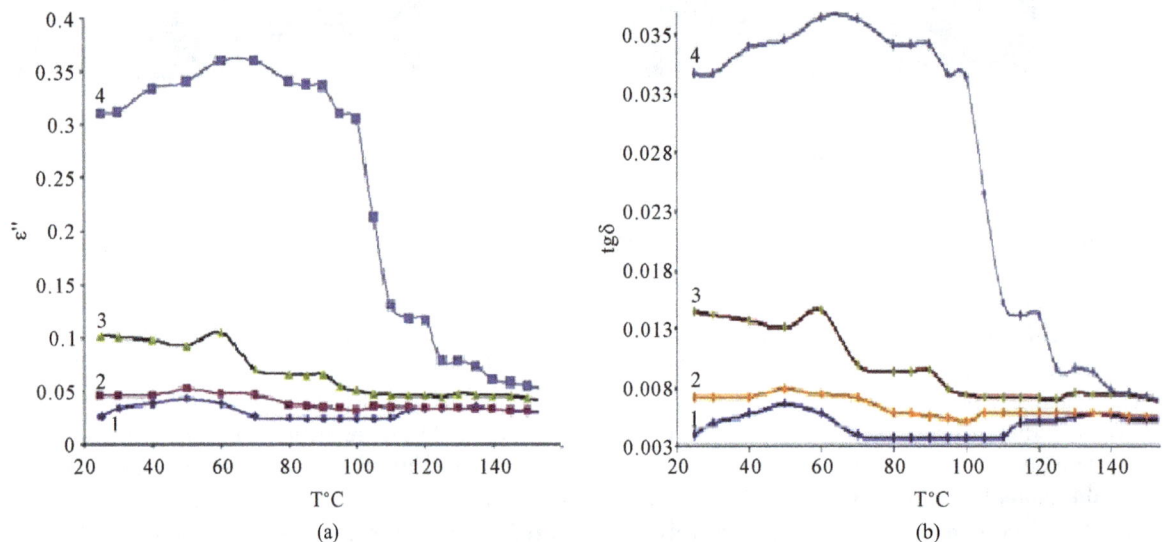

Figure 4. Temperature dependence of the imaginary part of dielectric permittivity (a) and tgδ (b) composite materials PE+xvol.%TlInSe$_2$, where 1) PE+1vol.%TlInSe$_2$, 2) PE+3vol.%TlInSe$_2$, 3) PE+5vol.%TlInSe$_2$, 4) PE+10vol.%TlInSe$_2$.

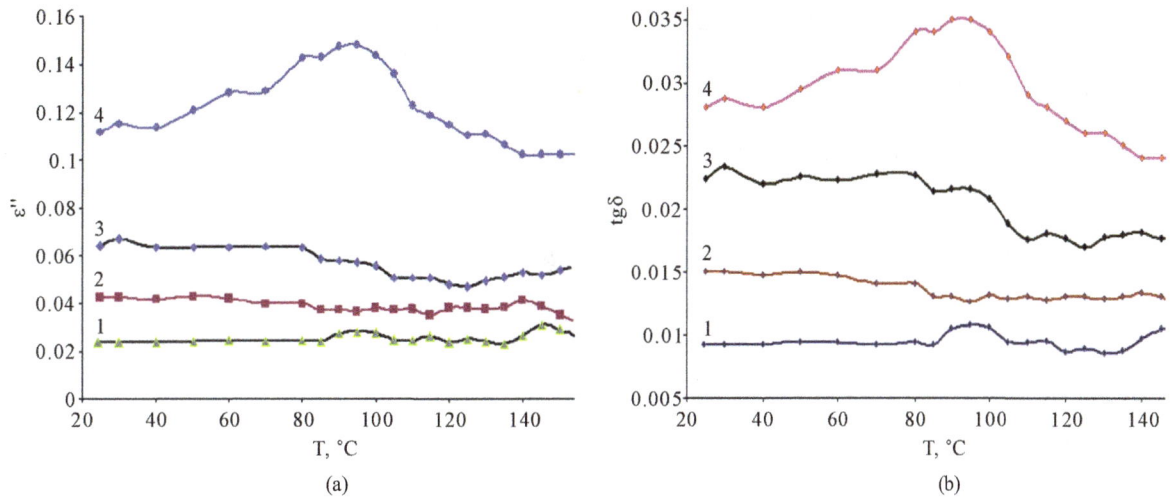

Figure 5. Temperature dependence of the imaginary part of dielectric permittivity (a) and tgδ (b) composite materials PE+xvol.%TlInSe$_2$<Al>, where 1) PE+3vol.%TlInSe$_2$+7vl.%Al, 2) PE+5vol.%TlInSe$_2$+5vl.%Al, 3) PE+7vol.%TlInSe$_2$+3vl.%Al, 4) PE+10vol.%TlInSe$_2$+10vl.%Al.

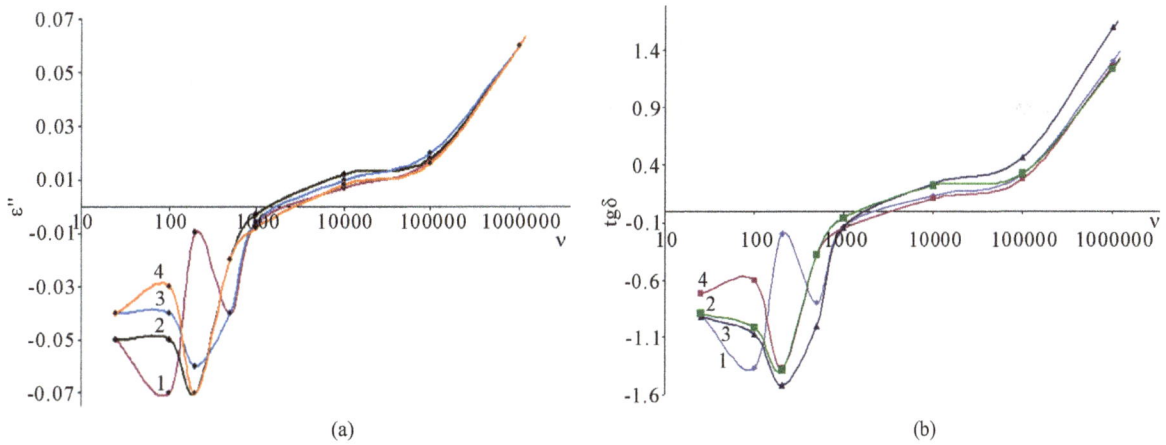

Figure 6. The frequency dependence of the imaginary part of dielectric permittivity (a) and tgδ (b) for composite materials PE+xvol.%TlInSe$_2$, where 1) PE+1vol.%TlInSe$_2$, 2) PE+10vol.%TlInSe$_2$, 3) PE+5vol.%TlInSe$_2$, 4) PE+3vol.%TlInSe$_2$.

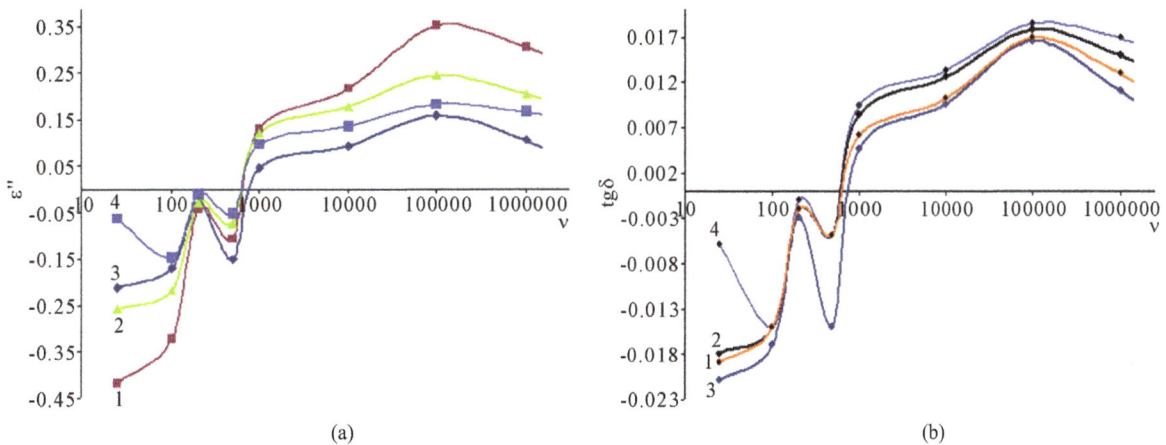

Figure 7. Frequency dependence of the imaginary part of dielectric permittivity (a) and tgδ (b) composite materials PE+xvol.%TlInSe$_2$<Al>, where 1) PE+5vol.%TlInSe$_2$+5%Al, 2) PE+3vol.%TlInSe$_2$+7%Al, 3) PE+7vol.%TlInSe$_2$+3%Al, 4) PE+10vol.%TlInSe$_2$10%Al.

As can be seen from **Figure 7(a)**, for these composite materials at high frequencies there is a slight change in ε''. Inversion of the sign occurs at a frequency of 1 kHz, a deep minimum at a frequency of 500 Hz and a pronounced maximum at a frequency of 200 Hz.

Figure 7(b) shows the results of a study of dielectric loss of composites with aluminum nano-particles depending on the frequency. Studies were carried out in the same frequency range. The results shown in **Figure 7(b)** imply that at the frequency of 800 Hz inversion of tgδ sign and at a frequency of 500 Hz deep minima, and at 200 Hz bright maximum is observed.

In general, the change in the imaginary part of the dielectric permittivity and dielectric loss frequency change occurs in a similar manner.

In a further increase in TlInSe$_2$ filler content to 10 vol% in the range 25°C - 100°C ε'' value increases for 6 times. In the range 100°C - 125°C ε'' decreases approximately four times and thereafter remains unchanged. For the investigated composites regular increase of ε'' with increasing filler content is observed.

Also the frequency dependence of ε'' has been investigated. The studies were conducted in the frequency range of 25 Hz - 1 MHz. Results of the study are shown in **Figure 4**.

As follows from figure, character of changes for ε'' depending on the frequency is the same for compositions with x = 3; 5 and 10. With an increase in frequency from 25 Hz to 200 Hz is revealed the deep minima, in the range of 200 Hz - 1 kHz sharp increase ε''. Further increase in frequency up to 1 MHz leads to increase in ε''. $\varepsilon''(v)$ for PE+1vol.% composite significantly different. At frequencies of 100 Hz and 500 Hz are observed deep minima, and between them at a frequency of 200 Hz bright maximum. In the frequency range of 500 Hz - 1 MHz character of change of ε'' is not different from other composites.

Analysis of the experimental results of the temperature dependence of dielectric permittivity of PE+xvol.% TlInSe$_2$ and PE+xvol.%TlInSe$_2$<Al> composites shows that in general, variation $\varepsilon''(t)$ in these materials are same. However, with increasing amounts of filler (**Figure 4(a)**) increasing ε'' values is observed. This behavior of the dielectric permittivity of the composite with semiconductor filler is largely determined by the Maxwell-Wagner polarization. The surface energy of the composite structure components becomes unstable and TlInSe$_2$ particles form clusters, the surface of which is less than the sum of the surface of their constituent particles. Increasing the number of clusters with increasing bulk filler content is accompanied by decrease in the dielectric layer between the particles. It leads to increase in electric capacity and accordingly ε''. In the composites obtained with the addition of aluminum nano-particles with size of 50 nm, the character of the dielectric permittivity variation does not change. However, with increasing of Al content in the composite the value of ε'' decreases throughout the temperature range studied. This is probably due to the fact that the aluminum nano-particles occupy vacancies-defects on the surface of the composites. This promotes the change in the electric resistance and the imaginary part of the dielectric permittivity. Using aluminum nano-particles allows obtaining composites having the desired dielectric permittivity and dielectric loss. Analysis of the frequency dependence of the dielectric permittivity and dielectric loss of PE+xvol.%TlInSe$_2$ composites and the same composite with aluminum nano-particles show that ε'' and tgδ significantly reduced. At introduction in the composites aluminum nano-particles isolated clusters is formed alongside with the semiconductor particles. Aluminum nano-particles—conductive clusters randomly distributed in the PE matrix. Increasing Al content in the composite leads to an increase in the number of nano-particles per cross section of the composite, which is equivalent to the proportion of Al in the total thickness of the composite—sample. Closed each other through-thickness sample clusters can be considered as active resistance connected between the electrodes. Since they have high compared with PE+xvol.%TlInSe$_2$ composites conductivity, it can be assumed that the resistance of the composite will be mainly determined by contacts between Al nano-particles. At the boundaries of clusters in the alternating electric field an accumulation and redistribution of free charges occurs, which changes the initial internal electric field. It is known that at low frequencies, internal electric fields are distributed accordingly conductivity and high frequencies—respectively the dielectric permittivity. Therefore, the decrease ε'' and tgδ with increasing content of aluminum nano-particles can be explained by the appearance of a relatively strong internal field in semiconductors and nano-clusters.

4. Conclusion

In conclusion, we study the materials at high frequencies and reveal that, the imaginary part of the permittivity is negative. In these frequencies, electronic and ionic polarization is in the main role. These materials have found application in the creation of new electronic devices.

References

[1]   Mironov, V. (2004) Fundamentals of the Scanning Probe Microscopy. Technosphere, Moscow, 197-201

[2]   Gojaev, E.M., Maharramov, A.M., Zeynalov, S.H.A., Osmanova, S.S. and Allakhyarov, E.A. (2010) Crown Electrets Based on High Density Polyethylene Composites with Semiconductor Filler $TlGaSe_2$. *Electronic Materials Processing*, No. 6, 266.

[3]   Zhigaeva, I.A. and Nikolaev, V.E. (2012) Study Based on Fluoropolymer Coronoelectrets-32L by Thermally Stimulated Relaxation of the Surface Potential. Education and Science Minister Modern Trends in Chemistry and Technology of Polymeric Materials. International Conference Abstracts St.-Petersburg, 38-39.

[4]   Mamedov, G.A., Godzhaev, E.M. and Magerramov, A.M. (2011) Study of Surface Topography by Atomic Force and Dielectric Properties of Compositions of High-Density Polyethylene with an Additive $TlGaSe_2$. *Electronic Materials Processing*, **47**, 94-98.

[5]   Gojayev, E.M., Safarova, S.S., Kafarova, D.M. Gulmammadov, K.D. and Ahmedova, J.R. (2013) Study of Surface Microrelief and Dielectric Properties of the Compositions of $PP+TlIn_0$, $98Ce_0$, $02Se_2$. *Electronic Materials Processing*, **49**, 267-271.

[6]   Godzhaev, E.M., Magerramov, A.M., Zeinalov, Sh.A., Osmanova, S.S. and Allakhyarov, E.A. (2011) Coronoelectrets Based on Composites of High Density Polyethylene with T_lGaSe_2 Semi-Conductor Filler. *Surface Engineering and Applied Electrochemistry*, **46**, 615-619. http://dx.doi.org/10.3103/S106837551006013X

17

Some Factors Influencing the Dielectric Properties of Natural Rubber Composites Containing Different Carbon Nanostructures

Ahmed A. Al-Ghamdi[1], Omar A. Al-Hartomy[1], Falleh R. Al-Solamy[2],
Nikolay Dishovsky[3*], Diana Zaimova[3], Rossitsa Shtarkova[4], Vladimir Iliev[5]

[1]Department of Physics, Faculty of Science, King Abdulaziz University, Jeddah, Saudi Arabia
[2]Department of Mathematics, Faculty of Science, King Abdulaziz University, Jeddah, Saudi Arabia
[3]Department of Polymer Engineering, University of Chemical Technology and Metallurgy, Sofia, Bulgaria
[4]Department of Chemistry, Technical University, Sofia, Bulgaria
[5]Department of Wireless Communications and Broadcasting, College of Telecommunications and Posts, Sofia, Bulgaria
Email: *dishov@uctm.edu

Abstract

Natural rubber based composites containing different carbon nanofillers (fullerenes, carbon nanotubes (CNTs) and graphene nanoplatelets (GNPs)) at different concentrations have been prepared. Their dielectric properties (dielectric permittivity, dielectric loss) have been studied in the 1 - 12 GHz frequency range. Some factors (electromagnetic field frequency, fillers concentration, fillers intrinsic structure) influencing the dielectric behavior of the composites have been investigated. The dielectric properties of the developed natural rubber composites containing conductive fillers (fullerenes, CNTs, GNPs) indicate that these composites can be used as broadband microwave absorbing materials.

Keywords

Carbon Nanostructure Materials, Composites, Natural Rubber, Dielectric Properties

1. Introduction

A growing number of demanding applications in electronics and telecommunications rely on the unique properties

---
*Corresponding author.

of carbon allotropes. The need for microwave absorbers and radar-absorbing materials is steadily increasing in military applications (reduction of radar signature of aircraft, ships, tanks, and targets) as well as in civilian applications (reduction of electromagnetic interference (EMI) among components and circuits, reduction of the back-radiation of microstrip radiators [1]).

In recent times researchers have tried three types of carbonaceous fillers, *i.e.*, carbon black (CB), carbon fiber (CF) and carbon nanotubes including multiwalled (CNTs/MWNTs) [2] with a suitable polymer matrix. In certain cases, pairs of fillers have been tried. The different structures and shapes of these conductive fillers and the morphologies of their dispersion will affect the ability to construct an effective conductive network, which is key to increasing the electrical conductivity of the polymer-filler composites [3] [4]. Some researchers [5] [6] have simultaneously introduced CB and CNTs into polymer matrices through conventional processing techniques.

A composite absorber that uses carbonaceous particles in combination with a polymer matrix offers a large flexibility for design and properties control, as the composite can be tuned and optimized via changes in both the carbonaceous inclusions (carbon black, carbon nanotube, carbon fiber, graphene) and the embedding matrix (rubber, thermoplastic) [1] [7]-[9].

Polymer composites containing conductive fillers have been developed in recent times as alternative EMI shielding materials since they possess the advantages like light weight, low cost, resistant to corrosion and processing advantages. However, in such materials, the EMI shielding effectiveness depends on many factors, including filler's intrinsic conductivity, filler loading, dielectric constant, aspect ratio and filler-polymer matrix interactions [10].

Dielectric properties of materials are those electrical characteristics of poorly conducting materials that determine their interaction with electric fields. Those properties determine how well energy from the high-frequency alternating electric fields can be absorbed and thus how rapidly the materials will be heated. The dielectric properties of the load materials are also important in the design of the radio frequency or microwave power equipment [11] [12].

The complex permittivity relative to free space may be represented as $\varepsilon = \varepsilon' - j\varepsilon''$, where ε' is the dielectric permittivity (dielectric constant) and ε'' is the dielectric loss factor. The real part of the permittivity represents the energy storage capability in the electric field in the dielectric material, and the imaginary part represents the energy dissipation capability of the dielectric by which energy from the electric field is converted into heat energy in the dielectric. Often, the loss angle of dielectrics is of interest, and the tangent of the loss angle δ is used (tan $\delta = \varepsilon''/\varepsilon'$) [13] [14].

In principle, the dielectric properties of most materials vary with several influencing factors. The dielectric properties depend on the frequency of the applied alternating electric field, the temperature and the water content of the material, its density, etc. The dielectric properties are dependent also on the presence of mobile ions and the permanent dipole moments. The distribution of the phases, including voids and cracks, has also a major influence on the dielectric properties of the composite materials [12] [15].

Among the fillers used in the last years, carbon-based nanostructures, such as fullerenes (Fs), carbon nanotubes (CNTs) and graphene nanoplatelets (GNPs) have been studied extensively. The reason for the interest in the above mentioned carbon nanostructures and their application in rubber composites can be explained with their unique properties (**Table 1**).

These materials have different aspect ratios and specific surface areas, and structures: CNTs are known to have high aspect ratio (several hundred to thousand), outstanding electrical and mechanical properties [19]-[21]. Also, GNPs are comprised of short stacks of platelet-shaped graphene sheets that are identical to those found in the walls of CNT, but in a planar form. GNPs can increase considerably the thermal conductivity and stability, and barrier properties [22] [23]. The fullerenes have a cage-like structure. Their major difference, compared to classical modification of carbon, is the molecule being a more or less punctiform, discrete unit instead of a structure repeating dimensionally through space [24]. It is obvious, that as Fs, CNTs, and GNPs have different characteristics, different geometries, aspect ratios, crystallography structure and physical properties, so their effects on the properties of polymer nanocomposites should be different, as shown in [25]-[28].

It is often difficult to explain most of the complex dielectric properties in the disordered composites [29]. Theoretical predictions of the fundamental physical data suffer from a lack of experimental data, especially dielectric data over a wide frequency range. We therefore conducted high resolution measurements of frequency-dependent dielectric properties in the carbonaceous fillers /natural rubber composites.

Table 1. Characteristics of some carbon nanostructures [16]-[18].

Material	Density g/cm^3	Specific surface area m^2/g	Length μm	Diameter or thickness nm	Aspect ratio	Electrical conductivity S/cm	Thermal conductivity W/mK
MWCNTs	2.6	50 - 1315	1 - 10	1 - 50	300 - 1000	10^2 - 10^6	2000 - 6000
GNPs	0.4	up to 2670	<50	<100	500	10^6	800 - 5300
Fs	1.7	50 - 500	>1	40 - 50	3000 - 5000	10^{-5}	0.4

The aim of this work is to prepare natural rubber based composites containing different amounts (2 - 10 phr) (phr = parts of filler by weight per hundred parts of rubber) of fullerenes, CNTs and GNPs; to analyze the role of some factors and specific structure features of these fillers on the microwave dielectric behavior of the composites with a view to propose lightweight, flexible materials suitable for microwave absorbers in wide frequency range (1 - 12 GHz).

2. Experimental

2.1. Materials

Natural rubber SMR 10 (Moony viscosity, ML$_{1+4}$ at 100˚C = 60; dielectric constant ε = 2.1) was purchased from North Special Rubber Corporation of Hengshui, Hebei Province, China.

Other ingredients such as zinc oxide (ZnO), stearic acid (SA), N-tert-Butyl-2-benzothiazolesulfenamide (TBBS) and sulphur (S) were commercial grade and used without further purification.

2.2. Characterization of the Fullerene Used

The investigations reported in this paper were on neat fullerene powder comprising 99.5% of C60 fullerene produced by Alfa Aesar (Johnson Matthey Company). Fullerene density was 1.65 g/cm^3. The micrograph in **Figure 1(a)** and the 50 nm marker on it show the fullerene particles to be aggregates of several hundred nm in size, e.g. 300 nm wide and 700 nm long, composed of initial particles about 50 - 60 nm large. The insert in **Figure 1(a)** is a micrograph taken in an electron diffraction regime of the aggregate which shows its having a crystal structure as the initial fullerene particles do.

2.3. Characterization of the Carbon Nanotubes Used

Multiwalled carbon nanotubes produced by Grafen Chemical Industries Co., Ankara, Turkey, were used in our investigation. The material's purity was more than 95%, density −2.6 g/cm^3. Carbon nanotubes were with average diameter about 15 nm and length 1 - 10 microns. **Figure 1(b)** shows the micrograph of the aggregate and the pattern taken in an electron diffraction regime showing that it has a crystal structure such as the primary CNTs particles have.

2.4. Characterization of the Graphene Used

Graphene used in the investigation was also produced by Grafen Chemical Industries Co., Ankara, Turkey. Graphene nanoplatelets (GNP) have a "platelet" morphology, meaning they have a very thin but wide aspect. Aspect ratios for this material can range into the thousands. Each particle consisted of several sheets of graphene with an overall thickness of 50 nm and average plate diameter 40 micron (**Figure 1(c)**), the density was 0.4 g/cm^3.

2.5. Preparation of Rubber Composites

Table 2 summarizes the formulation characteristics of the rubber compounds (in phr) used in the investigations.

The rubber compounds were prepared according to a specific recipe. The pre-characterized filler powder was incorporated into the natural rubber matrix at various loadings on an open two-roll laboratory mill (L/D 320 × 360 and friction 1.27). The speed of the slow roll was 25 min^{-1}. The formulations of the compounds prepared are shown in **Table 2**. The compounding was carried out as follows: the raw rubber was loaded into the mill;

Table 2. Compositions of rubber compounds (in phr).

	NR1	NR2	NR3
Natural rubber	100	100	100
Stearic acid	1	1	1
Zinc oxide	4	4	4
Processing oil	10	10	10
Filler (Fs, CNTs, GNPs)	2	6	10
MBTS[a]	2	2	2
TMTD[b]	1	1	1
Vulkanox 4020[c]	1	1	1
Sulphur	2	2	2

[a]Mercaptobenzothiazole sulphenamide; [b]Tetramethylthiuram disulphide; [c]Dimethyl butyl-phenyl-p-phenylendiamine.

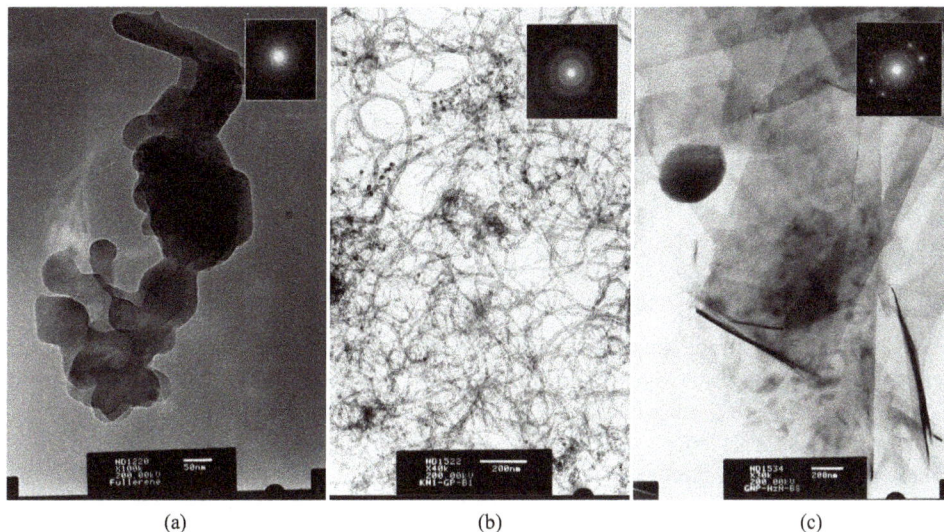

Figure 1. Micrographs of fullerene aggregates (a); CNT aggregates (b) and GNPs aggregates (c) in transmission regime and in selected area electron diffraction regime (inserts).

ZnO and stearic acid were added after 5 min. After 3 min of homogenization the filler was added. Following another homogenization for 7 min the accelerator and sulphur were added and the compound was rehomogenized for 4 min. The temperature of the rolls did not exceed 70°C. The experiments were repeated for verifying the statistical significance. The ready compounds in the form of sheets stayed 24 hours prior to their vulcanization. The optimal vulcanization time was determined by the vulcanization isotherms taken on an oscillating disc vulcameter MDR 2000 (Alpha Technologies) at 150°C, according to ISO 3417:2002. The vulcanization was performed on an electric hydraulic press which plate was 400 × 400 mm, at 10 MPa.

The properties of the composites obtained—dielectric constant and dielectric loss—were determined in the 1 - 12 GHz frequency range.

2.6. Measurements

2.6.1. Complex Permittivity

The determination of complex permittivity was carried out by the resonance method, based on the cavity perturbation technique [30]. The resonance frequency of an empty cavity resonator f_r was measured. After that the sample material was placed into the resonator and the shift in resonance frequency f_ε was registered. The real part of permittivity ε'_r (dielectric constant) was calculated from the shift in resonance frequency, cavity and the sample cross sections, S_r and S_ε, respectively:

$$\varepsilon'_r = 1 + \frac{S_r}{2S_\varepsilon} \cdot \frac{f_r - f_\varepsilon}{f_r} \tag{1}$$

The sample was in the form of a disc with a diameter of 11 mm and about 1.5 mm thick. Its location in the cavity was at the maximum electric field. Because the thickness of the sample was not equal to the height of the resonator, a dielectric occurred with an equivalent permittivity ε_e at the place of its inclusion. The parameter was determined by Equation (1) and instead ε'_r was saved ε_e. Then ε'_r was determined by

$$\varepsilon'_r \approx \varepsilon_e (\kappa + 1) - \kappa \left(\Delta \ll 1 \right) \tag{2}$$

where $k = l/\Delta$ and l is the distance from the disk to the top of the resonator.

2.6.2. Loss Factor tanδ

The loss factor tan δ was calculated from the quality factor of the cavity [31]

$$\tan \delta = \frac{1}{4\varepsilon_r} \cdot \frac{S_r}{S_\varepsilon} \left(\frac{1}{Q_\varepsilon} - \frac{1}{Q_r} \right) \tag{3}$$

where Q_ε—quality factor of the cavity with a sample and Q_r—quality factor of the cavity without a sample.

The measurement setup used several generators for the whole range: HP686A and G4-79 to 82, frequency meters: H 532A; FS-54, a cavity resonator. The scheme of the equipment used is shown on **Figure 2**.

The dielectric properties were measured in the frequency range from 1 GHz to 12 GHz.

3. Results and Discussion

3.1. Dielectric Properties Frequency Dependence

The frequency dependence of the dielectric permittivity of NR based composites, containing a different amount of fullerenes, CNTs and GNPs is shown in **Figures 3-5**. It is evident that the dielectric permittivity values increase with the increasing frequency in the case of any of the fillers used. In the 1 - 7 GHz range the increase is slightly pronounced while in the 7 - 12 GHz range it is more drastic. The specifics of the fillers used are expressed in the slope of plotted curves.

Figures 6-8 present the dielectric loss in the 1 - 12 GHz frequency range of the composites investigated. The changes in the imaginary part of the relative complex permittivity of the material also known as dielectric loss angle tangent—tan δ_ε—depend on the frequency. As expected, with the increasing frequency the dielectric loss decreases, more rapidly in the region 5 - 12 GHz. Evidently, when GNPs are used as filler, the frequency increase has a stronger effect on the dielectric losses of the samples at lower frequencies (1 - 4 GHz). Noteworthy is the fact that in the 11 - 12 GHz frequency range the values of dielectric losses of the samples examined become closer.

The results in **Figures 3-8** allow the conclusion that the dielectric properties of most materials vary considerably with the frequency of the applied electromagnetic field. An important phenomenon contributing to the frequency dependence of the dielectric properties is the polarization, arising from the orientation of molecules which have permanent dipole moments with the imposed electric field. The mathematical formulation developed by Debye to describe the permittivity for polar materials [32] can be expressed as

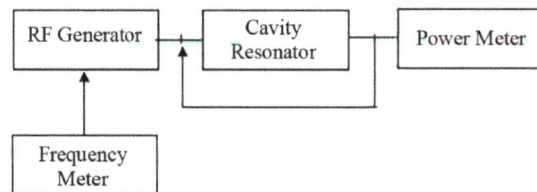

Figure 2. Scheme of the equipment for measuring the dielectric properties.

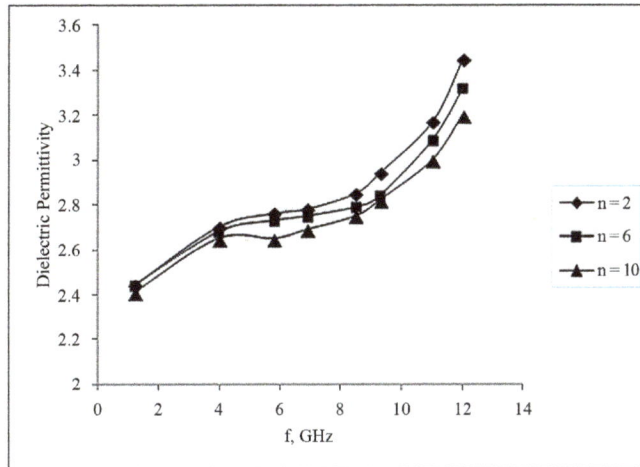

Figure 3. Frequency dependence of dielectric permittivity ε'_r at a various filler content (n-phr of fullerenes).

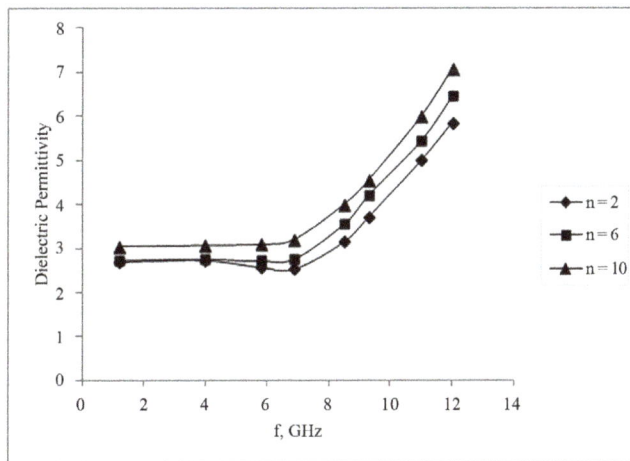

Figure 4. Frequency dependence of dielectric permittivity ε'_r at a various filler content (n-phr of CNTs).

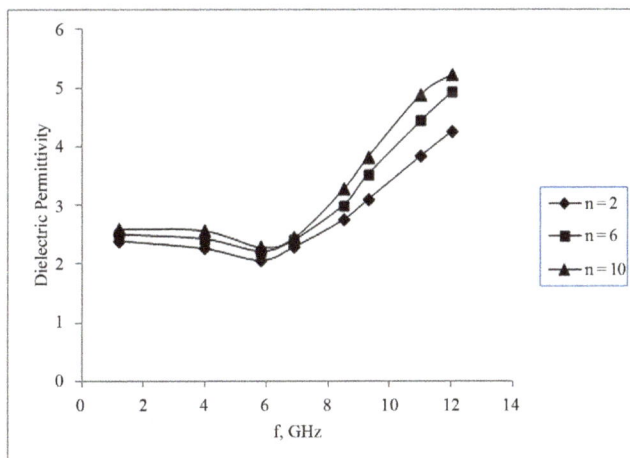

Figure 5. Frequency dependence of dielectric permittivity ε'_r at a various filler content (n-phr of GNPs).

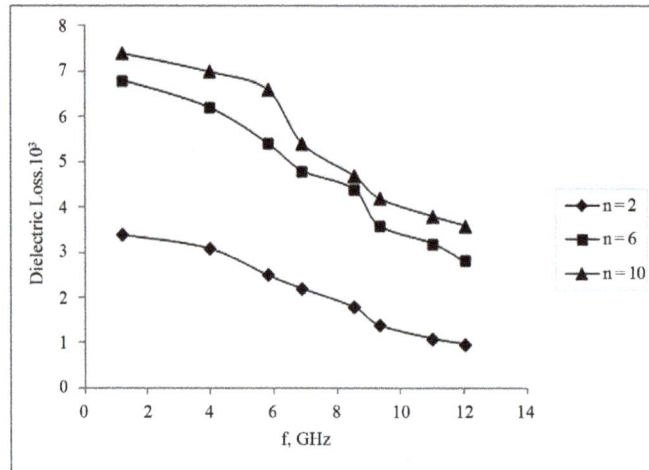

Figure 6. Frequency dependence of dielectric loss tanδ at a various filler content (n-phr of fullerenes).

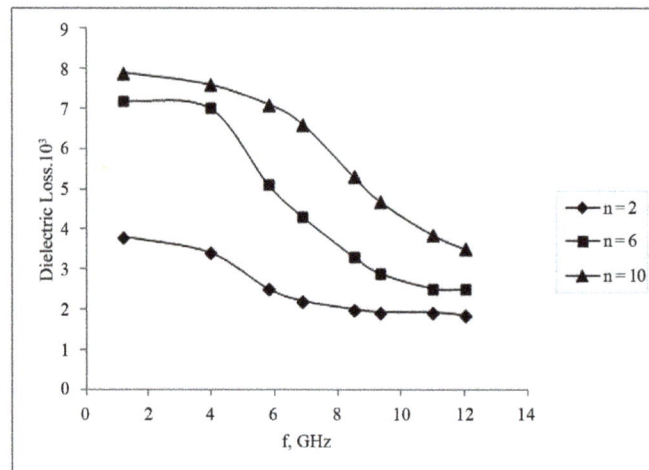

Figure 7. Frequency dependence of dielectric loss tanδ at a various filler content (n-phr of CNTs).

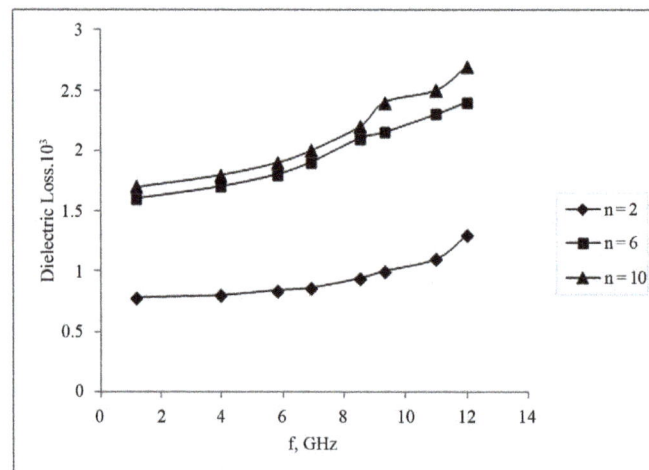

Figure 8. Frequency dependence of dielectric loss tanδ at a various filler content (n-phr of GNPs).

$$\varepsilon = \frac{\varepsilon_s - \varepsilon_\infty}{1 + j\omega\tau} \qquad (4)$$

where $\omega = 2\pi f$ is the angular frequency, ε_∞ represents the dielectric constant at frequencies so high that molecular orientation does not have time to contribute to the polarization, ε_s represents the static dielectric constant, *i.e.*, the value at zero frequency (dc value), and τ is the relaxation time in seconds, the period associated with the time for the dipoles to revert to random orientation when the electric field is removed. Separation of Equation (4) into its real and imaginary parts yields

$$\varepsilon' = \varepsilon_\infty + \frac{\varepsilon_s - \varepsilon_\infty}{1 + (\omega\tau)^2} \qquad (5)$$

$$\varepsilon'' = \frac{(\varepsilon_s - \varepsilon_\infty)\omega\tau}{1 + (\omega\tau)^2} \qquad (6)$$

Obviously, ε_∞, ε_s and in particular, τ are specific parameters for each of the fillers used, which depend on the fillers chemical and crystallographic nature. Therefore, the dielectric properties of the composites studied depend differently on the frequency, provided all other conditions are identical (the same polymer matrix and filling degree). The dependence of ε_∞, ε_s and τ on the fillers chemical nature and structure explains why in some frequency range (usually at lower frequencies) the dielectric properties change monotonously, while in some higher frequency range, when relaxation is hindered, the increase is drastic. According to the Debye theory of dielectric properties [31], ε'' is generally determined by relaxation and electrical conductivity losses. It is clear that both polarization relaxation and electrical conductance can affect ε''.

The Debye relaxation is one of the important dielectric loss mechanisms, which can be characterized by ε'' - ε' relationship. **Figure 9** presents the dependence of the dielectric loss ε'' on dielectric permittivity ε' for the NR based composites comprising different carbon nanostructures. As seen from this figure, for each of the fillers the dependence is strictly specific and differing from the rest. For instance, in the case of composites filled with GNPs the dielectric loss remains practically the same in a quite large interval of dielectric permittivity values (from 2.5 to 5.5). Moreover, there are regions of almost linear decrease of the dielectric loss with the increasing dielectric permittivity values for those composites. The dependence curve for the composites comprising

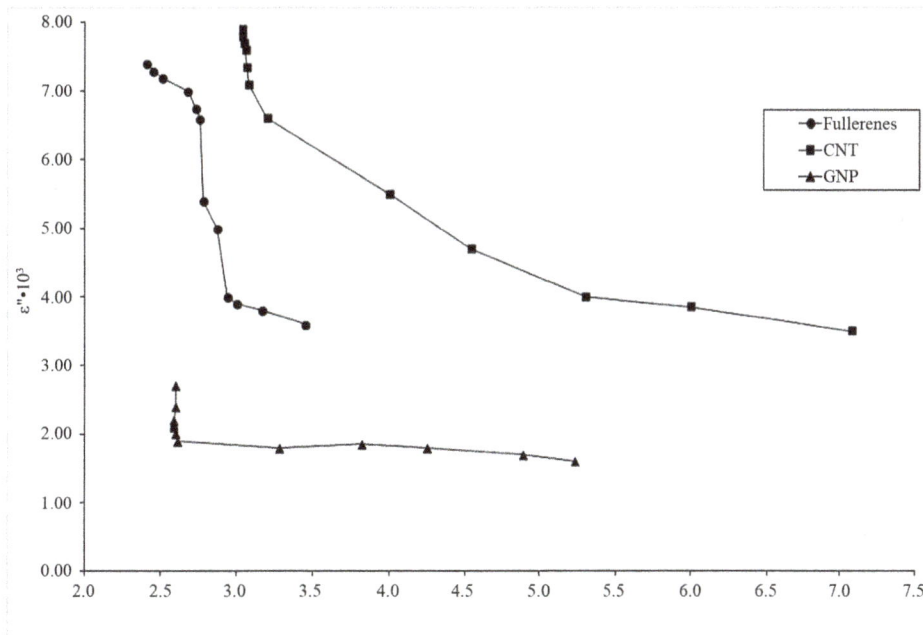

Figure 9. Dependence of dielectric loss on the dielectric permittivity values for composites filled with nanostructures of different chemical nature at 10 phr, studied in 1 - 12 GHz frequency range.

fullerenes has an S-shaped pattern with two regions wherein the dielectric loss values decrease almost linearly with the increasing dielectric permittivity, in the third region of curve that decrease is intermittent. As **Figure 9** shows, the relaxation proceeds in a different manner. For each of the studied fillers the real and imaginary parts of the complex dielectric permittivity change at different intervals. As all other conditions are identical, obviously, that peculiarity is due to the specific chemical nature and crystallographic structure of the fillers.

On the other hand, since the elastomer matrix is non-polar, apparently, the relaxation and polarization processes are greatly dependent on the fillers used and on their specific features. With regard to those specifics, the polarization may proceed according to three different mechanisms: electronic, ionic and orientational. All non-conducting materials are capable of electronic polarization. Therefore, we consider the polarization of the elastomer matrix used for the studied composites to proceed according to that mechanism. The ionic and orientational polarizations occur only in materials possessing ions and permanent dipoles, respectively. It might be assumed that the matrix and the fillers introduced into it polarize according to a different mechanism. The more polarization mechanisms of a composite are, the higher its dielectric constant will be. For example, a natural rubber based composite containing fillers with permanent dipoles have a dielectric permittivity higher than the one of non polar natural rubber. On the other hand, the more easily the various polarization mechanisms can act, the higher the dielectric permittivity will be. For example, among elastomers, the more mobile the chains are (*i.e.* the lower the degree of crystallinity), the higher the dielectric permittivity will be. It is important for the composites investigated, because the natural rubber crystallizes and the chemical nature of the fillers used and their amounts can change the degree of crystallinity and the dielectric permittivity. For polar structures the magnitude of the dipole also affects the magnitude of the occurring polarization. The fillers with non-centrosymmetric structure have especially strong spontaneous polarization, hence a high dielectric permittivity, respectively.

3.2. Dielectric Properties Filler Concentration Dependence

It is obvious from **Figures 3-8** that, in the investigated concentration interval, the values of the dielectric permittivity are too close to each other with a tendency to a slight increase with the increasing filler concentration. A similar tendency has been described by the authors [28] who have investigated the change in the dielectric permittivity of natural rubber based composites filled with fullerenes in the concentration interval from 0.065 to 0.75 phr. They have found an increase in those values from 4.8 to 5.2, respectively, with the increasing concentration; however, that investigation refers to a frequency of only 50 Hz.

When CNTs are used as fillers (**Figure 4**) the dielectric permittivity values are relatively close at lower frequencies (up to 7 GHz). At frequencies higher than 7 GHz the dielectric permittivity increases with the increasing filler concentration.

Of particular interest is the 9 - 12 GHz range wherein there is a relatively fast increase in the dielectric permittivity and its dependence on the amount of GNPs is the most prominent.

From the very beginning of that range the values of tanδ are clearly distinguished as dependent on the different degree of filling. As expected, the increase in filler amount leads to an increase in tanδ_ε values (**Figures 6-8**).

The imaginary part of the complex relative permittivity is more sensitive to changes in the filler amount than the real part is.

3.3. Dielectric Properties Physical Structure and Chemical Nature Dependence

Carbon nanofillers used have different geometries and exhibit different surface area/volume relations [24]. Some of their specific features are shown in **Table 1**.

The investigations carried out and the obtained results (**Figures 3-9**) show irrevocably the impact that the crystallographic structure and the chemical nature of the used fillers have upon the dielectric permittivity values and dielectric loss. The effect could be explained first of all by the polarization and by the mechanism according to which the polarization proceeds, as well as by the proceeding of relaxation and the time needed for the process. The aforementioned two processes are crucial for the dielectric properties of the materials when applying an electromagnetic field. The difference in the crystallographic structure and chemical nature of the fillers predetermine the differences in the dielectric permittivity values and dielectric losses of composites based on the same matrix. The effects of the two factors (crystallographic structure and chemical nature) sometimes intermingle and are hardly distinguishable. That could be the scope of future investigations.

4. Conclusion

Natural rubber (NR) based composites containing different carbon nanofillers (fullerenes, carbon nanotubes and graphene nanoplatelets) at concentrations from 2 to 10 phr have been prepared. Their dielectric properties (dielectric permittivity, dielectric loss) have been studied in the 1 - 12 GHz frequency range. It was found that the dielectric constant and dielectric loss of filled composites depend on amount and type of filler loading. It has been established that with increasing the frequency of the electromagnetic field the dielectric permittivity values increase, while those of the dielectric losses get lower. The higher the concentration of the fillers is (keeping identical all other conditions), the higher the values of dielectric permittivity and dielectric losses are. The observed effects are related first of all to the impact that the chemical nature and crystallographic structure of the fillers and elastomer matrix studied have upon the polarization time and mechanism as well as upon the relaxation time and mechanism. The latter processes determine the dielectric properties of the materials. The dielectric properties of the developed natural rubber composites containing conductive fillers (fullerenes, CNTs, GNPs) indicate that these composites can be used as broadband microwave absorbing materials.

Acknowledgements

The work is a part of a project funded by King Abdulaziz University, Saudi Arabia under grant number MB/11/12/436. The authors acknowledge the technical and financial support.

References

[1] Qin, F. and Brosseau, C. (2012) A Review and Analysis of Microwave Absorption in Polymer Composites Filled with Carbonaceous Particles. *Journal of Applied Physics*, **111**, 061301. http://dx.doi.org/10.1063/1.3688435

[2] Dinesh, P., Renukappa, N., Pasang, T., Dinesh, M. and Ranganathaiah, C. (2014) Effect of Nanofillers on Conductivity and Electromagnetic Interference Shielding Effectiveness of High Density Polyethylene and Polypropylene Nanocomposites. *European Journal of Advances in Engineering and Technology*, **1**, 16-28.

[3] Shang, S.M., Zeng, W. and Tao, X.M. (2010) Highly Stretchable Conductive Polymer Composited With Carbon Nanotubes and Nanospheres. *Advanced Materials Research*, **123**, 109-112. http://dx.doi.org/10.4028/www.scientific.net/AMR.123-125.109

[4] Spitalsky, Z., Tasis, D., Papagelis, K. and Galiotis, C. (2010) Carbon Nanotube-Polymer Composites: Chemistry, Processing, Mechanical and Electrical Properties. *Progress in Polymer Science*, **35**, 357-401. http://dx.doi.org/10.1016/j.progpolymsci.2009.09.003

[5] Sun, Y., Bao, H.D., Guo, Z.X. and Yu, J. (2009) Modeling of the Electrical Percolation of Mixed Carbon Fillers in Polymer-Based Composites. *Macromolecules*, **42**, 459-463. http://dx.doi.org/10.1021/ma8023188

[6] Jeevanand, T., Kim, N.H., Lee, J.H., Siddaramaiah, B. Deepa Urs, M.V. and Ranganathaiah, C. (2009) Investigation of Multi-Walled Carbon Nanotube Reinforced High-Density Polyethylene/Carbon Black Nanocomposites Using Electrical, DSC and Positron Lifetime Spectroscopy Technique. *Polymer International*, **58**, 775-780. http://dx.doi.org/10.1002/pi.2591

[7] Yang, C., Lin, Y. and Nan, C.W. (2009) Modified Carbon Nanotube Composites with High Dielectric Constant, Low Dielectric Loss and Large Energy Density. *Carbon*, **47**, 1096-1101. http://dx.doi.org/10.1016/j.carbon.2008.12.037

[8] Pierantoni, L., Mencarelli, D., Bozzi, M., Moro, R. and Bellucci, S. (2014) Graphene-Based Electronically Tuneable Microstrip Attenuator. *IEEE MTT-S International Microwave Symposium (IMS)*, Tampa, 1-6 June 2014, 1-3. http://dx.doi.org/10.1109/MWSYM.2014.6848645

[9] De Bellis, G., De Rosa, I.M., Dinescu, A., Sarto, M.S. and Tamburrano, A. (2010) Electromagnetic Absorbing Nanocomposites Including Carbon Fibers, Nanotubes and Graphene Nanoplatelets. *IEEE International Symposium on Electromagnetic Compatibility*, Fort Lauderdale, 25-30 July 2010, 202-207. http://dx.doi.org/10.1109/ISEMC.2010.5711272

[10] Sohi, N.J.S., Rahaman, M. and Khastgir, D. (2011) Dielectric Property and Electromagnetic Interference Shielding Effectiveness of Ethylene Vinyl Acetate-Based Conductive Composites: Effect of Different Type of Carbon Fillers. *Polymer Composites*, **32**, 1148-1154. http://dx.doi.org/10.1002/pc.21133

[11] Peng, Z.H., Peng, J.C., Peng, Y.F. and Wang, J.Y. (2008) Complex Conductivity and Permittivity of Single Wall Carbon Nanotubes/Polymer Composite at Microwave Frequencies: A Theoretical Estimation. *Chinese Science Bulletin*, **53**, 3497-3504. http://dx.doi.org/10.1007/s11434-008-0486-z

[12] Liu, L., Kong, L.B., Yin, W.Y., Chen, Y. and Matitsine, S. (2010) Microwave Dielectric Properties of Carbon Nanotube Composites. In: Marulanda, J.M., Ed., *Carbon Nanotubes*, InTech, Rijeka, 93-108.

http://dx.doi.org/10.5772/39420

[13] Pozar, D. (2012) Microwave Engineering. Fourth Edition, John Wiley &Sons, Hoboken.

[14] Dimiev, A., Zakhidov, D., Genorio, B., Oladimeji, K., Crowgey, B., Kempel, L., Rothwell, E.J. and Tour, J.M. (2013) Permittivity of Dielectric Composite Materials Comprising Graphene Nanoribbons. The Effect of Nanostructure. *ACS Applied Materials & Interfaces*, **5**, 7567-7573. http://dx.doi.org/10.1021/am401859j

[15] Raihan, R., Rabbi, F., Vadlamudi, V. and Reifsnider, K. (2015) Composite Materials Damage Modeling Based on Dielectric Properties. *Materials Sciences and Applications*, **6**, 1033-1053.

[16] Li, Y., Zhu, J., Wei, S., Ryu, J., Wang, Q., Sun, L. and Guo, Z. (2011) Poly(propylene) Nanocomposites Containing Various Carbon Nanostructures. *Macromolecular Chemistry and Physics*, **212**, 2429-2438.

[17] Zhou, Z., Wa, S., Zhang, Y. and Zhang, Y. (2006) Effect of Different Carbon Fillers on the Properties of PP Composites: Comparison of Carbon Black with Multiwalled Carbon Nanotubes. *Journal of Applied Polymer Science*, **102**, 4823-4830. http://dx.doi.org/10.1002/app.24722

[18] Thostenson, E., Li, C. and Chou, T. (2005) Nanocomposites in Context. *Composites Science and Technology*, **65**, 491-516. http://dx.doi.org/10.1016/j.compscitech.2004.11.003

[19] Guldi, D. and Martin, M. (Eds.) (2010) Carbon Nanotubes and Related Structures. Wiley-VCH Verlag, Weinheim. http://dx.doi.org/10.1002/9783527629930

[20] Tjong, S. (2009) Carbon Nanotube Reinforced Composites. Wiley-VCH Verlag, Weinheim. http://dx.doi.org/10.1002/9783527626991

[21] Harris, P. (2009) Carbon Nanotube Science-Synthesis, Properties and Application. Cambridge University Press, Cambridge. http://dx.doi.org/10.1017/CBO9780511609701

[22] Nieto, A., Lahiri, D. and Agarwal, A. (2012) Sythesis and Properties of Bulk Graphene Nanoplatelets Consolidated by Spark Plasma Sintering. *Carbon*, **50**, 4068-4077. http://dx.doi.org/10.1016/j.carbon.2012.04.054

[23] Yadav, S. and Cho, J. (2013) Functional Graphene Nanoplatelets for Enhanced Mechanical and Thermal Properties. *Applied Surface Science*, **266**, 360-367. http://dx.doi.org/10.1016/j.apsusc.2012.12.028

[24] Krueger, A. (2010) Carbon Materials and Nanotechnology. Wiley-VCH Verlag, Weinheim. http://dx.doi.org/10.1002/9783527629602

[25] Moniruzzaman, M. and Winey, K. (2006) Polymer Nanocomposites Containing Carbon Nanotubes. *Macromolecules*, **39**, 5194-5205. http://dx.doi.org/10.1021/ma060733p

[26] Li, Y., Zhu, J., Wei, S., Ryu, J., Sun, L. and Gu, Z. (2011) Poly(propylene)/Graphene Nanoplatelet Nanocomposites: Melt Rheological Behavior and Thermal, Electrical, and Electronic Properties. *Macromolecular Chemistry and Physics*, **212**, 1951-1959. http://dx.doi.org/10.1002/macp.201100263

[27] Potts, J., Dreyer, D., Bielawski, C. and Ruoff, R. (2011) Graphene-Based Polymer Nanocomposites. *Polymer*, **52**, 5-25. http://dx.doi.org/10.1016/j.polymer.2010.11.042

[28] Jurkovska, B., Jurkovski, B., Kamrovski, P., Pesetskii, S., Koval, V., Pinchuk, L. and Olkhov, Y. (2006) Properties of Fullerene-Containing Natural Rubber. *Journal of Applied Polymer Science*, **100**, 390-398. http://dx.doi.org/10.1002/app.22721

[29] Jiang, M.-J., Dang, Z.-M., Bozlar, M., Miomandre, F. and Bai, J. (2009) Broad-Frequency Dielectric Behaviors in Multi-Walled Carbon Nanotubes/Rubber Nanocomposites. *Journal of Applied Physics*, **106**, Article ID: 084902. http://dx.doi.org/10.1063/1.3238306

[30] Meng, B., Booske, J. and Cooper, R. (1995) Extended Cavity Perturbation Technique to Determine the Complex Permittivity of the Dielectric Materials. *IEEE Transactions on Microwave Theory and Techniques*, **43**, 2633-2636. http://dx.doi.org/10.1109/22.473190

[31] Kumar, A. and Sharma, S. (2007) Measurement of Dielectric Constant and Loss Factor of the Dielectric Material at Microwave Frequencies. *Progress in Electromagnetics Research*, **69**, 47-54. http://dx.doi.org/10.2528/PIER06111204

[32] Debye, P. (1929) Polar Molecules. The Chemical Catalogue Company, New York.

18

Structural and Dielectric Studies of Gd Doped ZnO Nanocrystals at Room Temperature

P. U. Aparna, N. K. Divya, P. P. Pradyumnan*

Department of Physics, University of Calicut, Malappuram, India
Email: *drpradyumnan@gmail.com

Abstract

Gadolinium doped Zinc oxide ($Zn_{1-x}Gd_xO$) nanocrystals with different percentage of Gd content (x = 0, 0.2, 0.4, 0.6, 0.8) have been prepared by the solid state reaction method. The structural, morphological and chemical studies of the samples were performed by X-ray diffraction (XRD), Scanning electron microscope (SEM) and Energy dispersive X-ray (EDX) analysis. The XRD spectra confirm that all the samples have hexagonal wurtzite structure. Decrease in average crystallite size with an increase in Gd concentration is observed in XRD. SEM images show that the grain size of undoped ZnO is larger than the Gd doped ZnO, specifying the hindrance of grain growth upon Gd doping. The chemical composition of the samples was confirmed using Energy dispersive X-ray (EDX) analysis. The variation of dielectric constant (ε_r), dielectric loss (tan δ) and AC conductivity as a function of frequency is studied at room temperature in a frequency which ranges from 100 Hz - 4.5 MHz by using LCR Hi TESTER. All the samples exhibit the normal dielectric behavior, *i.e.* decreases with increase in frequency which has been explained in the light of Maxwell-Wagner model. The dielectric constant and dielectric loss can be varied intensely by tuning Gd concentration in $Zn_{1-x}Gd_xO$ compounds.

Keywords

XRD, SEM, EDX, Dielectric Constant, AC Conductivity, Maxwell-Wagner Model

1. Introduction

ZnO is a versatile semiconductor having a wide band gap of 3.37 eV and large exciton energy of 60 meV which

---

*Corresponding author.

crystallizes in hexagonal wurtzite structure. Due to its unique physical and chemical properties, it has a wide spread application in solar cells, gas sensors, UV light emitters and surface acoustic wave (SAW) devices [1]-[3]. The lack of centre of symmetry in the wurtzite structure of ZnO crystals give rise to its piezoelectric and pyroelectric properties [2]. Therefore ZnO can be used for piezoelectric and pyroelectric applications such as transducer, actuator, IR sensors and energy generator. Some of the promising features of ZnO include its radiation hardness, biocompatibility and its high transparency in the visible region.

Doping ZnO with rare earth ions is of great interest for optoelectronics and spintronic applications [4]. The peculiar properties exhibited by rare earth ions are due to its intra f-shell transition. A good number of reports in the literature propose that doping with rare earth elements causes an enhancement in optical and magnetic properties of ZnO. K. Jayanthi et al. [5] observed a sharp and intense visible line emission from Nd^{3+} doped ZnO nanopowders. Achamma et al. [6] studied the luminescence properties in Ce doped ZnO nanocrystals. John and Rajakumari [7] reported the presence of ferromagnetic properties in Er doped ZnO nanocrystals. Reports on the dielectric and ferroelectric studies of rare earth ion doped ZnO are found limited. Nidhis Sinha et al. [8] reported about the dielectric and ferroelectric properties of Ce doped ZnO nanorods. Recently, Divya et al. [9] [10] studied the optical and dielectric properties of Er doped ZnO. The study of dielectric properties is concerned with the storage and dissipation of electric and magnetic energy in materials which helps to improve the design and quality of the devices. Many workers have studied the structural, optical and magnetic properties of Gd doped ZnO [11]-[16]. But dielectric studies on Gd doped ZnO are not found elsewhere. The aim of the present work is to synthesize Gd doped ZnO via the solid state reaction route and to examine whether the dielectric constant of these materials can be enhanced compared to pure ZnO. In this paper, we have investigated the effect of Gd doping in ZnO on its structural, morphological and dielectric properties.

2. Experimental Details

Gd doped ZnO were synthesized by solid state reaction route to study their structural and dielectric properties. The chemicals used in the experiment are ZnO (99.99% pure), Gd_2O_3 (99.99% pure) and $LiOH \cdot H_2O$ (99.99% pure). These chemicals were weighed using an electronic balance in accordance with the required stoichiometry. These materials were homogeneously mixed using an agate mortar for sufficient time to get fine powders. LiOH is an inorganic and water soluble compound used as a heat transfer medium for the synthesis of Gd doped ZnO. The prepared samples were mixed with ethanol and made into slurry. It is then dried in an oven for 1 hour at 100°C. After drying, the mixture was ground for 1 hour and made into pellets using hydraulic pelletizer. These pellets were sintered at 900°C for 4 hours in a high temperature furnace. The pellets were again ground and used as samples for the studies.

The XRD patterns of powder samples were attained by Rigaku Miniflex 600 X-ray diffractometer. Surface morphology and chemical composition of the samples were respectively examined by scanning electron microscopy (SEM) and energy dispersive X-ray analysis (EDX). For dielectric and AC conductivity measurements, the powder samples were made into pellets using hydraulic pelletizer of thickness 1 - 2 mm and of diameter 10 mm by applying pressure of 130 Kg-cm^2 for 1 minute. The dielectric constant, dielectric loss and AC conductivity were measured using HIOKI 3532-50 LCR Hi TESTER for a frequency range from 100 Hz to 4.5 MHz.

3. Result and Discussion

3.1. XRD Analysis

X-ray diffraction pattern of undoped and gadolinium doped ZnO, $Zn_{1-x}Gd_xO$ whereas ($x = 0$, $x = 0.2$, $x = 0.4$, $x = 0.6$, $x = 0.8$) are shown in **Figure 1**. These patterns have been compared with standard JCPDS data. All the XRD peaks can be indexed to a wurtzite structure of ZnO. It is found that up to 0.6 wt.% of Gd doping, there were no extra peaks which implies that Gd^{3+} ion perfectly replaced Zn^{2+} ion in the crystal matrix. Further increase of Gd^{3+} ($x \geq 0.8$), extra peaks arises and is shown in **Figure 2**.

It is noticeable that the XRD peaks of doped samples were found shifted towards the higher 2θ values. The peak shift observed in the XRD pattern for the peak corresponding to the plane (101) depicted in **Figure 3**. This shift is mainly due to the incorporation of larger sized Gd^{3+} ($r = 0.94$Å) than Zn^{2+} ($r = 0.74$Å) in the hexagonal wurtzite structure. This indicates that the doped Gd atoms substitute Zn atoms and the crystal structure remains unchanged.

Variation of lattice parameters with Gd concentration is shown in **Table 1**. Crystallite sizes were also estimated from XRD pattern using the Scherrer formula.

$$t = K\lambda/\beta\cos\theta , \tag{1}$$

where t is the crystallite size, K is the shape factor, λ is the wavelength of the incident X-ray radiation, θ is the bragg angle and β is the full width at half maximum (FWHM) in radian of the peak with given (hkl) value.

It was found that crystallite size decreases with increasing Gd concentration (up to 0.6 wt.% of Gd) as shown in **Table 2**. This is due to the distortion of host ZnO lattice by Gd^{3+} ions, which actually reduces the nucleation and subsequent growth rate of ZnO crystal. Similar results were reported by D. Mithal *et al.* [17], a decrease of crystallite size with increase in concentration of Gd in $Zn_{1-x}Gd_xO$.

It could be noted that the crystallite size of 0.8 wt.% Gd doped ZnO is larger when compared to other concentrations and it tending to the size of pure ZnO crystallites on higher doping percentage. This shows that the solu-

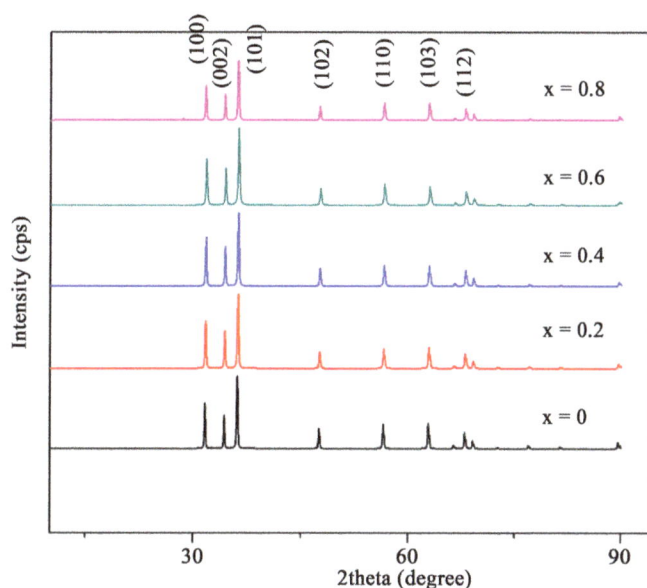

Figure 1. XRD pattern for Gd doped ZnO.

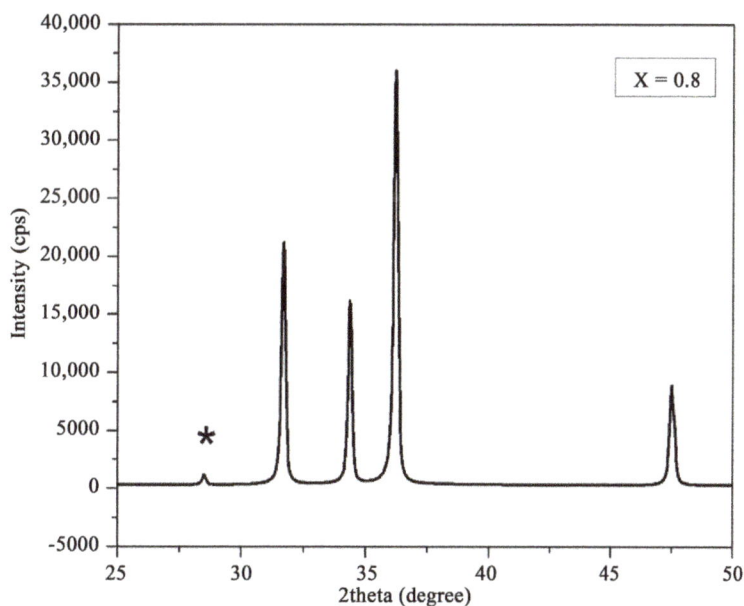

Figure 2. XRD pattern of 0.8 wt.% Gd doped ZnO.

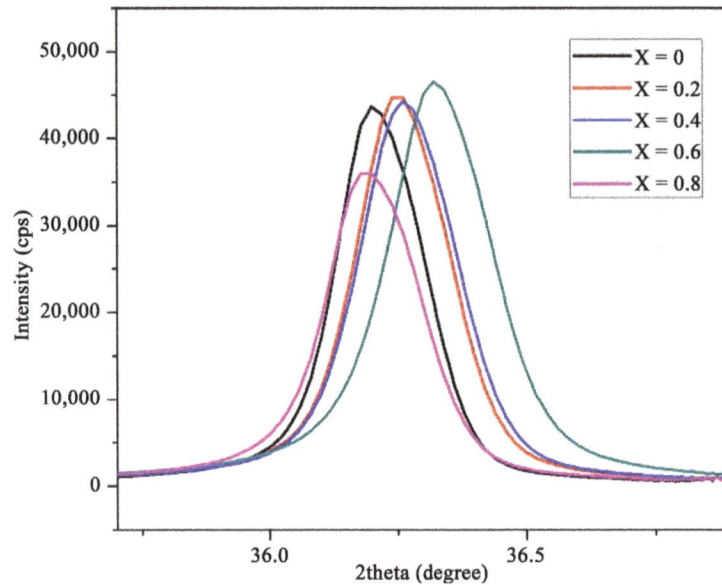

Figure 3. Peak shift observed in the XRD pattern for the peak corresponding to (101) plane.

Table 1. Variation of lattice parameters with Gd concentration.

Concentration of gadolinium (wt.%)	a (Å)	c (Å)	c/a
0	3.25331	5.21137	1.6018
0.2	3.25118	5.20893	1.6021
0.4	3.25012	5.20787	1.6023
0.6	3.24732	5.20111	1.6016
0.8	3.25422	5.21253	1.6017

Table 2. Variation of average crystalline size, volume and specific surface area with Gd concentration.

Concentration of gadolinium (wt.%)	Average crystalline size (*nm*)	Cell volume (Å³)	Specific surface area (m²/g) × 10³
0	64.5	47.79	16.435
0.2	52.1	47.71	20.346
0.4	51.2	47.67	20.704
0.6	47.1	47.52	22.506
0.8	56.0	47.83	18.929

bility limit of Gd ion in the ZnO crystal lattice is close to 0.6 wt.% and excess Gd ions may precipitate out on the particle surface.

The volume of the unit cell for hexagonal system has been calculated using the formula, $V = 0.866a^2c$. The unit cell volume is completely dependent on lattice constants. From the **Table 2**, it is clear that the volume of the unit cell decreases with small concentration of gadolinium. This is because of the perfect replacement of Zn^{2+} by Gd^{3+} in the crystal lattice. But as the concentration of doping increases, the gadolinium ion will also incorporate at the interstitial position, as a result lattice parameter changed and thus unit cell volume found increased.

The specific surface area of the crystallites of the samples was also determined using XRD. The specific surface area is a material property of solids which measures the total surface area of the crystallites present in per unit of mass. It is an important parameter that can be used to determine the type and properties of a material. It is particularly significant for adsorption, heterogeneous catalysis, and reactions on surfaces. The specific surface area can be calculated by Sauter formula,

$$S = 6 \times 10^3 / D_p \cdot \rho \qquad (2)$$

where S is the specific surface area, D_p is the size of the particle and ρ is the density of ZnO which equals to 5.606 g/cm^3. **Table 2** shows that, the specific surface area (S) of gadolinium doped ZnO nanocrystals was found to increase from 9.511 m$^2 \cdot$g^{-1} to 13.025 m$^2 \cdot$g^{-1} as the concentration of gadolinium in ZnO increased up to 0.6 wt.%. At 0.8 wt.% it becomes decreased due to the presence of gadolinium oxide in the sample.

3.2. Scanning Electron Microscopy

The SEM technique was employed to find the size and distribution of particles in the materials. **Figure 4** displays the surface morphology of undoped and Gd doped ZnO nanocrystalline powders. It can be seen that the samples prepared have smaller particle sizes.

Microstructural variation of Gd doped ZnO compared to pure ZnO is due to the significant difference in the ionic radius of Gd^{3+} related to the Zn^{2+} in ZnO. Ionic radius of Gd^{3+} is 0.938 Å, which is higher to Zn^{2+} (0.74 Å). Therefore higher radius Gd^{3+} may suppress the formation of larger nuclei during the crystallization process in ZnO. As a result of this, there is a reduction in the grain size happens. Thus Gd incorporation leads to a reduction

Figure 4. SEM images of (a) Pure ZnO (b) Zn$_{0.8}$Gd$_{0.2}$O (c) Zn$_{0.6}$Gd$_{0.4}$O (d) Zn$_{0.4}$Gd$_{0.6}$O (e) Zn$_{0.2}$Gd$_{0.8}$O.

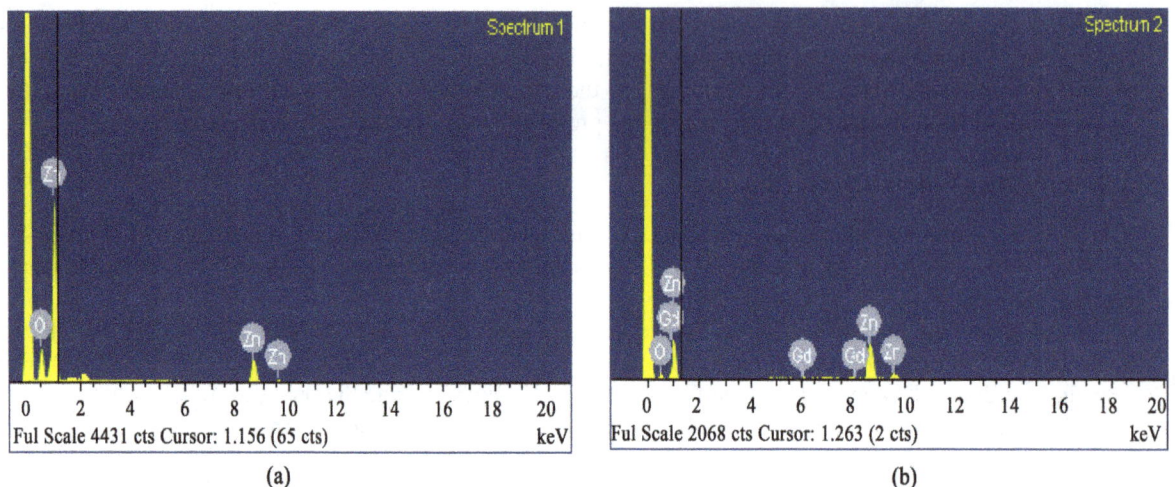

Figure 5. EDX spectrum of (a) ZnO (b) $Zn_{0.2}Gd_{0.8}O$.

in grain or crystallite size.

Another reason for the reduction of grain size is due to physiochemical effect. During the compound formation and sintering time, the larger radius Gd ion will enter into the lattice partially and the remaining will diffuse to the grain boundaries. This will lead to an isolated thin layer around the crystallites. There is a stress in the crystal when Gd ions get incorporated into the lattice. Those Gd ions which accumulated at the grain boundary act as a kinetic barrier for further grain displacement and thus hinder the grain growth [18]. Decrease in grain size with increasing Gd concentration in $(Bi_{1-x}Gd_x)_{0.5}Na_{0.5}TiO_3$ (BNGT) was reported by Vijayeta Pal *et al.* [19].

3.3. Energy Dispersive X-Ray Analysis

The Energy Dispersive X-ray analysis show peaks correspond to the element present in the sample. The higher a peak in a spectrum, the more concentrated the element is in the spectrum. The EDX image shown in **Figure 5** depicts the presence of Gd and Zn atom in the prepared samples. Absence of other elements in the spectra confirms the purity of the samples.

3.4. Dielectric Studies

The variation of the dielectric constant with log frequency at room temperatures is shown in **Figure 6**. The dielectric constant of all the samples found decreases with increasing frequency. This can be explained on the basis of Maxwell-Wagner model which is a result of the inhomogeneous medium of two-layer dielectric structure. In this model, dielectric structure is composed of well conducting grains, which are separated by the poorly conducting grain boundaries [20]. By hopping, electrons can accumulate at grain boundaries due to high resistance and produce polarization. As the frequency of the external electric field increases the hopping frequency of electrons cannot follow the alternating field. This decreases the probability of electron reaching the grain boundary and as a result polarization decreases. The results attained in this work and the explanation given above is in good agreement with the Koops phenomenological theory [21].

The observed higher value of dielectric constant at lower frequency is due to space charge polarization. While at higher frequency, polarization will lags behind the applied and hence decreases the value of dielectric constant.

Similar to dielectric constant, dielectric loss also decreases with increase in frequency and becomes constant at higher frequencies. Dielectric loss arises when the polarization lags behind the applied field and is caused by grain boundaries, impurities and imperfection in the crystal lattice [22]. **Figure 7** shows the variation of dielectric loss factor with frequency. When the frequency of the applied AC electric field is smaller than the hopping frequency of electrons between Zn^{2+} and Gd^{3+} ions, the electrons follow the field and hence the loss is maximum. At higher frequencies of the applied electric field, the hopping frequency of the electron exchange between these ions cannot follow the applied field beyond certain critical frequency and the loss is minimum. The value of tan δ is < 0.2 in the higher frequency range showing that the material is less lossy.

To find the effect of Gd substitution on the dielectric constant of the present samples, re-plot dielectric constant of $Zn_{1-x}Gd_xO$ as a function of Gd concentration is shown in **Figure 8**. It is found that dielectric constant increases slightly with small amount of Gd substitution ($x = 0.2$). Further increase in Gd concentration ($x = 0.4$) reduces the dielectric constant. Another maximum value of dielectric constant attained at $x = 0.6$ and additional Gd doping ($x = 0.8$) leads to another fall in the dielectric constant. The incorporation of small amounts of Gd^{3+} for Zn^{2+} would stabilize the wurtzite structure of ZnO and hence reduce the number of oxygen vacancies [23]. This leads to an increase in dielectric constant. Further increase of Gd content would result a unit cell contraction which is already discussed in the XRD section. Thus, the free volume available for the displacement of ions becomes smaller and this leads to a decrease in dielectric polarization. As Gd concentration increases from $x = 0.4$ to $x = 0.6$, the dielectric constant increases again due to the presence of more grain boundaries. A saturated level of Gd content is approaching when x is increased from 0.6 to 0.8. At this stage, the sample might be a composite of Gd_2O_3 and ZnO, with a lesser dielectric constant.

The effect of Gd substitution on the dielectric loss angle is shown in **Figure 9**. Compared to pure ZnO, mini-

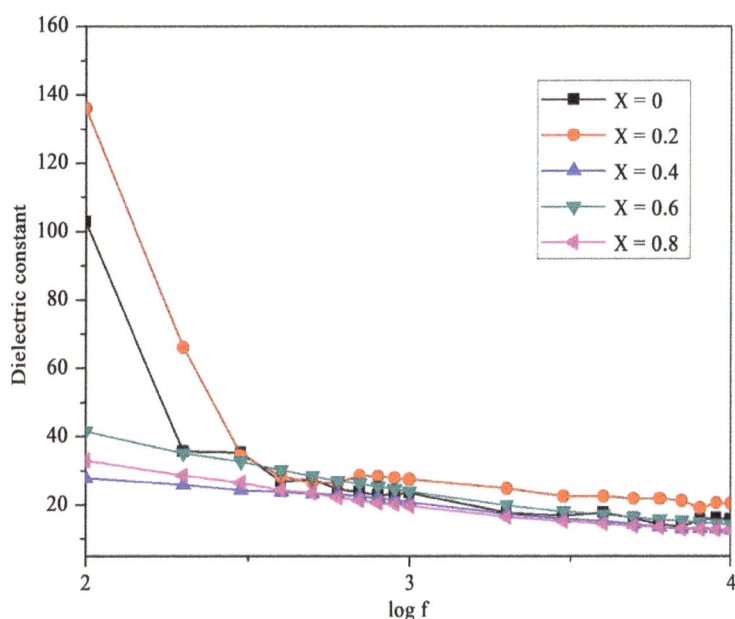

Figure 6. Variation of dielectric constant with frequency.

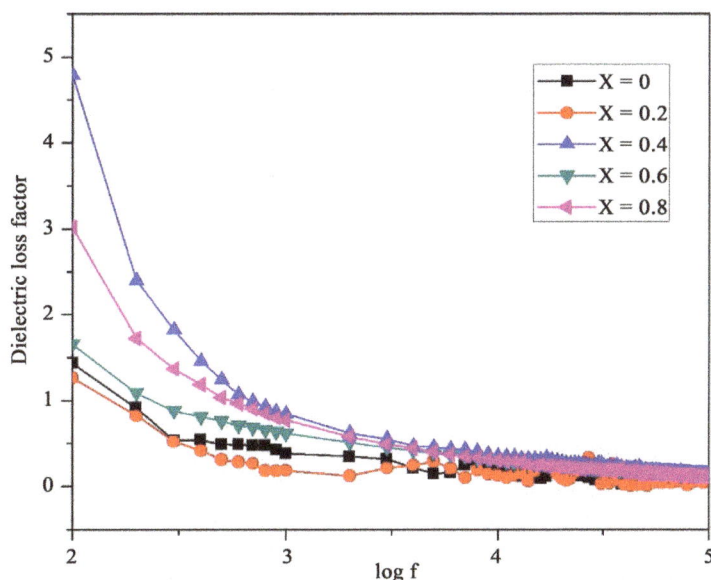

Figure 7. Variation of dielectric loss factor with frequency.

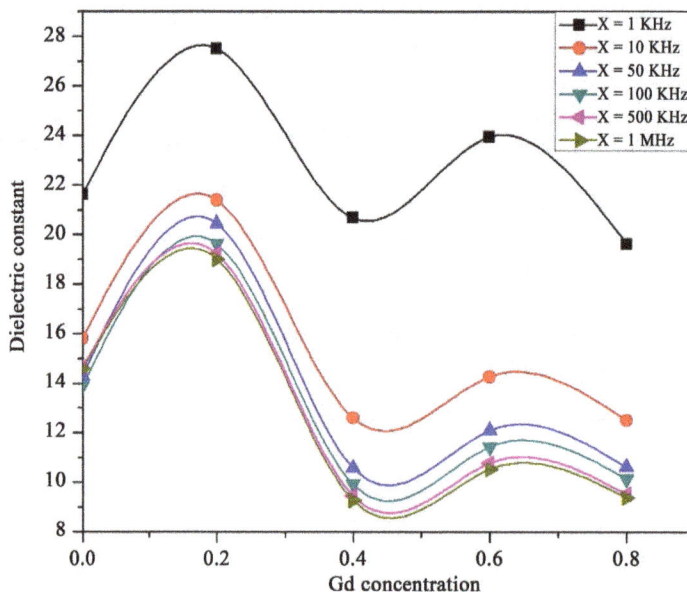

Figure 8. Variation of dielectric constant with Gd concentration for $Zn_{1-x}Gd_xO$.

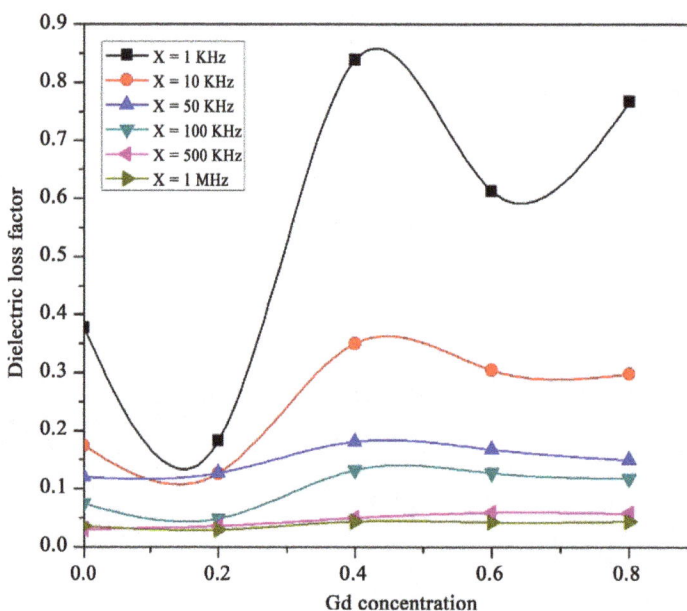

Figure 9. Variation of dielectric loss factor with Gd concentration for $Zn_{1-x}Gd_xO$.

mum dielectric loss is obtained at Gd content $x = 0.2$. It is also observed that all the samples have a less dielectric loss at higher frequency.

3.5. AC Conductivity

The conduction mechanism in the present samples was determined from the AC conductivity measurement. The variation of AC electrical conductivity (σ_{ac}) with frequency at room temperature is shown in **Figure 10**. At lower frequency, poorly conducting boundaries become more active. As a result hopping frequency between charge carriers decreases. This in turn decreases the conductivity value of the material. As frequency increases, the hopping frequency results in the increase of ac conductivity value of all the samples.

Figure 7 and **Figure 10** reveal a close relationship between tan δ and conductivity. The increase of conductivity is accompanied by an increase of the eddy current which in turn increases the energy loss tan δ [24].

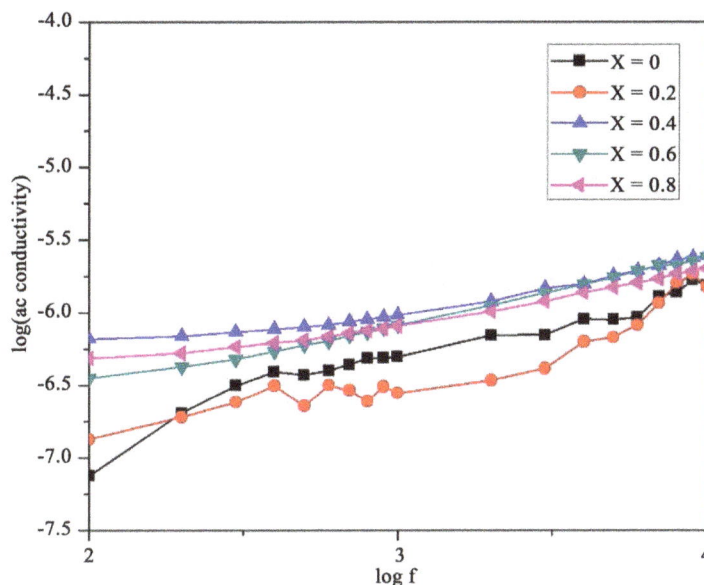

Figure 10. Variation of AC electrical conductivity with frequency.

4. Conclusion

Gadolinium (Gd) doped Zinc oxide (ZnO) nanocrystals were synthesized by the solid state reaction route by varying Gd concentration from 0 wt.% to 0.8 wt.%. Gd incorporation in the host lattice makes a structural distortion in ZnO due to the larger ionic radius of Gd compared to that of Zn and is evident from the structural studies. Increase the concentration of Gd hinders the growth of $Zn_{1-x}Gd_xO$ nanocrystals and multiphase growth is observed at higher concentrations. The frequency dependence of dielectric studies revealed that for all the samples studied, the dielectric constant and dielectric loss was found decreased with increase of frequency (between 100 Hz and 4.5 MHz), whereas AC conductivity was found increased. The gadolinium doping has an important effect on the dielectric properties of ZnO. At low Gd^{3+} concentrations, slightly higher value of dielectric constant is observed. The low dielectric loss at higher frequency makes this Gd doped ZnO nanocrystal as a candidate for high frequency applications.

Acknowledgements

Author (P. P. Pradyumnan) is thankful to SERB Govt. of India major research funding, DST-FIST Govt. of India, for projects sanctioned to Dept. of Physics, University of Calicut for the equipment facilities. One of the authors (Divya) acknowledges UGC-SAP for financial support.

References

[1] Zhang, Q.F., Dandeneau, C.S., Zhou, X.Y. and Cao, G.Z. (2009) ZnO Nanostructures for Dye Sensitized Solar Cells. *Advanced Materials*, **21**, 4087-4108. http://dx.doi.org/10.1002/adma.200803827

[2] Wang, Z.L. (2004) Zinc Oxide Nanostructures: Growth, Properties and Applications. *Journal of Physics*: *Condensed Matter*, **16**, R829-R858.

[3] Gurav, K.V., Fulari, V.J., Patil, U.M., Lokhande, C.D. and Joo, O. (2010) Room Temperature Soft Chemical Route for Nanofibrous Wurtzite ZnO Thin Film Synthesis. *Applied Surface Science*, **256**, 2680-2685. http://dx.doi.org/10.1016/j.apsusc.2009.09.080

[4] Kenyon, A.J. (2002) Recent Developments in Rare-Earth Doped Materials for Optoelectronics. *Progress in Quantum Electronics*, **26**, 225-284. http://dx.doi.org/10.1016/S0079-6727(02)00014-9

[5] Jayanthi, K., Manorama, S.V. and Chawla, S. (2013) Observation of Nd^{3+} Visible Line Emission in $ZnO:Nd^{3+}$ Prepared by a Controlled Reaction in the Solid State. *Journal of Physics D*: *Applied Physics*, **46**, 325101.

[6] Achamma, G., Sharma, S.K., Chawla, S., Malik, M.M. and Qureshi, M.S. (2011) Detailed of X-Ray Diffraction and Photoluminescence Studies of Ce Doped ZnO Nanocrystals. *Journal of Alloys and Compounds*, **509**, 5942-5946. http://dx.doi.org/10.1016/j.jallcom.2011.03.017

[7]   John, R. and Rajakumari, R. (2012) Synthesis and Characterization of Rare Earth Ion Doped Nano ZnO. *Nano-Micro Letters*, **4**, 65-72. http://dx.doi.org/10.1007/BF03353694

[8]   Sinha, N., Ray, G., Bhandari, S., Godara, S. and Kumar, B. (2014) Synthesis and Enhanced Properties of Cerium Doped ZnO Nanorods. *Ceramics International*, **40**, 12337-12342. http://dx.doi.org/10.1016/j.ceramint.2014.04.079

[9]   Divya, N.K., Aparna, P.U. and Pradyumnan, P.P. (2015) Dielectric Properties on Er^{3+} Doped ZnO Nanocrystals. *Advances in Materials Physics and Chemistry*, **5**, 287-294. http://dx.doi.org/10.4236/ampc.2015.58028

[10] Divya, N.K. and Pradyumnan, P.P. (2016) Solid State Synthesis of Erbium Doped ZnO with Excellent Photocatalytic Activity and Enhanced Visible Light Emission. *Materials Science in Semiconductor Processing*, **41**, 428-435. http://dx.doi.org/10.1016/j.mssp.2015.10.004

[11] Murmu, P.P., Kennedy, J., Ruck, B.J., Markwitz, A., Williams, G.V.M. and Rubanov, S. (2012) Structural and Magnetic Properties of Low Energy Gd Implanted ZnO Single Crystals. *Nuclear Instruments and Methods in Physics Research Section B*, **272**, 100-103. http://dx.doi.org/10.1016/j.nimb.2011.01.041

[12] Ma, X.Y. and Wang, Z. (2012) Optical Properties of Rare Earth Gd Doped ZnO nanocrystals. *Materials Science in Semiconductor Processing*, **15**, 227-231. http://dx.doi.org/10.1016/j.mssp.2011.05.013

[13] Gouri, M.I., Ahmed, E., Khalid, N.R., Ahmad, M., Ramzan, M., Shakoor, A. and Niaz, N.A. (2014) Gadolinium Doped ZnO Nanocrystalline Powders and Its Photocatalytic Performance for Degradation of Methyl Blue under Sunlight. *Journal of Ovonic Research*, **10**, 89-100.

[14] Dakhel, A.A. and El-Hilo, M. (2010) Ferromagnetic Nanocrystalline Gd Doped ZnO Powder Synthesized by Coprecipitation. *Journal of Applied Physics*, **107**, Article ID: 123905.

[15] Lin, W., Ma, R., Shao, W. and Liu, B. (2007) Structural, Electrical and Optical Properties of Gd Doped and Undoped ZnO: Al Thin Films Prepared by RF Magnetron Sputtering. *Applied Surface Science*, **253**, 5179-5183. http://dx.doi.org/10.1016/j.apsusc.2006.11.032

[16] Subramanian, M., Thakur, P., Tanemura, M., Hihara, T., Ganesan, V., Soga, T., *et al.* (2010) Intrinsic Ferromagnetism and Magnetic Anisotropy in Gd Doped ZnO Thin Films Synthesized by Pulsed Spray Pyrolysis Method. *Journal of Applied Physics*, **108**, Article ID: 053904. http://dx.doi.org/10.1063/1.3475992

[17] Mithal, D. and Kundu, T. (2013) Synthesis and Characterization of Gd Doped ZnO Nanocrystals. *Asian Journal of Chemistry*, **25**, 12-16.

[18] Rahman, M.T., Vargas, M. and Ramana, C.V. (2014) Structural Characteristics, Electrical Conduction and Dielectric Properties of Gadolinium Substituted Cobalt Ferrite. *Journal of Alloys and Compounds*, **617**, 547-562. http://dx.doi.org/10.1016/j.jallcom.2014.07.182

[19] Pal, V. and Dwivedi, R.K. (2012) Effect of Rare Earth Substitution on the Structural, Microstructure and Dielectric Properties of Lead Free BNT Ceramics. *Advanced Materials Research*, **585**, 200-204. http://dx.doi.org/10.4028/www.scientific.net/AMR.585.200

[20] Thakur, A., Mathur, P. and Singh, M. (2007) Study of Dielectric Behaviour of Mn-Zn Nano Ferrites. *Journal of Physics and Chemistry of Solids*, **68**, 378-381. http://dx.doi.org/10.1016/j.jpcs.2006.11.028

[21] Koops, C.G. (1951) On the Dispersion of Resistivity and Dielectric Constant of Some Semiconductors at Audio Frequencies. *Physical Review*, **83**, 121-124. http://dx.doi.org/10.1103/PhysRev.83.121

[22] Mangalaraja, R.V., Manohar, P. and Gnanam, F.D. (2004) Electrical and Magnetic Properties of $Ni_{0.8}Zn_{0.2}Fe_2O_4$/Silica Composite Prepared by Sol-Gel Method. *Journal of Materials Science*, **39**, 2037-2042. http://dx.doi.org/10.1023/B:JMSC.0000017766.07079.80

[23] Fanggao, C., Guilin, S., Kun, F., Ping, Q. and Qijun, Z. (2006) Effect of Gadolinium Substitution on Dielectric Properties of Bismuth Ferrite. *Journal of Rare Earths*, **24**, 273-276. http://dx.doi.org/10.1016/S1002-0721(07)60379-2

[24] Satter, A.A. and Samy, A.R. (2003) Dielectric Properties of Rare Earth Substituted Cu-Zn Ferrites. *Physica Status Solidi(a)*, **200**, 415-422. http://dx.doi.org/10.1002/pssa.200306663

19

Structural, Magnetic and Dielectric Properties of Fe-Co Co-Doped $Ba_{0.9}Sr_{0.1}TiO_3$ Prepared by Sol Gel Technique

Inas Kamal Batttisha[1*], Ibrahim Sayed Ahmed Farag[1], Mostafa Kamal[2],
Mohamed Ali Ahmed[3], Emad Girgis[1], Hesham Azmi El Meleegi[4], Fawzi Gooda El Desouky[1]

[1]National Research Center (NRC), Solid State Physics Department, Dokki, Giza, Egypt
[2]Metal Physics Laboratory, Physics Department, Faculty of Science, Mansoura University, Mansoura, Egypt
[3]Materials Science Lab (1), Physics Department, Faculty of Science, Cairo University, Giza, Egypt
[4]National Research Centre (NRC), Thin Film Lab, Electron Microscope Department, Dokki, Giza, Egypt
Email: [*]szbasha@yahoo.com, [*]ibattisha@gmail.com

Abstract

The structural, dielectric and magnetic properties of pure and Fe-Co co-doped $Ba_{0.9}Sr_{0.1}TiO_3$, $(Ba_{(1-x)}Sr_xTiO_3$, where $(x = 0.10)$ and $(Ba_{0.9}Sr_{0.1}Ti_{(1-x-y)}Fe_xCo_yO_3)$, where $(x = 0.1, y = 0)$ and $(x = 0$ and $y = 0.10)$ and $(x = 0.5, y = 0.5)$ in powder form, abbreviated as (BST) and (BST10FO), (BST10CO) and (BST5F5CO), respectively were prepared by a modified sol gel technique. Crystallization, surface morphology and electrical behavior of BST are improved by Fe^{3+} and Co^{2+} ions with optimized grain size. Phase identification by using X-ray diffraction and surface morphology will be studied by using transmission electron microscope (TEM) and scanning electron microscope imaging (SEM). Phase identification by using X-ray diffraction and surface morphology evaluation by using transmission electron microscope (TEM) and scanning electron microscope imaging (SEM) will be studied. The nano-scale presence and the formation of the tetragonal perovskite phase as well as the crystallinity were detected using the mentioned techniques. The dielectric properties of the prepared samples have been investigated as a function of temperature and frequency. The dielectric measurements are carried out in the frequency range of 42 Hz - 1 MHz, at temperature ranging between 25°C and 250°C. The results showed an abrupt decrease in the dielectric permittivity by increasing the frequency range. The magnetic hysteresis loop confirmed enhancement in the magnetization properties by co-doping with Fe^{3+}-Co^{2+} ions. An increase in the saturation of the magnetization at room temperature was detected by decreasing the crystallite sizes of the prepared samples.

---

[*]Corresponding author.

Keywords

Sol Gel, Nano-Structure, Magnetic Hysteresis Loop, Dielectric Permittivity, Nano-Composite
BaSrTiO₃ (BST), TEM, XRD, SEM

1. Introduction

Complex oxides are very appealing materials from a function point of view, due to their wide range of properties: such as ferroelectricity, ferromagnetism, ferroelasticity etc. They can be metals, insulator, semiconductors, superconductor etc. The coupling between some of these properties can give rise to new applications. Some of these oxides have the unique properties of both ferromagnetism and ferroelectricity in a single phase. This opens broader applications in transducers, magnetic field sensors and information storage industry and multifunctional devices such as memory devices [1]-[4]. The relationship between multiferroic and magnetoelectric materials is well reflected by **Figure 1** [5].

A single phase multiferroic material is the one that possesses two of the three "ferroic" properties *i.e.* ferroelectricity, ferromagnetism and ferroelasticity. Generally current trend is to exclude the requirement for ferroelastic property. Magnetoelectric coupling describes the coupling between magnetic and electric order parameters [6]-[10].

This paper aims to study the structural, dielectric and magnetic properties of pure BST and (Fe and Co) co-doping into ferroelectrics BST. The structure and phase identification will be evaluated by XRD. The TEM and SEM of pure barium strontium titinate (B10ST), B10ST10C and B10ST5C5F revealed the presence of the nanophase in the prepared samples. BST ceramics exhibit excellent single tetragonal phase by using sol gel technique as preparation process. A large amount of doping of Fe-Co which mainly acts as an acceptor to replace Ti in the B-site, leads to the appearance of lattice defects and vacancies. The dielectric properties of the samples will be studied in frequency range of 42 Hz up to 1 MHz. Ferroelectricity and ferromagnetism of samples are simultaneously observed. The magnetic measurements were carried out at room temperature using lakeshore vibrating sample magnetometer (VSM 7410) model lakeshore 7110.

2. Materials and Methods

2.1. Samples Preparation

$Ba_{0.9}Sr_{0.1}TiO_3$, (BST) and $Ba_{0.9}Sr_{0.1}Ti_{(1-x-y)}Fe_xCo_yO_3$, where $(x = 0.1, y = 0)$ (BST10F), $(x = 0, y = 0.10)$, (BST10C) and $(x = 0.5, y = 0.5)$, (BST5F5C), respectively in powder forms were prepared by a modified sol gel method. **Table 1** shows dopant concentration, sample abbreviation, chemical formula and oxygen vacancies of Fe^{3+} and Co^{2+} ions concentration doped nano-composite B10ST powders.

The mentioned samples have been prepared using barium acetate (Ba(Ac)₂) (99%, Sisco Research Laboratories PVT.LTD, India) and titanium butoxide (Ti(C₄H₉O)₄), (97%, Sigma-Aldrich, Germany) are used as the starting materials; acetyl acetone (AcAc, C₅H₈O₂), (98%, Fluka, Switzerland) acetic acid (HAc)-H₂O mixture (96%, Adwic, Egypt) were adopted as solvents of (Ti(C₄H₉O)₄), and Ba(Ac)₂, respectively. Strontium bromide is

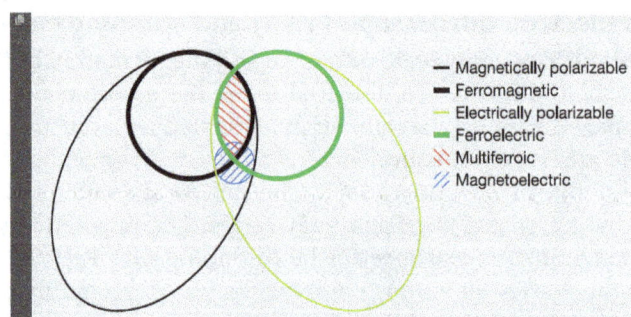

Figure 1. The relationship between multiferroic and magnetoelectric materials [5].

Table 1. Dopant concentration, sample abbreviation, chemical formula and oxygen vacancies of Fe^{3+} and Co^{2+} ions concentration doped nano-composite B10ST powders.

Dopant Concentration	Sample abbreviations	Chemical Formula	δ
$x = 0.1, y = 0, z = 0$	B10ST	$Ba_{0.9}Sr_{0.1}Ti_1O_3$	$\delta = 0.000$
$x = 0.1, y = 0.1, z = 0$	B10ST10F	$Ba_{0.9}Sr_{0.1}Ti_{0.9}Fe_{0.1}O_{2.95}$	$\delta = 0.050$
$x = 0, y = 0, z = 0.1$	B10ST10C	$Ba_{0.9}Sr_{0.1}Ti_{0.9}Co_{0.1}O_{2.9}$	$\delta = 0.100$
$x = 0.1, y = 0.05, z = 0.05$	B10ST5F5C	$Ba_{0.9}Sr_{0.1}Ti_{0.9}Fe_{0.05}Co_{0.05}O_{2.925}$	$\delta = 0.075$

added to the precursor with constant 0.1 mol % molar ratios. Iron and cobalt nitrates are added to the final solution and its content has been changed. Densification of the gel is achieved by sintering in air for one hour at heat treatment temperature 750°C, in a muffle furnace type (Carbolite CWF 1200).

2.2. Characterization

The phases of the obtained samples are characterized by X-ray diffraction (XRD) (BRUKUR D8 ADVANCED TARGET Cu Kα with Secondary monochromatic KV = 40, mA = 40 Germany) in a wide range of Bragg angle from 10° - 80° using Cu Ka (1.5406 Å) radiation with a step size of 0.02 at room temperature. The crystallite size (G) is determined from the Scherrer's equation;

$$G = K\lambda / D \cos\theta \tag{1}$$

where K is the Scherer constant, in the present case $K = (0.9)$, λ is the wavelength and D is the full width (in radians) of the peak at half maximum (FWHM) intensity. The microstructure and surface morphology of the samples were observed by (TEM) transmission electron microscope (using JEOL JEM-1230 equipment operat- ing at 120 kV with attached CCD camera) and (SEM) scanning electron microscope (Quanta 250 FEG (Field emission Gun) was used to determine grain size and uniformity of the sample analysis. The phase transitions above room temperature have been investigated by DTA (SDT Q 600 V 20.9 Build 20) measurements.

Magnetization hysteresis (M-H) measurements were carried out at room temperature using lakeshore vibrating sample magnetometer (VSM 7410) model lakeshore 7110.

The relative dielectric permittivity was calculated using the relations:

$$\varepsilon' = Cd / \varepsilon_o A \tag{1}$$

where C is the capacitance of the measured sample in Farad, d is the thickness of the sample in meters, A is the cross section area of the sample and ε_o is the permittivity of free space (8.854×10^{-12} Fm^{-1}).

$$\varepsilon'' = \varepsilon' \times \tan\delta \tag{2}$$

where, ε'' is the dielectric loss and $\tan\delta$ is the loss tangent.

3. Results and Discussion

3.1. XRD Investigation

For comparison between the un-doped and co-doped samples with Fe^{3+} and Co^{2+} ions calcinated at constant temperature 850°C for four hours, the XRD patterns of pure B10ST (a) and the doped powder samples with (10 mol% Fe^{3+}) (B10ST10F) (b), (10 mol% Co^{2+}) (B10ST10C) (c) and doped with both (Fe^{3+} & Co^{2+} at 5 mol%) (B10ST5F5C) (d) were plotted in **Figure 2** (a)-(d), respectively. All the diffraction peaks are indexed and the tetragonal structure phase is identified [11]. The XRD patterns of the prepared samples are in good agreement with the tetragonal BST phase (ICCD card number 44-0093,) as shown in **Figure 2** (b)-(d). Very weak line corresponding to the residual carbonates phases, such as $BaCO_3$, $SrCO_3$ and (Ba, Sr) CO_3 were appeared in the doped samples with no other observable iron and cobalt oxide phases in the systems [12] [13]. So the Fe^{3+} and Co^{2+} ions are embedded in the BST crystal lattice and substitute the Ti^{4+} ions where no obvious secondary phase from Fe^{3+}, Co^{2+} and Ti^{4+} ions are observed in the diffraction patterns as shown in **Figure 2** (b)-(d).

From the **Table 2** we can see that the lattice parameters of the B10ST and B10ST10F samples exhibit a slight variation, but the crystal structures remain almost the same. The values of lattice parameter (a) and (c) increase

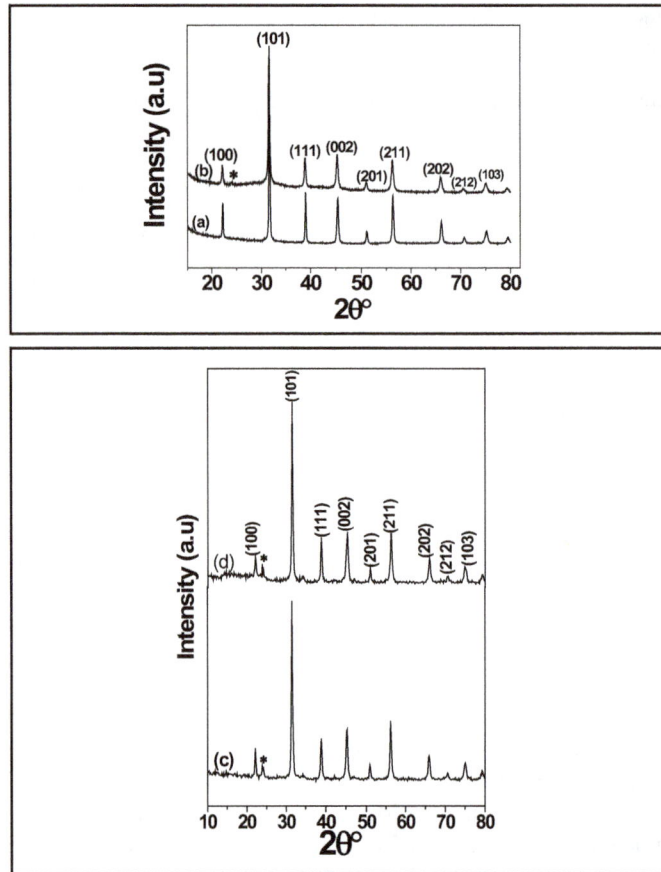

Figure 2. XRD patterns of (B10ST) (a), (B10ST10F) (b), (B10ST10C) (c) and (B10ST5F5CO) (d) in powder form calcinated at 850°C for 4 h (Ba, Sr(Co$_3$) is marked by *).

by doping with the Fe^{3+} ions, indicating a lattice expansion of BSTF systems. For instance, the unit cell volume increases due to the different sizes between Fe^{3+} and Ti^{4+} ions (0.645 Å for Fe^{3+} and 0.68 Å for Ti^{4+}) [13] and the possible ion vacancies presented in the lattices as shown in **Table 1**. In the sample doped with cobalt oxide some Co^{2+} cation proportions exist and this cation can be considered as acceptor dopants as it has a lower valence than Ti^{4+}. The substitution of Ti^{4+} by Co^{2+} would give rise vacancy in the oxygen sub-lattice as in **Figure 2** (c). The lattice parameter of the doped samples is carefully determined, as listed in **Table 1**. The values of lattice parameter (a) and (c) is increased by doping with the Co^{2+} ions, indicating a lattice expansion of B10ST10C system. The unit cell volume increases as shown in **Table 1** which can be attributed to the different sizes between Co^{2+} and Ti^{4+} ions (0.65 Å for Co^{2+} and 0.68 Å for Ti^{4+}). These vacancies may be favorable for stabilizing the tetragonal structure of BSTC systems.

The average crystallite sizes of our samples calculated using Scherrer's formula were decreased by doping with both Fe^{3+} and Co^{2+} ions to be equal to 52, 31, 31.3, 27.45 for B10ST, (B10ST10F), (B10ST10C) and (B10ST5F5C), respectively as detected in **Table 1**.

3.2. Transmission Electron Microscope (TEM) Study

Figure 3 shows the representative TEM of B10ST5F5C ≈ (27 nm), thermally synthesized in air for 4 h at 850°C. Some degree of agglomerates has been found in the clusters consisting of many small particles. The calculated average particle size from TEM was about 27 nm for B10ST5F5C, which is nearly agree with the value obtained from XRD for the same sample 27.45 nm. The TEM was used to confirm the data calculated from XRD patterns and that the sample is in nano-scale. Images in **Figure 3**, clearly shows that the additive of cobalt and iron oxides have leads to a grains refining [14] [15].

Table 2. Lattice parameters and unit cell volume of B10ST, B10STF, B10ST10C and B10ST5F5C.

Sample abbreviations	Crystal size (XRD) nm	a, b (Å)	c (Å)	c/a	V (Å)³
B10ST	52.00	3.9996	3.9995	0.9999	63.97879
B10ST10F	31.00	4.0066	3.9980	0.99875	64.17926
B10ST10C	31.30	4.00725	4.00194587	0.9986	64.26316
B10ST5F5C	27.45	4.00139	4.001985	1.00015	64.07627

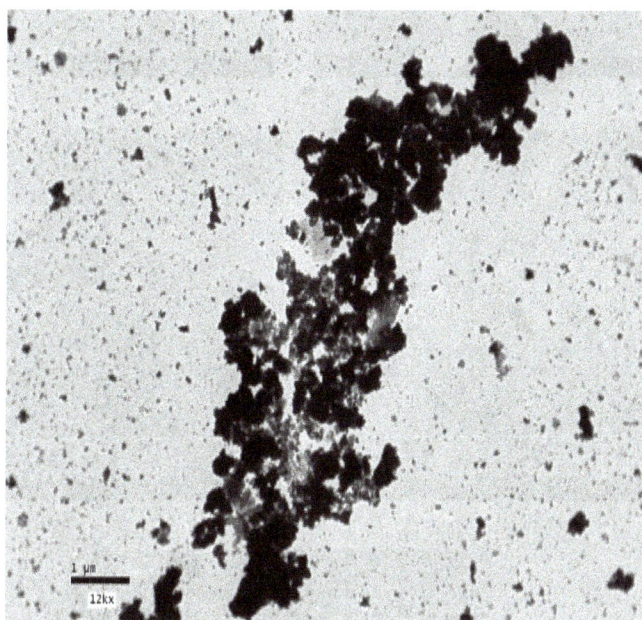

Figure 3. The TEM micrograph of (B10ST5F5C) powder sample calcinated at 850°C for 4 h.

3.3. Scanning Electron Microscope (SEM) Study

Figure 4, shows the surface morphologies obtained through Scanning Electron Microscope (SEM) with different magnifications for nanostructure pure (B10ST) and co-doped with both Fe^{3+} and Co^{2+} ions (B10ST10F) (B10ST10C) and (B10ST5F5C) powder samples calcined for 4 hours at 850°C are illustrated in **Figures 4(a)-(e)**, respectively. These images show grains typical of pure B10ST powders, which consist of a granular microstructure with irregular shaped grains, with two different magnifications and in two different areas, as shown in **Figure 4(a)** and **Figure 4(b)**. The images in **Figures 4(c)-(e)** show comparatively more accumulated particles with higher density showing an increase in grain growth by doping with both Fe^{3+} and Co^{2+} ions and the particles have a well-defined shape.

Fe^{3+} and Co^{2+} dopant contents are effective to increase the density and improve the microstructure homogeneity of the prepared powder samples. The particles are nearly tetragonal in nature and less agglomerated in the pure B10ST sample and indicate the well-distributed crystallites and the dense nanoparticles surfaces. While by doping with Fe^{3+} and Co^{2+} the particles have a well-defined shape, and they are still tetragonal and highly dispersed. This is might be due to some aggregates of particles together.

3.4. Dielectric Properties of B10ST, B10ST10F, B10ST10C and B10ST5F5C, Respectively

Figure 5 shows the variation of dielectric constant, (ε') measured at room temperature as a function of frequency (f) in the frequency range (10 KHz - 5 MHz) for pure B10ST powder sample and doped samples B10ST10F, B10ST10C and B10ST5F5C, respectively. We can see a sharp fall in the value of (ε') for the pure BS10T nanoparticles that is followed by a rise in the (ε') value for B10ST10F and B10ST5F5C, respectively.

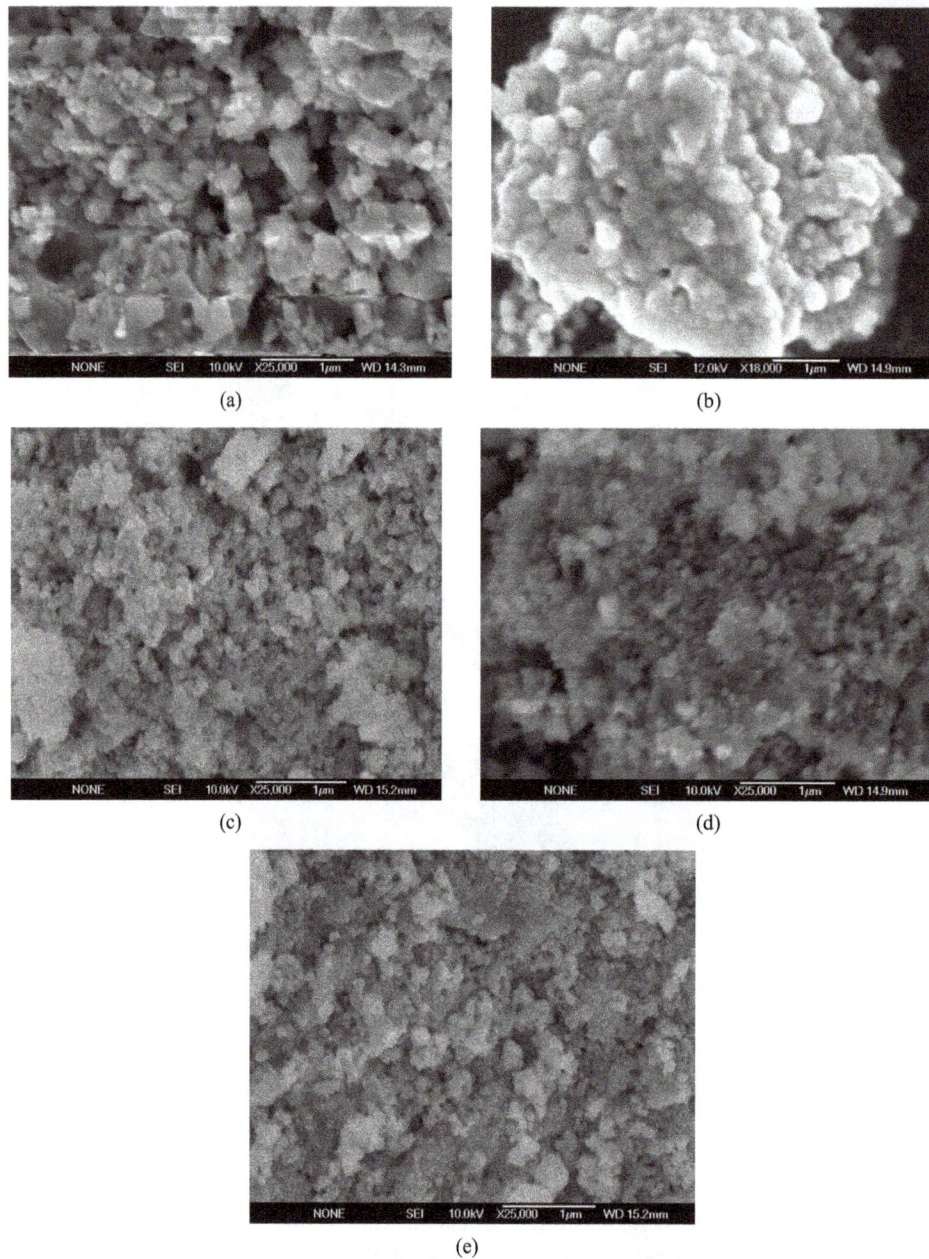

Figure 4. The SEM micrograph of pure (B10ST) (a) calcined at 750°C for one hour [13], pure (B10ST) (b) calcined at 850°C for 4 hours, (B10ST10C) (c), (B10ST10F) (d) and (B10ST5F5C) (e), all calcinated for 4 h at 850°C.

It is evident that the dielectric constant (ε') decreases with increasing frequency and becomes nearly constant at the higher frequencies range. It follows the dipole relaxation where at low frequencies the dipoles follow the frequency of the applied field. The value of dielectric constant is highest for the B10ST10C. The decrease in (ε') value as a function of frequency is due to the electric dipole response in the prepared samples which, decreased at higher frequencies, where it need some time for realignment and cannot follow electric-field changes. Also the periodic reversal of the electric field occurred so fast that there was no excess charge carrier diffusion in the direction of the field. As the frequency increased, the dipoles were less able to rotate and maintain phase with the field. Thus, they reduced their contribution to the polarization field, and hence an observed reduction in the real part (ε') is appeared. Where the low frequency dispersion region is attributed to the charge accumulation at the electrode—sample interface.

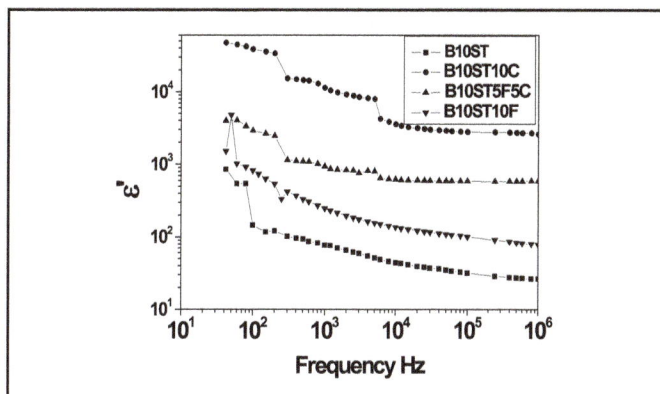

Figure 5. Relative permittivity (ε') of B10ST, and doped samples, B10ST10F, B10ST10C and B10ST5F5C, respectively.

For the dissipation factor $(\tan \delta)$ shown in **Figure 6**, it was noticed that this attenuation of $(\tan \delta)$ by increasing frequency might be attributed to the phonon dipole interaction which, led to a lowering of the energy transferred to the dielectric medium. Also, as the frequency increased, the dipoles polarization tended to zero and tan δ depended only on electronic polarization.

The low value of dissipation factor $(\tan \delta)$ indicated low conversion of electrical energy to heat energy and reduced power loss for the network. The dissipation factor, as well as other electrical parameters, depended on the frequency, humidity and purity of the sample.

The cationic size are (0.65 °A for Co^{2+}, 0.645 °A for Fe^{3+} and 0.68 °A for Ti^{4+}) Therefore, cationic substitution causes expansion in the ABO_3 structure, and modifies the dielectric properties as mentioned in the XRD section in this work. Addition of Fe^{3+} and Co^{2+} ions increases the relative permittivity of the ceramics compared to pure B10ST thermally treated at the same calcination temperature. These changes in electrical properties are believed to be associated with both improving microstructure and fired density [16] [17].

3.5. Magnetic Properties of B10ST10F and B10ST5F5C

Figure 7 shows the M-H curve of nano-structure B10ST, B10ST10F and B10ST5F5C, respectively. Measurement of magnetization in the nanocrystalline B10ST, with C.S. equal to (52 nm) sample was measured at room temperature and the results are as previously published by our team work [18] and as detected in **Table 3**. It is clear that a weak ferromagnetism at room temperature with a coercive field and saturation magnetization equal to 74.219000 Oe and 0.041556 emu/g, respectively, for pure B10ST was obtained. It may be mentioned that the 52 nm B10ST sample showed higher saturation magnetization. In contrast, the bulk B10ST sample failed to exhibit magnetic hysteresis, as expected it exhibits diamagnetic behavior at room temperature. It is therefore, understandable that an increase in the particle size eliminates the magnetism due to the decrease in the surface to volume ratio.

Both the B10ST10F and the B10ST5F5C samples exhibit ferromagnetism at room temperature with clear hysteresis loops. **Figure 7** shows M-H curves of B10ST10F and B10S5F5C respectively, in powder form calcined at 850°C for 4 h. It is known that magnetism in oxide nanoparticles arises from vacancies. From **Table 3** we can see that oxygen vacancies for B10ST, B10ST10F and B10S5F5C increase by doping with different transition elements, respectively, as a result of different ionic radius, where Fe^{3+} and Co^{2+} ions mainly acts as an acceptor to replace Ti in the B-site, leading to the further ionization of Fe^{3+} and Co^2 ions and the appearance of lattice defects and vacancies, which are favorable for single-phase. From Table 3 the saturated magnetization M_S increases by decreasing the B10ST, B10ST10F and B10S5F5C particle sizes, this result is in agreement with the previously reported [18]-[20], as the size of the grain decreases, the material reaches a single domain state where there is no domain wall and all the magnetic domains are completely separated from each other, the enhancement of magnetization also come from the magnetic moment per Fe^{3+} and Co^{2+} ions due to the combined contributions of magnetic Fe^{3+} and Co^{2+} ions distributed over pentahedral and octahedral Ti sites. The Co^{2+} ions presence led to the increase of anisotropy property, hence an increase in oxygen vacancies as a result of different size between Ti^{4+} and Co^{2+} was appeared.

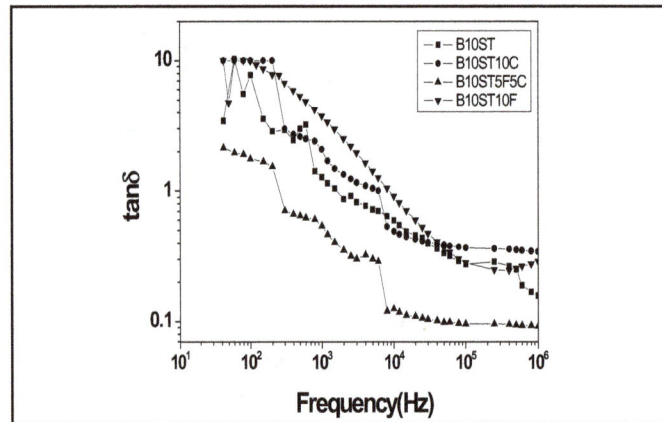

Figure 6. The dissipation factor (tan δ) of B10ST, and doped samples, B10ST10F, B10ST10C and B10ST5F5C, respectively.

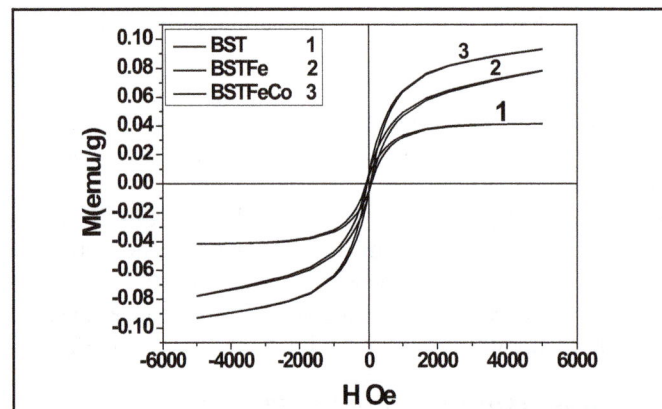

Figure 7. M-H curve of nano-structure (1) B10ST [18], (2) B10ST10F and (3) B10ST5F5C, respectively.

Table 3. Magnetic parameters at room temperature and particle size.

Samples	C.S nm	Hc (Coerc. Field)	Mr (Rem. Mag.)	Ms (Sat. Mag.)	Anisotropy
Abr.		Oe	emu/g	emu/g	Const.
B10ST	52.0	74.219	0.0049275	0.041556	1.5421
B10ST10F	31.00	72.196	0.0065871	0.07802	2.8163
B10ST5F5C	27.45	46.335	0.0059189	0.09298	2.15391

4. Conclusions

Nano-structure B10ST, B10ST10F, B10ST10C and B10S5F5C, have been successfully synthesized by a modified sol gel technique. XRD patterns confirm the tetragonal structure phase presence of the prepared samples. The prepared samples exhibit ferromagnetism at room temperature with clear hysteresis loops. It can be seen that the oxygen vacancies for B10ST , B10ST10F , B10ST10C and B10S5F5C increase respectively, as a result of different ionic radius, where Fe^{3+} and Co^{2+} ions, which mainly act as an acceptor to replace Ti in the B-site, leading to the further ionization of Fe^{3+} and Co^{2+} ions and the appearance of lattice defects and vacancies.

The results showed the achievement of Fe-doped BST with single phase perovskite structure and improved magnetic properties. The significant increase in magnetization by doping the BST with both Fe^{3+} and Co^{2+} ions increases the possibility of use for BST10F and BST5F5C in memory based applications.

Acknowledgements

The authors acknowledge 1) The team work of both projects—Imhotep project, 2009-2011, entitled; New nano-structured magnetoelectric materials based on doped BaTiO3: from ceramics to sol-gel deposited nano-films, for supporting—Imhotep project 2011-2013, entitled; Multiferroic nanostructure for multifunction sensors application, 2) Team work of Mena Sweden-Egyptian sharing project 2009-2015; entitled; Synthesis and investigation of the physical and electrical properties of BaSrTiO3 (BST) nano-structures prepared by sol-gel methods. For their support and helps in measuring all the Scanning Electron microscope of BST samples in this work.

References

[1]   Li, B., Wanga, C., Liu, W., Ye, M. and Wang, N.G. (2013) Multiferroic Properties of La and Mn Co-Doped $BiFeO_3$ Nanofibers by Sol-Gel and Electrospinning Technique. *Materials Letters*, **90**, 45-48. http://dx.doi.org/10.1016/j.matlet.2012.09.012

[2]   Bhushan, B., Das, D., Priyamc, A., Vasanthacharya, N.Y. and Kumar, S. (2012) Enhancing the Magnetic Characteristics of $BiFeO_3$ Nanoparticles by Ca,Ba Co-Doping Centre for Applied Physics. *Materials Chemistry and Physics*, **135**, 144-149. http://dx.doi.org/10.1016/j.matchemphys.2012.04.037

[3]   Annapu Reddy, V., Patha, N.P. and Nath, R. (2012) Particle Size Dependent Magnetic Properties and Phase Transitions in Multiferroic $BiFeO_3$ Nano-Particles. *Journal of Alloys and Compounds*, **543**, 206-212. http://dx.doi.org/10.1016/j.jallcom.2012.07.098

[4]   Wu, W.W., Cai, J.C., Wu, X.H., Liao, S. and Huang, A.G. (2012) $Co_{0.35}Mn_{0.65}Fe_2O_4$ Magnetic Particles: Preparation and Kinetics Research of Thermal Process of the Precursor. *Powder Technology*, **215-216**, 200-205. http://dx.doi.org/10.1016/j.powtec.2011.09.048

[5]   Ahadi, K., Nemati, A., Mahdavi, S.M. and Vaezi, A. (2013) Effect of Simultaneous Chemical Substitution of A and B Sites on the Electronic Structure of $BiFeO_3$ Films Grown on $BaTiO_3/SiO_2/Si$ Substrate. *Journal of Materials Science: Materials in Electronics* **24**, 2128-2134. http://dx.doi.org/10.1007/s10854-013-1069-6

[6]   Zhou, J., Wang, P., Qiu, Z., Zhu, G. and Liu, P. (2008) Flower-Like $Pb(Zr_{0.52}Ti_{0.48})O_3$ Nanoparticles on the $CoFe_2O_4$ Seeds. *Journal of Crystal Growth*, **310**, 508-512. http://dx.doi.org/10.1016/j.jcrysgro.2007.10.066

[7]   Li, B., Wang, C., Liu, W., Zhong, Y. and An, R. (2012) Synthesis of Co-Doped Barium Strontium Titanate Nanofibers by Sol-Gel Electrospinning Process. *Materials Letters*, **75**, 207-210. http://dx.doi.org/10.1016/j.matlet.2012.02.035

[8]   Bao, D. (2008) Multilayered Dielectric/Ferroelectric Thin Films and Superlattices. *Current Opinion in Solid State and Materials Science*, **12**, 55-61. http://dx.doi.org/10.1016/j.cossms.2009.01.006

[9]   Gao, L., Zhai, J. and Yao, X. (2009) The Influence of Co Doping on the Dielectric, Ferroelectric and Ferromagnetic Properties of $Ba_{0.70}Sr_{0.30}TiO_3$ Thin Films. *Applied Surface Science*, **255**, 4521-4525. http://dx.doi.org/10.1016/j.apsusc.2008.11.064

[10] Solopan, S.A., Vyunov, O.I., Belous, A.G., Tovstolytkin, A.I. and Kovalenko, L.L. (2010) Magnetoelectric Effect in Composite Structures Based on Ferroelectric.Ferromagnetic Perovskites. *Journal of the European Ceramic Society*, **30**, 259-263. http://dx.doi.org/10.1016/j.jeurceramsoc.2009.05.043

[11] Deshpandea, S.B., Khollamb, Y.B., Bhoraskarb, S.V., Dateb, S.K., Sainkara, S.R. and Potdara, H.S. (2005) Synthesis and Characterization of Microwave-Hydrothermally Derived $Ba1-xSrxTiO_3$ Powders. *Materials Letters*, **59**, 293-296. http://dx.doi.org/10.1016/j.matlet.2004.10.006

[12] Zuo, X.H., Deng, X.Y., Chen, Y., Ruan, M., Li, W., Liu, B., Qu, Y. and Xu, B. (2010) A Novel Method for Preparation of Barium Strontium Titanate Nanopowders. *Materials Letters*, **64**, 1150-1153. http://dx.doi.org/10.1016/j.matlet.2010.02.034

[13] Mahani, R.M., Battisha, I.K., Salem, M.A. and Abou Hamad, A.B. (2010) Structure and Dielectric Behavior of Nano-Structure Ferroelectric $Ba_xSr_{1-x}TiO_3$ Prepared by Sol-Gel Method. *Journal of Alloys and Compounds*, **508**, 354-358. http://dx.doi.org/10.1016/j.jallcom.2010.05.060

[14] Wei, X., Xu, G., Ren, Z., Wang, Y., Shen, G. and Han, G. (2008) Size-Controlled Synthesis of $BaTiO_3$ Nanocrystals via a Hydrothermal Route. *Materials Letters*, **62**, 3666-3669. http://dx.doi.org/10.1016/j.matlet.2008.04.022

[15] El-Naggar, M.Y., Dayal, K., Goodwin, D.G. and Bhattacharya, K. (2006) Graded Ferroelectric Capacitors with Robust Temperature Characteristics. *Journal of Applied Physics*, **100**, 114115. http://dx.doi.org/10.1063/1.2369650

[16] Tsai, K.-C., Wu, W.-F., Chao, C.-G., Lee, J.-T. and Shen, S.-W. (2006) Improving Electrical Properties and Thermal Stability of (Ba,Sr)TiO3 Thin Films on Cu(Mg) Bottom Electrodes. *Japan Journal of Applied Physics*, **45**, 5495-5500. http://dx.doi.org/10.1143/JJAP.45.5495

[17] Subramanyam, G., Ahamed, F. and Biggers, R. (2005) A SiMMIC Compatible Ferroelectric Varactor Shunt Switch for Microwave Applications. *IEEE Microwave and Wireless Components Letters*, **15**, 739-741. http://dx.doi.org/10.1109/LMWC.2005.858992

[18] Ahmed, M.A., Kamal, M., El Desouky, F.G., Girgis, E., Farag, I.S.A. and Batttisha, I.K. (2013) Synthesis and Characterization of Cobalt-Doped Nano-Structure Barium Strontium Titanate Prepared by Sol-Gel Process. *Journal of Applied Sciences Research*, **9**, 2432-2438.

[19] Culity, B.D. and Graham C.D. (2009) Introduction to Magnetic Materials. John Wiley, Sons, Inc., Hoboken, New Jersey.

[20] Guo, Z., Pan, L., Qiu, C.H., Zhao, X., Yang, L. and Rafique, M.Y. (2013) Structural and Multiferroic Properties of Fe-Doped $Ba_{0.5}Sr_{0.5}TiO_3$ Solids. *Journal of MMM*, **325**, 24-28.

20

Improving the Dielectric Properties of High Density Polyethylene by Incorporating Clay-Nanofiller

Ossama E. Gouda[1], Sohair F. Mahmoud[2], Ahmed A. El-Gendy[3], Ahmed S. Haiba[2]

[1]Electrical Department, Faculty of Engineering, Cairo University, Giza, Egypt
[2]High Voltage Metrology Lab, National Institute of Standards (NIS), Giza, Egypt
[3]Nanotechnology and Nanometrology Lab, National Institute of Standards (NIS), Giza, Egypt
Email: Prof_ossama11@yahoo.com, sohairfakhry@hotmail.com, dr_aboad2000@yahoo.com, eng_haiba@yahoo.com

Abstract

Polymer nanocomposites have been used for various important industrial applications. The preparation of high density polyethylene composed with Na-montmorillonite nanofiller using melt compounding method for different concentrations of clay-nanofiller of 0%, 2%, 6%, 10%, and 15% has been successfully done. The morphology of the obtained samples was optimized and characterized by scanning electron microscope showing the formation of the polymer nanocomposites. The thermal stability and dielectric properties were measured for the prepared samples. Thermal gravimetric analysis results show that thermal stability in polymer nanocomposites is more than that in the base polymer. It has been shown that the polymer nanocomposites exhibit some very different dielectric characteristics when compared to the base polymer. The dielectric breakdown strength is enhanced by the addition of clay-nanofiller. The dielectric constant (ε_r) and dissipation factor (Tan δ) have been studied in the frequency range 200 Hz to 2 MHz at room temperature indicating that enhancements have been occurred in ε_r and Tan δ by the addition of clay-nanofiller in the polymer material when compared with the pure material.

Keywords

Polymer Nanocomposites, High Density Polyethylene, Dielectric Breakdown, Dielectric Constant, Dissipation Factor

1. Introduction

Polymers play an important role for many applications due to their unique properties which can be classified as heat sensitive, flexible, electrically insulating, amorphous, or semi-crystalline materials. For that reason, polymers are the most commonly used dielectrics because of their reliability, availability, ease of fabrications, and low cost. The selection of the proper dielectric polymer for a desired application depends on the requirements and operating conditions of the applied system [1]. The electrical properties of polymers can be improved by the addition of inorganic nano-fillers to the polymers forming new materials called polymer nanocomposites (PNC). Polymer nanocomposites are composite materials having several wt% of inorganic particles of nanometer dimensions homogeneously dispersed into their polymer matrix. PNC with better dielectric and electrical insulation properties are slowly emerging as excellent functional materials for dielectrics and electrical insulation application and the term "nanodielectrics" for such materials is increasingly becoming popular. Although the technology of addition of fillers to polymers to enhance a particular dielectric property has been in existence for several decades [2]-[4], the effect of filler size on the dielectric behaviour of the polymer composites has not been understood fully. It is with the advent of nanotechnology leading to the availability and commercialization of nanoparticles that polymer nanocomposite technology started to gain momentum. Polymer nanocomposites have been found to exhibit enhanced physical, thermal, mechanical, and dielectric properties when compared to the traditional polymer materials especially at low nano-filler concentrations (1% - 10%) [5]-[7]. However, it is only recently that the dielectric properties of such polymer nanocomposites were looked into and limited research results demonstrate very encouraging dielectric properties for these materials. Irrespective of the type of base polymer material (thermoplastic or thermoset), significant enhancements in several physical properties, like thermal conductivity (with conducting fillers) or dielectric properties like resistivity, permittivity, dissipation factor, dielectric strength, tracking and partial discharge resistant characteristics (with insulating fillers) were observed when compared to similar properties in traditional neat polymers [8]-[10]. These observations were mainly attributed to the unique properties of nanoparticles and the large interfacial area in polymer nanocomposites [11]-[13].

The present work focuses on the dielectric properties of polyethylene (PE) nanocomposites. Polyethylene is one of the thermoplastic polyolefin which is traditionally one of the most widely used polymer classes with applications in structural, textile, and packaging industries, and their nanocomposites have found multiple applications for the same uses. This paper shows the preparation and characterization of high density polyethylene composed with Na-montmorillonite clay-nanofiller (HDPE/clay) with different concentrations of clay-nanofiller as 0%, 2%, 6%, 10% and 15%. Then the dielectric properties, such as dielectric constant, dissipation factor, dielectric breakdown, and insulation resistance, of the prepared samples will be discussed and compared to the base polymer material.

2. Experimental Work

2.1. Materials

HDPE with melt flow rate of 0.75 g/min and density of 960 kg/m^3 is chosen as the base polymer material for the current study. It was manufactured by the International Company for Manufacturing Plastic Products. Sodium montmorillonite clay K10 (MMT) was acquired from fluka chemika company. Hexadecyl Trimethyl Ammonium Bromide, modifier or surfactant material, was obtained from Merck KGaA, Darmstadt, Germany.

2.2. Modification of Clay

The preparation of polymer/clay nanocomposites with good dispersion of clay layers within the polymer matrix is not possible by physical mixing of polymer and clay particles. It is not easy to disperse nanolayers in most polymers due to the high face to face stacking of layers in agglomerated tactoids and their intrinsic hydrophilisity which make them incompatible with hydrophobic polymers. The intrinsic incompatibility of hydrophilic clay layers with hydrophobic polymer chains prevents the dispersion of clay nanolayers within polymer matrix and causes to the weak interfacial interactions. Modification of clay layers with hydrophobic agents is necessary in order to render the clay layers more compatible with polymer chains, and result in a larger interlayer spacing. In addition, modification process improves the strength of the interface between the inorganic clay and the polymer matrix. So, sodium montmorillonite (Na-MMT) clay was modified with the compatiblizer of Hexadecyl Trimethyl Ammonium Bromide [14].

100 g of clay was dispersed into 1000 ml of methanol solvent and placed on hot plate with magnetic stirrer to allow continuous stirring for 2 hours. On the other hand, 100 g of hexadecyl trimethyl ammonium bromide was dissolved in 500 ml of methanol. Then the solution was added to clay dispersion. The stirring continued for 72 hours. After that, the modified clay was filtered and collected. Finally the filtrate was dried in a vacuum oven at 70°C for 6 hours [15].

2.3. Preparation of HDPE/Clay Composites

The concentrations of modified clay-nanofiller were added as 0%, 2%, 6%, 10%, and 15% into the base polymer material. HDPE/clay nanocomposites were prepared by melt compounding method (master batch method) using twin screw extruder (TSE) at zones temperature 163°C, 167°C, and 167°C, for Zone 1, Zone 2, and Zone 3 respectively. The screw speed was maintained at 30 rpm. After extrusion, the dried pellets of nanocomposites were preheated using Morgan Press Injection unit at 160°C for 30 min and injected to produce test samples with dimensions 7.5 cm × 7.5 cm × 0.25 cm for dielectric measurements [15]. The prepared samples are referred to in this paper as HDPE 0% (pure material), HDPE 2%, HDPE 6%, HDPE 10%, and HDPE 15%.

2.4. Characterization of HDPE/Clay Composites

The prepared samples were characterized by the scanning electron microscope (SEM). SEM is a type of electron microscope that produces images of a sample by scanning it with a focused beam of electrons. The electrons interact with atoms in the sample, producing various signals that can be detected and that contain information about the sample's surface topography and composition. The scanning electron microscope images were carried out by using SEM, model Quanta 250 FEG (Field Emission Gun) attached with EDX unit (Energy Dispersive X-ray Analyses), with accelerating voltage 30 kV, magnification 14× up to 1,000,000×, and a resolution of 1 nm.

Thermal stability was measured by using thermo gravimetric analysis (TGA). TGA experiments were done by a shimadzu TA-50 thermal analyzer using scanning rate of 5°C/min under N2 with 20 ml/min flow rate, from room temperature to 600°C.

2.5. Dielectric Properties

Dielectric breakdown refers to a rapid reduction in the resistance of an electrical insulator when the voltage applied across it exceeds the breakdown voltage. Dielectric breakdown measurements were performed using AC Dielectric Test Set. The samples were sandwiched between two electrodes and tested at room temperature under an ac voltage ramp of 750 V/sec. The ac voltage was increased with a rate of 750 V/Sec until breakdown occurred.

Dielectric constant is called relative permittivity which is a parameter that indicates the relative charge storage capability of dielectrics in the presence of an electric field. The used instrument is an Agilent E4980A LCR meter with dielectric sample holder. The equivalent parallel capacitance (C_p) was measured directly by the LCR meter, then the dielectric constant is calculated as shown below in the results section.

Dissipation factor is called loss tangent or Tan δ. It represents the energy loss in the dielectrics and it is preferred to be smaller for insulation materials. It was measured directly by an Agilent E4980A LCR meter with dielectric sample holder in the frequency range 200 Hz to 2 MHz at room temperature.

Also, insulation resistance was measured directly by LCR meter at the same conditions.

3. Results and Discussion

3.1. Scanning Electron Microscopy (SEM)

The morphology of the SEM images for HDPE with 2% clay, 6% clay, 10% clay, and 15% clay composites is shown in **Figures 1-4** respectively. Each sample has two images with different magnifications. All SEM images for all samples revealed that, clay was dispersed in polymer matrix very well and there wasn't any accumulation of clay-nanofiller in it. An important observation is that the thickness of clay content is still in nano-size range (1 - 100 nm). This means that the samples were successfully prepared.

3.2. Thermal Analysis

The thermal stability of the prepared samples was measured using thermo-gravimetric analyzer (TGA). In this

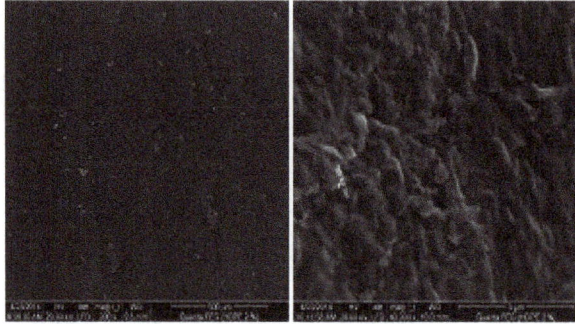

Figure 1. SEM images for HDPE 2% sample at (200× & 40,000×) magnifications.

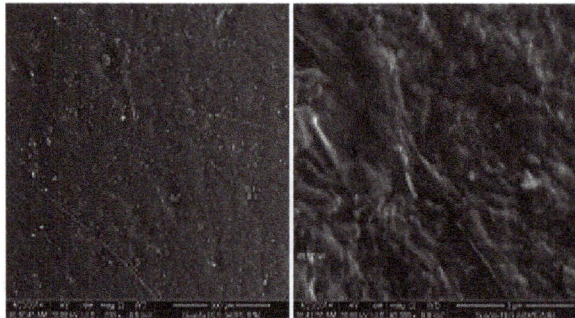

Figure 2. SEM images for HDPE 6% sample at (200× & 40,000×) magnifications.

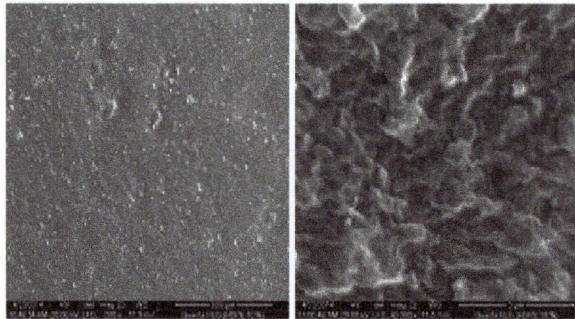

Figure 3. SEM images for HDPE 10% sample at (200× & 40,000×) magnifications.

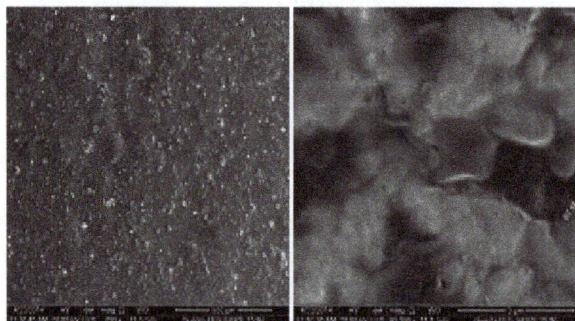

Figure 4. SEM images for HDPE 15% sample at (200× & 40,000×) magnifications.

technique, the weight loss of the material due to the formation of volatile compounds under degradation because of the heating and temperature rise is monitored.

The data available from TGA is tabulated in **Table 1** and graphed in **Figure 5** including $T_{10\%}$ (onset temperature), the temperature at which 10% degradation from the sample occurs, $T_{50\%}$, the temperature at which 50% degradation occurs, T_{max}, the temperature at which maximum degradation occurs, and residual loss at 600˚C.

According to TGA results as shown in **Figure 5**, the incorporation of MMT to HDPE improved the thermal stability at higher degradation temperature ranges compared to pure HDPE. The temperature of the 10% degradation of HDPE 2%, HDPE 6% and HDPE 10% has been shifted to lower temperatures relative to HDPE 0%, while the 10% degradation temperature of HDPE 15% shifted to higher temperatures compared to HDPE 0%. The 50% and maximum degradation temperatures have been shifted to higher temperatures compared to HDPE 0%. This means that, thermal stability has been occurred with increasing the concentration of MMT composed to HDPE. The residual weight of the samples at 600˚C increased with increasing the concentration of clay composed to HDPE. Thus, thermal stability of HDPE/clay has been improved compared to pure HDPE.

3.3. Dielectric Properties

3.3.1. Dielectric Breakdown Strength

The dielectric breakdown strength of the composites is analyzed using an AC dielectric test set at room temperature. The test was repeated 5 times for each sample and the average value was recorded and plotted as shown in **Figure 6**.

Figure 6 shows the behavior of the dielectric strength for HDPE/clay composites. Results show that the dielectric breakdown voltage increases with increasing the concentrations of clay-nanofiller to HDPE when compared to the pure material having the same dimensions until reaching to an optimum value (36.1 kV) at HDPE 6% then, the breakdown voltage starts to decrease at 10%, and 15% clay-nanofillers. Although the breakdown voltage decreases at 10% clay-nanofiller, its value is larger than the value of pure material. It is observed that the breakdown voltage value of the sample HDPE 15% is lower than that of pure material. As a result, the dielectric breakdown strength has been improved at all concentrations of clay-nanofiller except 15% clay concentration when compared to the unfilled material. The optimum enhancement occurred at HDPE 6%.

Table 1. TGA results for HDPE and HDPE/clay composites.

Samples	$T_{10\%}$ (˚C)	$T_{50\%}$ (˚C)	T_{max} (˚C)	Residual Weight (mg) at 600˚C
HDPE 0%	403.6	451.4	478.3	0.32
HDPE 2%	403.5	451.6	479.3	0.42
HDPE 6%	396.7	459.0	481.2	1.24
HDPE 10%	399.1	463.1	484.2	1.77
HDPE 15%	405.3	464.8	485.1	4.34

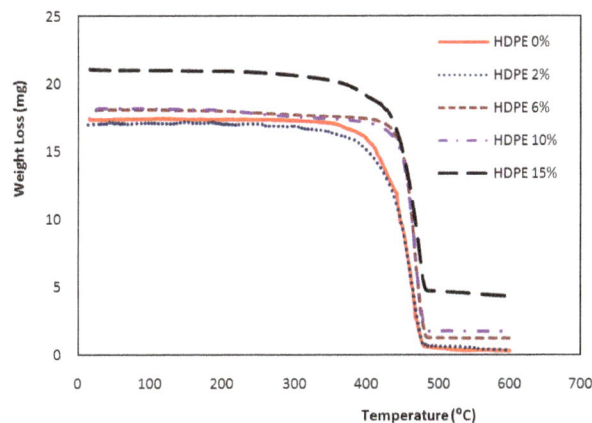

Figure 5. TGA curves for HDPE and HDPE/clay composites.

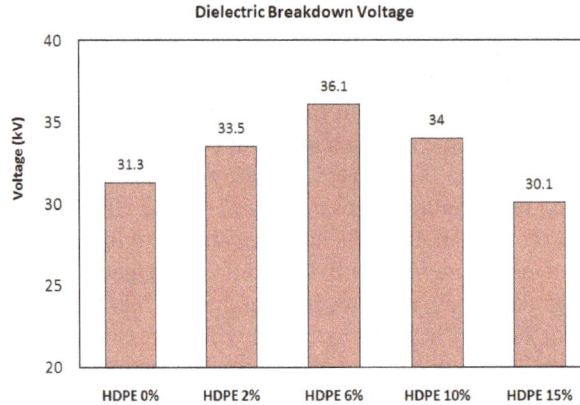

Figure 6. Dielectric breakdown strength measurement for HDPE/clay composites.

3.3.2. Dielectric Constant (ε_r)

Measured quantity was the equivalent parallel capacitance (C_p) of the samples in the frequency range of 200 Hz to 2 MHz, then the dielectric constant (ε_r) was calculated by the following equations [16] and plotted as shown in **Figure 7**.

$$C_p = \frac{\varepsilon_0 \varepsilon_r A}{t} \tag{1}$$

$$\varepsilon_r = C_p \frac{t}{\varepsilon_0 A} \tag{2}$$

where ε_0 = 8.854 × 10 - 12 F/m is the permittivity of free space, (A) is the area of electrodes, and (t) is the thickness of the samples.

Figure 7 shows the variation of the dielectric constant (ε_r) with frequency at room temperature for all samples. Observed differences were found in dielectric constant between pure HDPE and HDPE composites with different concentrations of clay-nanofiller. It is seen that, ε_r decreases with increasing frequency for all samples. An important observation is that dielectric constant decreases considerably with the addition of clay-nanofiller up to 6% filler concentration, and then it starts to increase at 10% and 15% filler concentrations. The value of ε_r at HDPE 10% is still lower than that of pure HDPE and its value at HDPE 15%, is higher than the pure material. HDPE 6% has the lowest dielectric constant and HDPE 15% has the highest dielectric constant. The increasing of ε_r at 10% and 15% filler concentrations may be due to the effect of ε_r of composites (inclusions + matrix) on the resultant permittivity [17] [18]. This means that an enhancement occurred in dielectric constant at 2%, 6%, and 10% filler concentrations when these composites used as insulating materials.

3.3.3. Dissipation Factor (Tan δ)

Figure 8 shows the variation of the dissipation factor (Tan δ) with frequency at room temperature for all samples. As shown in the figure, Tan δ decreases with increasing frequency for all samples. Also an important observation is that Tan δ decreases with increasing the concentrations of clay-nanofiller incorporated in polymeric material up to 6% filler concentration, then it further increases at 10% and 15% filler concentrations. This may be due to the increasing of conductivity according to increasing of nano-filler concentration [17] [18]. HDPE 6% has the lowest dissipation factor and HDPE 15% has the highest dissipation factor. In addition, the values of Tan δ for IIDPE 2%, HDPE 6% and HDPE 10% samples are less than that of pure material. On the other hand, the values of Tan δ for 15% filler concentrations are higher than that of pure material. This means that clay-nanofiller improves the dissipation factor for HDPE polymeric material when used for insulating purpose.

3.3.4. Insulation Resistance (R)

Figure 9 shows the variation of the insulation resistance (R) with frequency at room temperature for all samples. Marked differences were found in the insulation resistance between pure HDPE and HDPE composites with different concentrations of clay-nanofiller. It is seen that, the insulation resistance decreases with increasing frequency for all samples. An important observation is that the insulation resistance increases with the addition of clay-nanofiller up to 6% filler concentration, and then it starts to decrease at 10% and 15% filler concentrations. Although that, the insulation resistance value at 10% filler concentration is still higher than that of pure HDPE. On the other hand, the insulation resistance value at 15% filler concentrations is lower than that of pure sample. The decreasing of the insulation resistance at 10 % and 15% filler concentrations may be due to the increasing of conductivity of composites at high concentrations. This means that an enhancement occurred in the insulation resistance up to 10% filler concentration when compared to pure material.

Figure 10 and **Figure 11** show the instruments which used for dielectric and breakdown measurements respectively.

4. Conclusion

HDPE/clay composites are prepared by melt compounding method (Master Batch method). Morphology structure (clay dispersion in polymer matrix and the thickness of clay-nanofiller) of the prepared samples is investigated by SEM. SEM images show that clay content is well dispersed in the polymer matrix indicating samples are successfully prepared. Thermal stability and dielectric properties are investigated for the prepared samples. TGA results show that HDPE nanocomposites have thermal stability more than unfilled polymer material. Dielectric breakdown strength is improved by the addition of clay-nanofillers. Dielectric constant and dissipation factor are studied at room temperature in the frequency range 200 Hz to 2 MHz. The experimental results show that there is an enhancement in both ε_r and Tan δ due to the unique behavior of clay-nanofiller when incorpo-

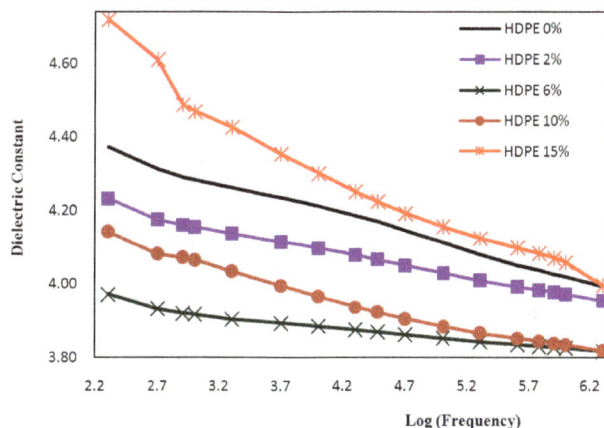

Figure 7. Frequency dependence of dielectric constant at room temperature.

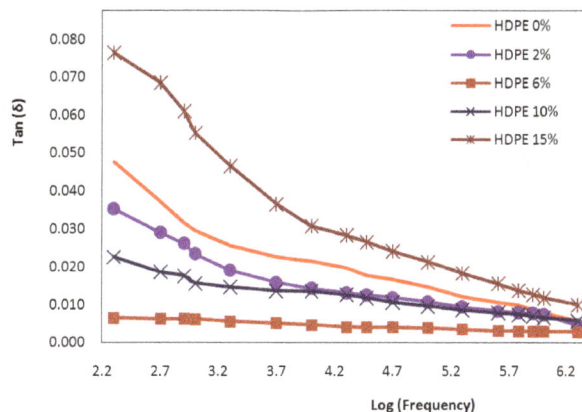

Figure 8. Frequency dependence of dissipation factor at room temperature.

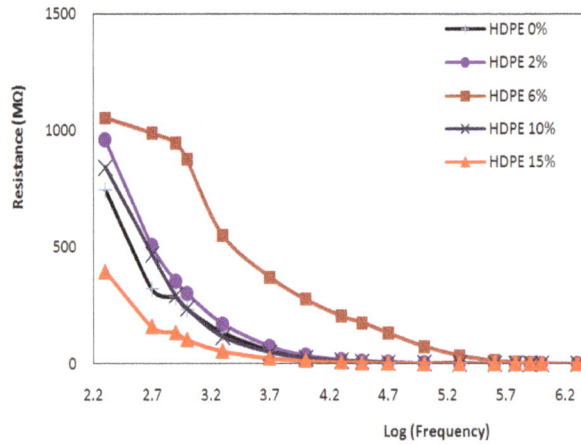

Figure 9. The insulation resistance variation with variable frequencies at room temperature.

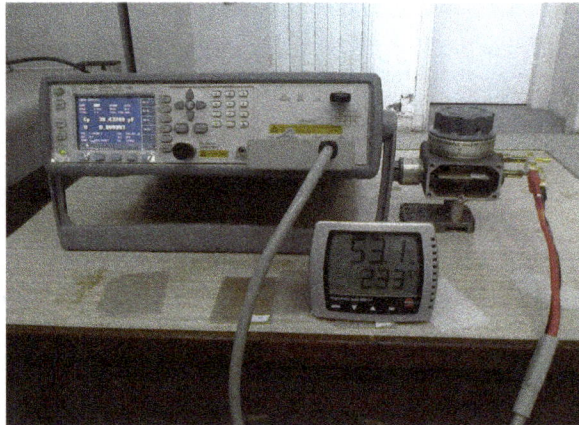

Figure 10. Agilent E4980A LCR meter with dielectric sample holder.

Figure 11. Dielectric test set for dielectric breakdown measurement.

rated into the polymer base matrix HDPE. Also insulation resistance has been improved by the addition of clay-nanofiller. From all results, it can be noticed that 6% filler concentration is the optimum clay content for HDPE/clay system.

Acknowledgements

The authors would like to thank the Fire and Explosion Protection Lab at NIS for giving access to the TSE facility used in this research work. The authors are also grateful to Prof. Dr. Mostafa, in Thermometry Lab at NIS, for his help in the TGA measurements.

References

[1] Du, B.X. and Liu, H.J. (2010) Effects of Atmospheric Pressure on Tracking Failure of Gamma-Ray Irradiated Polymer Insulating Materials. *IEEE Transactions on Dielectrics and Electrical Insulation*, **17**, 541-547. http://dx.doi.org/10.1109/TDEI.2010.5448110

[2] Carmona, F. (1989) Conducting Filled Polymers. *Physica A*, **157**, 461-469. http://dx.doi.org/10.1016/0378-4371(89)90344-0

[3] Bai, Y., Cheng, Z.-Y., Bharti, V., Xu, H.S. and Zhang, Q.M. (2000) High Dielectric-Constant Ceramic-Powder Polymer Composites. *Applied Physics Letters*, **76**, 3804-3806. http://dx.doi.org/10.1063/1.126787

[4] Ueki, M.M. and Zanin, M. (1999) Influence of Additives on the Dielectric Strength of High-Density Polyethylene. *IEEE Transactions on Dielectrics and Electrical Insulation*, **6**, 876-881. http://dx.doi.org/10.1109/94.822030

[5] Messersmith, P.B. and Giannelis, E.P. (1994) Synthesis and Characterization of Layered Silicate-Epoxy Nanocomposites. *Chemistry of Material*, **6**, 1719-1725. http://dx.doi.org/10.1021/cm00046a026

[6] Ray, S.S. and Okamoto, M. (2003) Polymer/Layered Silicate Nanocomposites: A Review from Preparation to Processing. *Progress in Polymer Science*, **28**, 1539-1641. http://dx.doi.org/10.1016/j.progpolymsci.2003.08.002

[7] Gensler, R., Groppel, P., Muhrer, V. and Muller, N. (2002) Applications of Nanoparticles in Polymers for Electronic and Electrical Engineering. *Particle and Particle Systems Characterization*, **19**, 293-299. http://dx.doi.org/10.1002/1521-4117(200211)19:5<293::AID-PPSC293>3.0.CO;2-N

[8] Tanaka, T. (2005) Dielectric Nanocomposites with Insulating Properties. *IEEE Transactions on Dielectrics and Electrical Insulation*, **12**, 914-928. http://dx.doi.org/10.1109/TDEI.2005.1522186

[9] Cao, Y., Irwin, P.C. and Younsi, K. (2004) The Future of Nanodielectrics in the Electrical Power Industry. *IEEE Transactions on Dielectrics and Electrical Insulation*, **11**, 797-807. http://dx.doi.org/10.1109/TDEI.2004.1349785

[10] Imai, T., Sawa, F., Ozaki, T., Inoue, Y., Shimizu, T. and Tanaka, T. (2006) Comparison of Insulation Breakdown Properties of Epoxy Nanocomposites under Homogeneous and Divergent Electric Fields. 2006 *IEEE Conference on Electrical Insulation and Dielectric Phenomena*, Kansas, 15-18 October 2006, 306-309. http://140.98.202.196/xpl/login.jsp?tp=&arnumber=4105431&url=http%3A%2F%2F140.98.202.196%2Fstamp%2Fstamp.jsp%3Ftp%3D%26arnumber%3D4105431

[11] Lewis, T.J. (2004) Interfaces Are the Dominant Feature of Dielectrics at the Nanometric Level. *IEEE Transactions on Dielectrics and Electrical Insulation*, **11**, 739-753. http://dx.doi.org/10.1109/TDEI.2004.1349779

[12] Roy, M., Nelson, J.K., MacCrone, R.K. and Schadler, L.S. (2005) Polymer Nanocomposite Dielectrics—The Role of the Interface. *IEEE Transactions on Dielectrics and Electrical Insulation*, **12**, 629-643. http://dx.doi.org/10.1109/TDEI.2005.1511089

[13] Ajayan, P.M., Schadler, L.S. and Braun, P.V. (2003) Nanocomposite Science and Technology. John Wiley & Sons Inc., New York. http://dx.doi.org/10.1002/3527602127

[14] Liu, X.H. and Wu, Q.J. (2001) PP/Clay Nanocomposites Prepared by Grafting-Melt Intercalation. *Elsevier Polymer*, **42**, 10013-10019. http://dx.doi.org/10.1016/S0032-3861(01)00561-4

[15] Hassan, M.A., El-Sayed, I., Nour, M.A. and Mohamed, A.A. (2012) Flammability Properties of HDPE Nanocomposites Based on Modification of Na-MMT with Organo Silane and Ammonium Phosphate Mono and Di Basic. *Elixir Applied Chemistry*, **46**, 8328-8333. http://www.elixirpublishers.com/index.php?route=articles/category&path=305_306&page=9

[16] Li, L. (2011) Dielectric Properties of Aged Polymers and Nanocomposites. Ph.D. Thesis, Iowa State University, Iowa.

[17] Singha, S. and Thomas, M.J. (2008) Permittivity and Tan Delta Characteristics of Epoxy Nanocomposites. *IEEE Transactions on Dielectrics and Electrical Insulation*, **15**, 2-11. http://dx.doi.org/10.1109/T-DEI.2008.4446731

[18] Singha, S. and Thomas, M.J. (2008) Dielectric Properties of Epoxy Nanocomposites. *IEEE Transactions on Dielectrics and Electrical Insulation*, **15**, 12-23. http://dx.doi.org/10.1109/T-DEI.2008.4446732

21

Measurement of Error in Electrostriction Based Dielectrics

Anjani Kumar Singh

Bharati Vidyapeeth's College of Engineering, New Delhi, India
Email: anjaninsit@gmail.com

Abstract

While correlating the various components of mechanical stress tensor due to elastic response with the corresponding components of electrically induced stress tensor pertaining to quadratic electrostriction, proper precautions are to be observed for higher order electromechanical coupling. Contributions from lateral stresses, electrical and mechanical boundary conditions are to be considered for the correct estimation of induced strain in elastic dielectrics. The knowledge of dependence of Maxwell's electrostatic stresses on dielectric constants and on the orientation of dielectric material with respect to the electric field vector is necessary for the exact estimation of electrically induced strains. The contributions from the variation in transverse components of dielectric tensor produced by the variation in lateral stresses are to be incorporated into the expression for the correct estimation of electrostrictive coefficients. The electromechanical behavior of elastic dielectric is discussed and the errors often committed in using incorrect formulae for electrostriction are reported.

Keywords

Elastic Dielectric, Electrostriction, Induced Strain, Lateral and Shear Stresses

1. Introduction

The analysis that has been presented in this paper gives a guideline indicating errors committed in the estimation of the electrically induced strain in elastic dielectrics. The Maxwell stress effect occurs due to variation in electric field distribution with strain and the phenomena of electrostriction occur due to variation in dielectric properties of the material with strain. The linear elastic response governed by Hooke's law, the Maxwell electrostatic stress governed by coulomb's law and the electrostriction stress (dielectric response) are illustrated in **Figure 1**.

The deformed material (generally non piezoelectric) is no longer isotropic, and the scalar permittivity (ε) becomes dielectric tensor $\left(\varepsilon_{ij}\right)$ due to its anisotropy behavior.

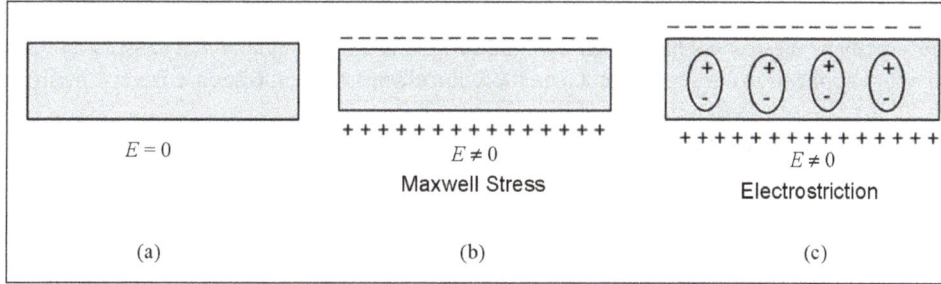

Figure 1. Schematic diagrams illustrating (a) pure linear elastic response governed by Hooke's law for $E = 0$ (b) Maxwell stress produced by the surface charge and governed by coulomb law for $E \neq 0$ & (c) Electrostriction stress produced due to alignment of dipoles inside the dielectric material for $E \neq 0$.

As the deformation is extremely small, only the first order terms in strain tensor S_{ij} have been considered and the dielectric tensor ε_{ij} in terms of the strain tensor is given [1] by

$$\varepsilon_{ij} = \varepsilon^0 \delta_{ij} + a_1 S_{ij} + a_2 S_{kk} \delta_{ij}, \tag{1}$$

where ε^0 is the permittivity of the undeformed body and a_1 & a_2 are two parameters describing the variation in dielectric properties of the material in shear and bulk deformation respectively.

The Maxwell stress tensor [1], derived on the fundamental assumptions for the linear electrostriction, is a result of force produced by the electric field and is given as

$$T_{ij} = T_{ij}^0 + \frac{\varepsilon_0}{2}\left(2\varepsilon^0 - a_1\right)E_i E_j - \frac{\varepsilon_0}{2}\left(\varepsilon^0 + a_2\right)E^2 \delta_{ij}, \tag{2}$$

where T_{ij}^0 is stress tensor in the absence of an external electric field. Generally we use to neglect T_{ij}^0 in isotropic dielectrics but in case of composite dielectric material or dielectric interface, the presence of significant number of point defects introduces distortion. In case of size difference of two particles from two phases, elastic stress strains are created. A larger atom introduces compressive stress and corresponding strain around it, while a smaller interacting atom creates a tensile stress-strain field. An interstitial atom also produces strain around the void it is occupying. However, it is very difficult to evaluate T_{ij}^0 precisely [2].

$$S_{ij} = \frac{1}{Y}\left[(1+\sigma)T_{ij} - \sigma T_{kk}\delta_{ij}\right], \tag{3}$$

conversely,

$$T_{ij} = \frac{Y}{1+\sigma}\left(S_{ij} + \frac{\sigma}{1-2\sigma}S_{kk}\delta_{ij}\right). \tag{4}$$

It is very important to note that the above Equations (2) and (4) have been correlated in the electromechanical coupling on taking certain approximations, however these equations have been formulated certainly on different assumptions particularly in respect of range of interaction for the body forces and surface forces in continuum mechanics, and polarization effect in dielectrics. For the measurement of electrically induced strains, the dependence of traction vector (at the boundary/interface between the dielectric material and electrodes) on the orientation of dielectric sample with the electric field is of practical interest [3]. However the boundary conditions [4] at the interface between electrode and the dielectrics are the points of main concern as it is very difficult to predict exact mechanical and electrical boundary conditions in the deformed dielectrics.

The objective of this work is to provide a clear understanding of the possible mechanism involving various dielectric and mechanical parameters for electric field induced strains in elastic dielectrics. Experimental data on the subject are not abundant and the researchers [5] [6] frequently used incorrect formulae in the derivation of elastic strain with respect to the Maxwell stress effect, particularly in case of polyurethane elastomer (a crossed linked polymer), assuming various unrealistic approximations as observed by [7]. A linear electromechanical effect does not exist in case of elastomers and the Hooke's law based on thermodynamics consideration (Helmholtz free energy and Gibbs free energy concept) should not be applied for elastomers up to a large extent due to its nonlinear elastic behavior. Electromechanical coupling effects in case of non-piezoelectric material such as polyurethane elastomers have been exploited in the areas of fundamental sensors and actuators [8] [9]. Due to potential applications in sensing and actuation, the electrostriction response is very significant [10].

2. Maxwell Electrostatic Stresses

In electromagnetic field theory, the Maxwell stress tensor T_{ij} has been formulated [11] using classical electrostatic and assuming uniform electric field as

$$
\begin{aligned}
T_{ij} &= E_i D_j - \frac{1}{2}\delta_{ij}\sum_{k=1}^{3} E_k D_k \\
&= E_i \sum_{k=1}^{3} \varepsilon_{jk} E_k - \frac{1}{2}\delta_{ij}\sum_{m=1}^{3}\sum_{k=1}^{3}\varepsilon_{mk} E_k E_m
\end{aligned}
\tag{5}
$$

where E_i is the electric field strength vector, D_i is the electric displacement vector, and ε_{ij} is the dielectric tensor. A symmetric dielectric tensor ε_{ij} in case of triclinic crystal is given by

$$
\varepsilon_{ij} = \begin{bmatrix} \varepsilon_{11} & \varepsilon_{12} & \varepsilon_{13} \\ \varepsilon_{21} & \varepsilon_{22} & \varepsilon_{23} \\ \varepsilon_{31} & \varepsilon_{32} & \varepsilon_{33} \end{bmatrix}
\tag{6}
$$

If an electric field is parallel to the Z-axis (unidirectional field as in case of parallel plate capacitor), the tensor T_{ij} is reduced [3] to

$$
\begin{aligned}
T_{ij}(0,0,E_z) &= \begin{bmatrix} -\frac{1}{2}\varepsilon_{33}E_z^2 & 0 & 0 \\ 0 & -\frac{1}{2}\varepsilon_{33}E_z^2 & 0 \\ \varepsilon_{13}E_z^2 & \varepsilon_{23}E_z^2 & \frac{1}{2}\varepsilon_{33}E_z^2 \end{bmatrix} \\
&= \begin{bmatrix} 0 & 0 & 0 \\ 0 & 0 & 0 \\ \varepsilon_{13}E_z^2 & \varepsilon_{23}E_z^2 & \frac{1}{2}\varepsilon_{33}E_z^2 \end{bmatrix} - \frac{1}{2}\varepsilon_{33}E_z^2 \begin{bmatrix} 1 & 0 & 0 \\ 0 & 1 & 0 \\ 0 & 0 & 1 \end{bmatrix}
\end{aligned}
\tag{7}
$$

the first term of this equation represents tension along X & Y direction, the second term implies a compression along Z-direction. Generally compression along longitudinal axis (Z-axis) is always accompanied by extension along transverse plane (X-Y plane) as in (**Figure 2**) and vice-versa. It is also clear from the above Equation (7) that the stress tensor with electric field in Z-direction has affected only the components $\varepsilon_{33}, \varepsilon_{13}$ & ε_{23} of dielectric tensor ε_{ij} corresponding to the strain component S_{33}, S_{13} & S_{23} respectively and other components become zero. The lateral strains $(S_{13}$ & $S_{23})$ exist [11] in the capacitor electrode even if the dielectric medium is empty space. So the contribution from lateral strains should not be neglected for the correct estimation of results.

3. Coupling of Dielectric and Elastic Response Parameters

The direction of electric field is assumed to be along Z-axis $(E_z \neq 0, E_x = E_y = 0)$ in the cartesian co-ordinate system. For the mechanically isotropic material $(T_{ij}^0 = 0)$, the electrically induced stresses given by Equation (2) become

$$
T_{33} = +\frac{1}{2}\varepsilon_0\varepsilon^0 E^2 \left(1 - \frac{a_1 + a_2}{\varepsilon^0}\right),
\tag{8}
$$

$$
T_{11} = T_{22} = -\frac{1}{2}\varepsilon_0\varepsilon^0 E^2 \left(1 + \frac{a_2}{\varepsilon^0}\right),
\tag{9}
$$

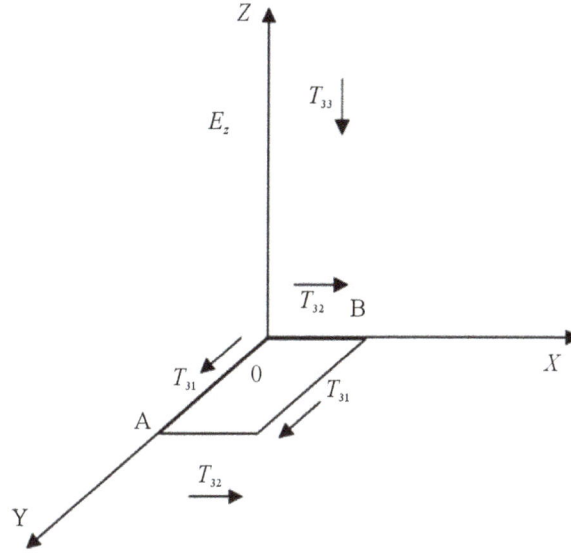

Figure 2. Schematic diagram indicating compression due to normal stress along Z-directions and is accompanied by the extension due to shear stresses along X-Y plane.

$$T_{13} = T_{23} = 0. \tag{10}$$

Equation (8) represents tension along longitudinal (Z-axis) axis and the Equation (9) represents compression along transverse plane. However the researchers [5] [7] [12] have indicated the opposite characteristic of Equations (8) and (9).

The electrically induced stresses will generate elastic strains in equilibrium condition by the method of superposition. On putting the principal value of Maxwell stress Tensor from Equation (8) and Equation (9) into the corresponding components of Equation (3), we get Sacerdote's formula for the relative change in thickness of an elastic dielectric material as the coupling expressions between two phenomena and are given by

$$S_{11} = S_{22} = -\frac{1}{2}\varepsilon_0\varepsilon^0 E_z^2 \left[1 - \frac{\sigma a_1}{\varepsilon^0} + \frac{(1-2\sigma)a_2}{\varepsilon^0} \right] Y^{-1}, \tag{11}$$

$$S_{33} = +\frac{1}{2}\varepsilon_0\varepsilon^0 E_z^2 \left[(1+2\sigma) - \frac{a_1 + (1-2\sigma)a_2}{\varepsilon^0} \right] Y^{-1} \tag{12}$$

In unilateral mechanical deformations described in continuum mechanics, the components of elastic stress tensor (elastic response) of Equation (4) have been correlated with the corresponding components of electrostriction stress tensor (dielectric response) of Equation (2). Sufficient precautions have to be taken while coupling the dielectric response (represented by Equation (2)) with the elastic response (represented by Equations (3) and (4)) to derive the Equations (11) and (12) and correlating higher order components of two phenomena [11]. Because assumptions made for the derivation of mechanical stress tensor (Equation (4)) and electrostriction stress tensor (Equation (2)) are almost inconsistent and both of them can be endowed with a different physical meaning [13]. It has also been pointed out [14] [15] that the body forces (described by a body force density) and the surface force (described by a stress tensor) cannot be applied to all continuous media.

The principal elastic strain, S_{33}, represents the relative change in thickness of dielectric slab where as the principal strains, S_{11} or S_{22}, represent the relative change in diameter of the capacitor as described by [7] which is not correct in case of a parallel plate capacitor in which S_{31} & S_{32} actually represents the lateral change in dimension as the electric field E has only one component along Z direction (**Figure 3**) *i.e.* only one normal component of stress, T_{33} along the direction of electric field and two shearing components of stress tensor, T_{31} & T_{32} along X-Y plane.

On comparing Equations (11) & (12) with the general equation, $S_{ij} = \gamma_{ijkl}E_kE_l$, for electrostriction, we get

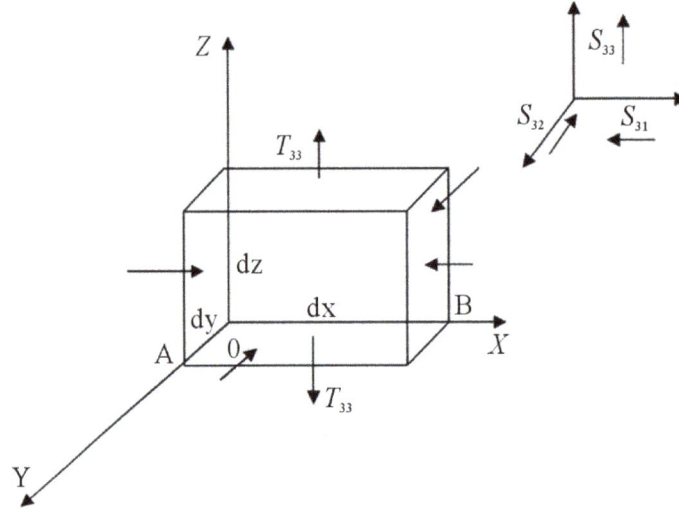

Figure 3. Schematic diagram illustrating extension due to electrostriction stress tensor along Z-direction and compression due to shear stress tensor along X-Y plane.

$$S_{11} = S_{22} = -\gamma_{31}E^2, \ S_{33} = +\gamma_{33}E^2 \tag{13}$$

where γ_{31} & γ_{33} are electrostrictive coefficients. The negative sign for strains S_{11} & S_{22} indicates contraction along X & Y direction, and the positive sign for S_{33} indicates expansion along Z direction as indicated in **Figure 3**.

If the electric field E is changed to $E + \Delta E$, then the change in S_{33} becomes,

$$\Delta S_{33} = +2\lambda_{33}E_0\Delta E = \mu_{33}\Delta E , \tag{14}$$

where μ_{33} is a piezoelectric coefficient. Similarly, the coefficient μ_{31} & μ_{32} can be found along X & Y directions. The piezoelectric transverse coefficients have been frequently used [16] in piezoelectric thin film devices. The average value of compressive elastic strain induced along Z-direction is

$$K\int_0^d \Delta S_{33}\mathrm{d}z , \tag{15}$$

where d is the longitudinal thickness of elastic dielectric slab. Similarly, the expression for the compressive strains (ΔS_{32} & ΔS_{31}) along X & Y direction can be obtained. For nonuniform field as in case of cylindrical capacitor or interface, average value of E^2 can be calculated as

$$\overline{E^2} = \frac{1}{d}\int_0^d E^2(z)\mathrm{d}z , \tag{16}$$

where the integration is over the thickness d of dielectric slab of a capacitor. In the presence of a non-uniform electric field with high magnitude, the phenomenon of dielectrophoresis [4] takes place, however it is in general a weak effect particularly in case of a solid dielectric.

4. Measurement of Electrostrictive Coefficients and the Induced Strain

On differentiating Equation (1) with respect to S_{33}, we get

$$\frac{\mathrm{d}\varepsilon_{33}}{\mathrm{d}S_{33}} = a_1 + a_2\frac{\mathrm{d}S_{kk}}{\mathrm{d}S_{33}} = a_1 + a_2(1 - 2\sigma), \tag{17}$$

Assuming, if the normal force is acting only along Z direction *i.e.* along the direction of electric field E, then we have $S_{11} = S_{22} = 0$ and if the sides of dielectric material are also fixed by the rigid wall (**Figure 4**) then we have $S_{32} = S_{31} = 0$. We have, for the constrained situations as in **Figure 4**,

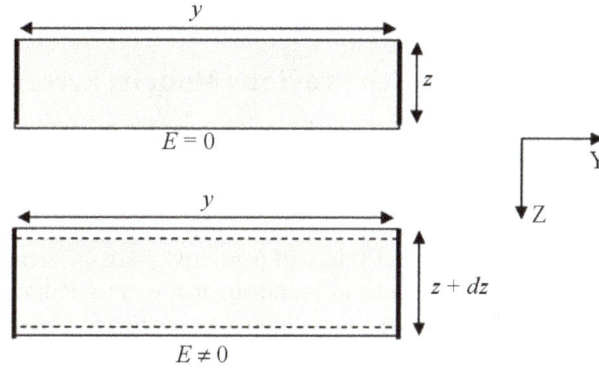

Figure 4. Schematic diagrams illustrating dielectric sample with side of capacitor rigid and fixed (in a constrained situation with $S_{32} = S_{31} = 0$) for (a) $E = 0$ & (b) $E \neq 0$.

$$\frac{\mathrm{d}\varepsilon_{33}}{\mathrm{d}S_{33}} = a_1 + a_2. \tag{18}$$

The second term under bracket of Sacerdate formula Equation (12) can be compared with Equation (17) and the Equation (12) is given as

$$S_{33} = \frac{\nabla z}{z} = -\frac{1}{2}\varepsilon_0\varepsilon^0 E^2\left[(1 + 2\sigma) - \frac{1}{\varepsilon_0}\frac{\mathrm{d}\varepsilon_{33}}{\mathrm{d}S_{33}}\right]. \tag{19}$$

Several researchers [5] [6] have measured second term under bracket by various methods.
For a parallel plate capacitor, the capacity is

$$C = \frac{\varepsilon_0\varepsilon^0 A}{h}, \tag{20}$$

where A is area of plate, h is separation between plates, ε^0 is permittivity. Differentiating above equation, we get

$$\mathrm{d}C = \frac{\varepsilon_0 A}{h}\mathrm{d}\varepsilon^0 + \frac{\varepsilon_0\varepsilon^0}{h}\mathrm{d}A - \left(\varepsilon_0\varepsilon^0 A\right)\mathrm{d}h. \tag{21}$$

From (20) & (21), we get

$$\frac{\mathrm{d}C}{C} = \frac{\mathrm{d}\varepsilon^0}{\varepsilon^0} + \frac{\mathrm{d}A}{A} - \frac{\mathrm{d}h}{h} \tag{22}$$

If the electrodes are rigid and incompressible, then $\mathrm{d}A = 0$ and Equation (22) becomes

$$\left(\frac{\mathrm{d}C}{C}\right)_{die} = \frac{\mathrm{d}\varepsilon^0}{\varepsilon^0} - \frac{\mathrm{d}h}{h}, \tag{23}$$

and

$$\left(\frac{\mathrm{d}C}{C}\right)_{air} = -\frac{\mathrm{d}h}{h} \approx S_{33}. \tag{24}$$

From Equation (19), the electrostrictive term is

$$\frac{1}{\varepsilon_0}\frac{\mathrm{d}\varepsilon_{33}}{\mathrm{d}S_{33}} = \frac{\mathrm{d}\varepsilon_{33}/\varepsilon_0}{\mathrm{d}h/h}. \tag{25}$$

The relative change in thickness and the dielectric parameters, a_1 & a_2, are experimentally obtained [17] by determining $\mathrm{d}\varepsilon_{33}/\mathrm{d}S_{33}$ in constrained and unconstrained situation, with and without dielectric material, in a capacitor. However the linearity of the theory for electrostriction fails if the deformations are not quadratic in the applied field particularly at higher field strength. But very few reliable measurements of electrostrictive coefficients (a_1 & a_2) for solids have been reported in the literature.

5. Discussion and Comparison with Previous Models: Errors in Estimation

Errors are often committed in the derivation of suitable expression for the interpretation of experimental results. The above expressions (Equations (23)-(25)) have been derived on the sole assumption that permittivity ε_{33}^0 changes with longitudinal strain only but the permittivity also changes with respect to the lateral strain/stress in the capacitor depending on the crystal structure of elastic dielectric medium. In triclinic, monoclinic & orthorhombic crystal systems, all the three principal values of permittivity are different, whereas, in the cubic system, these values are same [1]. So the correction due to lateral strain & corresponding change in lateral permittivities ε_{13} & ε_{23} (Equation (6)) must be incorporated to get correct mathematical modeling. The researcher [17] has stopped the change in lateral strain by considering constrained situation (considering sides of capacitor fixed & rigid) but fail to understand that the change in lateral permittivity is also due to lateral electrostatic stresses (Equation (7)).

5.1. Shear Stresses

Zhang *et al.* [6] use non-tensor form of stresses (T) and strains (S) for experimental determination of induced strains in the case of Polyurethane elastomers (DOW 2103-80AE) and the relative change in thickness of the dielectric slab of a parallel plate capacitor is considered as

$$S_{33} = -\frac{1}{2Y}\varepsilon_0 \varepsilon^0 E_z^2 \tag{26}$$

which is quite different from the Equation (12) particularly in case of an elastomer or an elastic dielectric material with large Poisson's ratio $(\sigma \approx 0.5)$. So the neglect of shear stresses/strains leads to incorrect estimation [7] of induced strain. Similarly, another researcher [5] used Equation (8) without considering contributions from shear stresses, edge effect and orientation of electric field with respect to sample.

Errors in the estimations of induced strain can be pointed out for the case of Polycarbonate (PC). The following data for Polycarbonate are given [17] as:

$$|a_1| = 1.404, \ |a_2| = 2.6 \ .$$

Substituting above data in Equations (12) and (26), an overestimation 179% has been found. Similarly comparing the Equation (8) divided by Y as considered by [5] with Equation (12), an overestimation of 137% has been observed.

The use of correct Equation (12) is necessary particularly for the material having high value of Poisson's ratio. Using dielectric constant and the elastic compliance data the contribution of the Maxwell stress to the total strain response can be determined. For accurate estimation of induced strains, the contributions from coordinate axis or orientation (of sample with respect to the normal principal stress) dependent of Young modulus, Poisson's ratio and generalized moduli should also be incorporated into the expressions. A comprehensive investigation and computations [18] carried out for BCC crystal under hydrostatic stresses show their dependency on coordinate axis and orientation of crystal with respect to the principal stress. However due to viscoelastic behavior of polymeric materials, most of the assumptions made for linear dielectric material are not valid for elastomer up to large extent for large deformation. Despite of these limitations, the researchers are frequently using the formula derived on assumptions made for a linear elastic and linear dielectric material.

5.2. Boundary Conditions: Dependence of Stress Tensor on the Orientation of Dielectric Slab with Respect to the Electric Field Vector

The expression (27) is valid if the direction of electric field $\underline{E}_z$ is making an angle θ with respect to the normal to the surface of sample (**Figure 5**), then the expression [3] for the stress tensor in triclinic case would be

$$T_{ij} = \begin{bmatrix} -\dfrac{1}{2}\varepsilon_{33}E_z^2 & 0 & 0 \\[2mm] \varepsilon_{13}E_z^2\sin\theta & \dfrac{1}{2}\varepsilon_{33}E_z^2\left(\sin^2\theta - \cos^2\theta\right) + \varepsilon_{23}E_z^2\sin\theta\cos\theta & \varepsilon_{33}E_z^2\sin\theta\cos\theta - \varepsilon_{23}E_z^2\sin\theta \\[2mm] \varepsilon_{13}E_z^2\cos\theta & \varepsilon_{33}E_z^2\sin\theta\cos\theta + \varepsilon_{23}E_z^2\cos^2\theta & \dfrac{1}{2}\varepsilon_{33}E_z^2\left(\cos^2\theta - \sin^2\theta\right) - \varepsilon_{23}E_z^2\sin\theta\cos\theta \end{bmatrix} \quad (27)$$

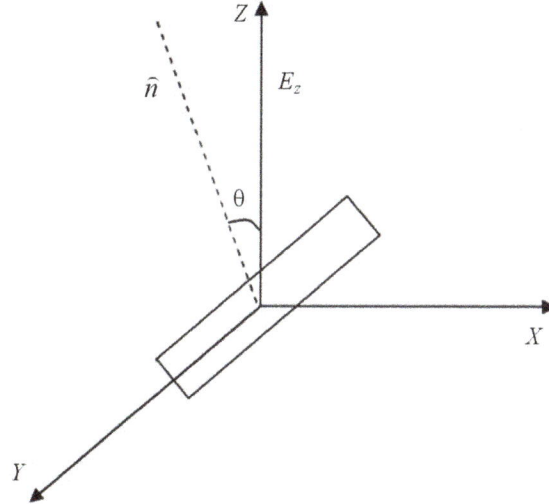

Figure 5. Schematic diagram illustrating the normal to the surface of a sample makes an angle θ with the direction of applied electric field E i.e. orientation of sample with electric field vector.

For experimental study [6] [17], a parallel plate capacitor with dielectric film/slab has been considered due to its simple symmetrical geometry. However, due to electrostrictive deformation, the permittivity of different regions is different so the solution to the field or potential must be different although having the same general form of solution (to the Laplace equation). The appropriate electrical boundary conditions must be satisfied at the interface between two regions.

For mechanical boundary conditions, the external forces on the boundary may be regarded as a continuation of the internal stress distribution and in equilibrium, the total forces consisting of total internal body forces and total external forces, like forces due to gravitational field, in every volume element of elastic dielectrics must be balanced to a zero value [4].

However it is very difficult to predict the exact boundary conditions taking place at the interface between the electrodes and the deformed dielectrics or between two regions with different permittivities and non-uniform field. The interface in Nanometric dielectrics with non-uniform electric fields also exhibits electromechanical properties [19].

6. Conclusion

The idea of the dependence of electrostatic stresses on the orientation of dielectric material slab with respect to electric field vector is necessary for the correct measurement of electrically induced strains. Most of the recent works fail to correlate correctly the Maxwell stress tensor, the electrostrictive stress tensor and the mechanical stress tensor, as the assumptions made for their formulations are different. As the details of exact boundary conditions at the electrode-dielectric interface of a deformed elastic material are unknown, the correct estimation of electrically induced deformation is very difficult. Contribution due to lateral/shear components of dielectric constant as in Equation (27), in addition to the longitudinal deformation, must be incorporated into the mathematical expression to get correct estimation of results. In elastic dielectrics, the shear and lateral stresses play a major role and hence neglect of the contribution from lateral and shear stresses leads to over estimation or under estimation of results. Experimental data on the subject are not abundant, and the researchers [5] [6] frequently used an incorrect formula in the derivation of elastic strain with respect to the Maxwell stress effect, particularly in the case of polyurethane elastomer (a cross-linked polymer), assuming various unrealistic approximations.

References

[1]   Landau, L.D. and Lifshitz, L.M. (1984) Electrodynamics of Continuous Media. 2nd Edition, Pergamon Press, Oxford.

[2]   Thakur, O.P. and Singh, A.K. (2009) Electrostriction and Electromechanical Coupling in Elastic Dielectrics at Nanometric Interfaces. *Materials Science—Poland*, **27**, 839-850.

[3]   Kloos, G. (1995) The Dependence of Electrostatic Stresses at the Surface of a Dielectric on Its Orientation in an Electric Field. *Journal of Physics D: Applied Physics*, **28**, 2424-2429. http://dx.doi.org/10.1088/0022-3727/28/12/006

[4]   Stratton, J.A. (1941) Electromagnetic Theory. McGraw-Hill, New York.

[5]   Shkel, Y.M. and Klingenberg, D.J. (1996) Material Parameter for Electrostriction. *Journal of Applied Physics*, **80**, 4566. http://dx.doi.org/10.1063/1.363439

[6]   Zhang, Q.M., Su, J., Kim, C.H., Ting, R. and Capps, R. (1997) An Experimental Investigation of Electromechanical Responses in a Polyurethane Elastomer. *Journal of Applied Physics*, **81**, 2770. http://dx.doi.org/10.1063/1.363981

[7]   Krakovský, I., Romijn, T. and Posthuma de Boer, A. (1999) A Few Remarks on the Electrostriction of Elastomers. *Journal of Applied Physics*, **85**, 628. http://dx.doi.org/10.1063/1.369418

[8]   Ladabaum, I., Khuri-Yakub, B.T. and Spoliansky, D. (1996) Micromachined Ultrasonic Transducers: 11.4 MHz Transmission in Air and More. *Applied Physics Letters*, **68**, 7. http://dx.doi.org/10.1063/1.116764

[9]   Herbert, J. M. (1982) Ferroelectric Transducers and Sensors. Gordon and Breach, New York.

[10]   Bar-Cohen, Y. (Ed.) (2004) Electroactive Polymer (EAP) Actuators as Artificial Muscles: Reality, Potential and Challenges. 2nd Edition, SPIE Press, Bellingham, Vol. PM136.

[11]   Juretschke, H.J. (1977) Simple Derivation of Maxwell Stress Tensor and Electrostrictive Effect in Crystal. *American Journal of Physics*, **45**, 277-280. http://dx.doi.org/10.1119/1.10642

[12]   Lewis, T.J. (2005) Interfaces: Nanometric Dielectrics. *Journal of Physics D: Applied Physics*, **38**, 202-212. http://dx.doi.org/10.1088/0022-3727/38/2/004

[13]   Barletta, A. and Zanchini, E. (1994) Can the Definition of Mechanical Stress Tensor Be Applied to a Dielectric Fluid in an Electrostatic and Magneto Static Field? *IL Nuovo Cimento*, **16D**, 177-187.

[14]   Wang, C.C. (1979) Mathematical principles of Mechanics and Electromagnetism. Plenum, New York.

[15]   Truesdell, C. (1966) The Elements of Continuum Mechanics. Springer-Verlag, Berlin.

[16]   Kanno, I., Kotera, H. And Wasa, K. (2003) Measurement of Transverse Piezoelectric Properties of PZT Thin Films. *Sensors and Actuators Actuators* A, **107**, 68-74.

[17]   Lee, H.Y., Peng, Y. and Shkel, Y.M. (2005) Strain-Dielectric Response of Dielectrics as Foundation for Electrostriction Stresses. *Journal of Applied Physics*, **98**, Article ID: 074104. http://dx.doi.org/10.1063/1.2073977

[18]   Thakur, O.P. and Singh, A.K. (2008) Modeling of Capacitive Sensor Filled with Elastic Dielectrics and Its Advantages. 3*rd International Conference on Sensing Technology*, 30 November-3 December 2008, Taiwan, 467-471.

[19]   Thakur, O.P. and Singh, A.K. (2007) Electromechanical Phenomena at the Interface in Nanometric Dielectrics. *Proceedings of* 2*nd International Conference on Sensing Technology*, 26-28 November 2007, Palmerston North, 188-192.

Permissions

All chapters in this book were first published by Scientific Research Publishing; hereby published with permission under the Creative Commons Attribution License or equivalent. Every chapter published in this book has been scrutinized by our experts. Their significance has been extensively debated. The topics covered herein carry significant findings which will fuel the growth of the discipline. They may even be implemented as practical applications or may be referred to as a beginning point for another development.

The contributors of this book come from diverse backgrounds, making this book a truly international effort. This book will bring forth new frontiers with its revolutionizing research information and detailed analysis of the nascent developments around the world.

We would like to thank all the contributing authors for lending their expertise to make the book truly unique. They have played a crucial role in the development of this book. Without their invaluable contributions this book wouldn't have been possible. They have made vital efforts to compile up to date information on the varied aspects of this subject to make this book a valuable addition to the collection of many professionals and students.

This book was conceptualized with the vision of imparting up-to-date information and advanced data in this field. To ensure the same, a matchless editorial board was set up. Every individual on the board went through rigorous rounds of assessment to prove their worth. After which they invested a large part of their time researching and compiling the most relevant data for our readers.

The editorial board has been involved in producing this book since its inception. They have spent rigorous hours researching and exploring the diverse topics which have resulted in the successful publishing of this book. They have passed on their knowledge of decades through this book. To expedite this challenging task, the publisher supported the team at every step. A small team of assistant editors was also appointed to further simplify the editing procedure and attain best results for the readers.

Apart from the editorial board, the designing team has also invested a significant amount of their time in understanding the subject and creating the most relevant covers. They scrutinized every image to scout for the most suitable representation of the subject and create an appropriate cover for the book.

The publishing team has been an ardent support to the editorial, designing and production team. Their endless efforts to recruit the best for this project, has resulted in the accomplishment of this book. They are a veteran in the field of academics and their pool of knowledge is as vast as their experience in printing. Their expertise and guidance has proved useful at every step. Their uncompromising quality standards have made this book an exceptional effort. Their encouragement from time to time has been an inspiration for everyone.

The publisher and the editorial board hope that this book will prove to be a valuable piece of knowledge for researchers, students, practitioners and scholars across the globe.

List of Contributors

N. K. Divya, P. U. Aparna and P. P. Pradyumnan
Department of Physics, University of Calicut, Malappuram, India

E. M. Gojayev and A. G. Hasanova
Azerbaijan Technical University, Baku, Azerbaijan

E. M. Gojaev, Sh. V. Alieva, K. C. Gulmammadov and S. S. Osmanova
Azerbaijan Technical University, Baku, Azerbaijan

Alabur Manjunath, Tegginakeri Deepa, Naraganahalli Karibasappa Supreetha and Mohammed Irfan
Department of P.G. Studies in Physics, Government Science College, Chitradurga, India

Robert E. Baier
Industry/University Center for Biosurfaces, 110 Parker Hall, State University of New York at Buffalo, Buffalo, NY, USA

Sankararao Gattu
Department of Physics, MVJ College of Engineering, Bangalore, India

Venuturupalli Durga Prasadu and Kocharlakota Venkata Ramesh
Department of Physics, GITAM Institute of Technology, GITAM University, Visakhapatnam, India

E. M. Gojayev, S. I. Mammadova and G. S. Djafarova
Azerbaijan Technical University, Baku, Azerbaijan

A. Y. Ismailova
Ganja State University, Ganja, Azerbaijan

R. K. Raju and H. S. Jayanna
Department of Physics, Kuvempu University, Shankaraghatta, India

S. M. Dharamaprakash
Department of Physics, Mangalore University, Mangalore, India

Salima Hennani, Lahcen Ez-zariy and Abdelmajid Belafhal
Laboratory of Nuclear, Atomic and Molecular Physics Department of Physics, Faculty of Sciences, Chouaïb Doukkali University, El Jadida, Morocco

Huanyou Wang, Yaqi Chen and Xiangyan He
Department of Physics and Electronic Information Engineering, Xiangnan University, Chenzhou, China

Yalan Li
Department of Physics and Electronic Information Engineering, Xiangnan University, Chenzhou, China
College of Physical Science and Technology, Huazhong Normal University, Wuhan, China

Kumar Brajesh
Materials Engineering, Indian Institute of Science, Bangalore, India

Kiran Kumari
P G Department of Physics, R N College Hajipur (Vaishali), Bihar, India

Shuna Hou, Jianfei Xie, Ye Kuang, Xianhong Zheng, Lan Yao and Yiping Qiu
Key Laboratory of Textile Science and Technology, Ministry of Education, College of Textiles, Donghua University, Shanghai, China

Rassel Raihan and Kenneth Reifsnider
University of Texas at Arlington, Arlington, USA

Fazle Rabbi and Vamsee Vadlamudi
University of South Carolina, Columbia, USA

E. M. Gojayev, Kh. R. Ahmadova, S. I. Safarova, G. S. Djafarova and Sh. M. Mextiyeva
Azerbaijan Technical University, Baku, Azerbaijan

P. U. Aparna, N. K. Divya and P. P. Pradyumnan
Department of Physics, University of Calicut, Malappuram, India

Ossama E. Gouda
Electrical Department, Faculty of Engineering, Cairo University, Giza, Egypt

Sohair F. Mahmoud and Ahmed S. Haiba
High Voltage Metrology Lab, National Institute of Standards (NIS), Giza, Egypt

Ahmed A. El-Gendy
Nanotechnology and Nanometrology Lab, National Institute of Standards (NIS), Giza, Egypt

Anjani Kumar Singh
Bharati Vidyapeeth's College of Engineering, New Delhi, India